LUMINESCENCE AND NONLINEAR OPTICS

LYUMINESTSENTSIYA I NELINEINAYA OPTIKA

ЛЮМИНЕСЦЕНЦИЯ И НЕЛИНЕЙНАЯ ОПТИКА

The Lebedev Physics Institute Series

Editor: Academician D. V. Skobel'tsyn

Director, P. N. Lebedev Physics Institute, Academy of Sciences of the USSR

Proceedings (Trudy) of the P. N. Lebedev Physics Institute

Volume 59

LUMINESCENCE AND NONLINEAR OPTICS

Edited by
Academician D. V. Skobel'tsyn
Director, P. N. Lebedev Physics Institute
Academy of Sciences of the USSR, Moscow

Translated from Russian by
Albin Tybulewicz
Editor: *Soviet Physics–Semiconductors*

Springer Science+Business Media, LLC

Library of Congress Cataloging in Publication Data

Main entry under title:

Luminescence and nonlinear optics.

(Proceedings (Trudy) of the P. N. Lebedev Physics Institute, v. 59)
Translation of Lîuminesfsenfsiîa i nelineĭnaia optika.
Includes bibliographical references.
1. Luminescence—Addresses, essays, lectures. 2. Nonlinear optics—Addresses, essays,
lectures. I. Skobel'fsyn, Dmitriĭ Vladimirovich, 1892-   ed. II. Series: Akademiia
nauk SSSR. Fizicheskiĭ institut. Proceedings, v. 59.
QC476.5.L8513   535'.35   73-83897
ISBN 978-1-4899-2652-4

ISBN 978-1-4899-2652-4   ISBN 978-1-4899-2650-0 (eBook)
DOI 10.1007/978-1-4899-2650-0

The original Russian text was published by Nauka Press in Moscow in 1971 for the
Academy of Sciences of the USSR as Volume 59 of the Proceedings of the P. N.
Lebedev Institute. The present translation is published under an agreement with
Mezhdunarodnaya Kniga, the Soviet book export agency.

United Kingdom edition published by Consultants Bureau, London
A Division of Plenum Publishing Company, Ltd.
Davis House (4th Floor), 8 Scrubs Lane, Harlesden, London NW10 6SE, England

CONTENTS

SEPARATION OF COMPLEX SPECTRA INTO INDIVIDUAL BANDS BY THE GENERALIZED ALENTSEV METHOD

M. V. Fok

The paper deals with the theory of the Alentsev method for separation of complex spectra into individual bands. The method is generalized to the many-band case. An estimate is given of the precision of the results obtained and the conditions of validity of the method are discussed. Numerical examples are considered in detail and the recommended sequence of operations is given.

The need to separate complex absorption or luminescence spectra into their individual components arises in many cases such as luminescence analysis of complex mixtures or investigations of luminescence centers in crystal phosphors known to contain many types of luminescence center. Whenever the spectral bands belonging to centers or molecules of different types overlap one another we are dealing with some form of a complex spectrum. Usually, a spectrum of this kind is separated into individual bands on the assumption that the bands have Gaussian profiles and the problem usually discussed is that of the simplest method for finding the half-widths and the positions of the maxima of such bands. This is typical of the approach used, for example, in [1, 2].

However, since we are always dealing with spectra measured with finite precision, an analysis of the kind just mentioned does not give unambiguous results 3, 4]. This is clear from Fig. 1 which shows a Gaussian curve and which coincides with a sum of two narrower Gaussian curves that are shifted quite strongly relative to one another. The scale of the figure is insufficient to indicate the difference between the two main curves.

Moreover, it is not theoretically self-evident why an elementary profile should be of the Gaussian type and not some other bell-shaped curve. It is well known that the band profile depends on the system of coordinates in which it is plotted. Different workers use different quantities along the abscissa (these quantities are proportional to the wavelength or frequency). The ordinate usually represents the spectral energy density E_λ or E_ν and in some cases N_ν is used; here, $E_\lambda d\lambda$ is the power emitted by a given body in the spectral interval $d\lambda$, $E_\nu d\nu$ is the power emitted in the interval $d\nu$, $N_\nu d\nu$ is the number of quanta emitted per unit time in the interval $d\nu$. The conversion from E_λ to E_ν requires multiplication of the function describing the spectral profile by a factor which is proportional to λ^2, whereas the conversion to N_ν involves a factor which is proportional to λ^3. In spite of these important differences many workers fail to indicate what they are plotting along the ordinate and call the quantity employed the "relative intensity." It is quite obvious that the factor proportional to λ^2 or even λ^3 (the reverse conversions require factors proportional to ν^2 or ν^3) must displace the maximum of the spectral curve and distort its shape. An even greater distortion results from the replacement of λ with ν or ν with λ along the abscissa.

1

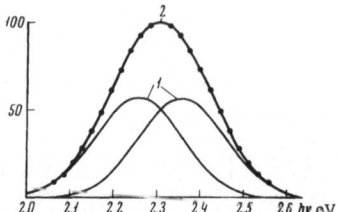

Figure 1. Spectrum consisting of two
Gaussian bands: 1) Gaussian bands with
half-widths of 0.232 eV shifted by 0.1 eV
relative to one another; 2) sum of the
curves denoted by 1. The points represent a Gaussian curve of half-width 0.27
eV.

Figure 2 shows, in two systems of coordinates, three spectral curves which are strictly
Gaussian in the $(h\nu, N_\nu)$ system and which differ only in their half-widths. We can see that in
the (λ, E_ν) system the maxima of the two curves are now located at different points, the half-
widths are smaller, and the profiles are asymmetrical. Similar changes occur when the curves
of Gaussian shape in the (λ, E_λ) coordinates are replotted in the $(h\nu, N_\nu)$ system. The situation
is only slightly better in the case of the absorption spectra. Since the quantity which we know
as the absorption coefficient does not change on conversion from one system of coordinates to
another, the factor λ^2 or λ^3 is not needed. Therefore, the position of an absorption maximum is
independent of the coordinate system. However, the profile of an absorption band changes (on
transition from one system of coordinates to another) almost as much as the profile of a luminescence curve.

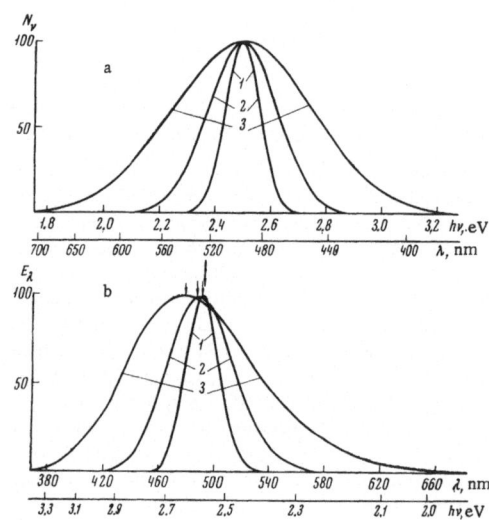

Fig. 2. Spectral curves of Gaussian shape
with their maxima all located at 2.5 eV (494
nm) are plotted in the coordinates $(h\nu, N_\nu)$
in Fig. 2a and in the coordinates (λ, E_λ) in
Fig. 2b. The half-widths of curves 1, 2,
and 3 in Fig. 2a are 0.15 eV (30 nm), 0.30
eV (60 nm), and 0.60 eV (118 nm), respectively; the corresponding half-widths of
curves 1, 2, and 3 in Fig. 2b are 0.14 eV
(28 nm), 0.29 eV (57 nm), and 0.58 eV (108
nm), respectively. The short arrows indicate the positions of the maxima in Fig. 2b.
The long arrow shows the position of the
same maxima in the coordinates $(h\nu, N_\nu)$.

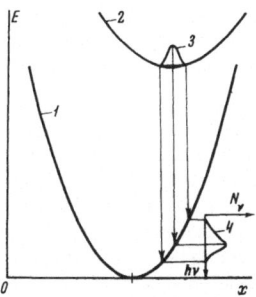

Fig. 3. Construction of a luminescence spectrum with the
aid of the simplest possible potential diagram of a luminescence center. Here, E is the
energy of the system; $h\nu$ is the
energy of the emitted quanta; x
is the configurational coordinate.
1) Potential energy of a center
in its ground state; 2) the same
center in an excited state; 3) distribution of probabilities of the
configurational coordinate in the
excited state; 4) luminescence
spectrum in the coordinates ($h\nu$,
N_ν). The vertical arrows represent changes in the state of the
system as a result of electron
transitions.

It follows that if there is no additional information on the profiles of individual bands, it
is not clear what coordinates should be used to plot the combined band before it is separated
into its Gaussian components. This is particularly important in the case of wide bands because
their profiles change greatly when the coordinates are altered. The asymmetry of a band plotted in a particular coordinate system and its large half-width may suggest that a given band is
complex whereas in other coordinates the same band may have the exact Gaussian profile, i.e.,
such a band may be regarded as elementary by those who consider the Gaussian shape to be
fundamental.

Moreover, the theory of spectra predicts a Gaussian profile only when very stringent conditions are satisfied. The first condition is that electron transitions must not have any phonon
replicas. However, even in this case the Gaussian profile is not always obtained. Let us illu-
strate it by considering a simple example. We shall assume that in order to calculate the
luminescence spectrum in the ($h\nu$, N_ν) coordinate system we can use a single-coordinate potential diagram of the type shown in Fig. 3. We shall also postulate that the temperature is sufficiently low so that all the luminescence centers are in the ground vibrational state. Then, the
distribution of the centers along the configurational coordinate will be described by a Gaussian
curve. According to the Franck-Condon principle (provided it is applicable) the atoms are not
displaced during the short time required for an electron transition. Therefore, the electron
transitions (including radiative transitions) are represented by vertical lines in the potential
diagram. If the probability of a radiative electron transition is independent of the configura-

tional coordinate, the luminescence spectrum may be Gaussian if the ground-state potential below the excited-state minimum can be replaced by a rectilinear segment. In all other cases the luminescence spectrum will not be Gaussian. Similar considerations apply also to absorption spectra.

Thus, there are many conditions which must be satisfied if the spectral profile is to be Gaussian. Therefore, there is no justification for the a priori assumption that a given band will be Gaussian. The fact that the experimentally obtained spectral curves can frequently be approximated by Gaussian shapes is simply the result of low precision of measurements. Moreover, it is usual to consider only a part of the spectra curve extending over a range slightly greater than its half-width. This approach does not always reveal any asymmetry and only asymmetric bands are regarded as nonelementary.

A correct analysis of a complex spectrum into its individual components must include also the band "wings" since that is where the influence of the neighboring bands is greatest and errors might be committed in the determination of the position of the maximum and the width of the neighbouring band if the wings are ignore. The deviations from the Gaussian profile and the asymmetry of the bands are strongest in the wings. Therefore, one must not be satisfied with a good agreement between the central part of a band and the theoretical Gaussian curve.

M. N. Alentsev suggested a method for separating two overlapping luminescence bands in which no assumptions are made about their profiles or the positions of the maxima. This method has been used for many years in our laboratory but it is unknown outside because it was not published due to the premature death of the author.

The importance of the analysis of complex spectra is such that it has become necessary to generalize the Alentsev method to the many-band case and to develop the theory of this method in detail.

The method is based on the following considerations. We shall assume that the luminescence spectrum of a substance consists of two partly overlapping bands whose maxima are located at different wavelengths. We shall denote the profile of the first band by $\varphi_1(\lambda)$ and that of the second band by $\varphi_2(\lambda)$. We shall assume that the luminescence of the substance in question is excited by two different methods in such a way that the luminescence spectra obtained are different in the two cases. If it can be assumed that the profiles of the bands are not affected by the excitation method, the differences between the two spectra must be due to the differences between the relative contributions of the first and second bands to the combined luminescence spectrum.* In this case the experimentally determined luminescence spectra are described by the functions $f_1(\lambda)$ and $f_2(\lambda)$ defined as follows:

$$f_1(\lambda) = \varphi_1(\lambda) + \varphi_2(\lambda),$$
$$f_2(\lambda) = a_1\varphi_1(\lambda) + a_2\varphi_2(\lambda),$$

$$(1)$$

where the unknowns are not only φ_1 and φ_2 but also the constant coefficients a_1 and a_2. We have

---

* The excitation methods can be varied in different ways. For example, we can alter the wavelength of the exciting radiation in order to change the proportion of the excitation quanta absorbed by the two types of center involved. We can also compare the luminescence and the afterglow spectra or we can use the differences between the emission delay times of the centers. In the case of crystal phosphors emitting recombination luminescence we can employ the nonlinear properties of these substances and later the excitation intensity or subject a crystal to de-excitation with long-wavelength light. It is undesirable to alter the temperature of a sample because this may change the profiles of the two bands in question.

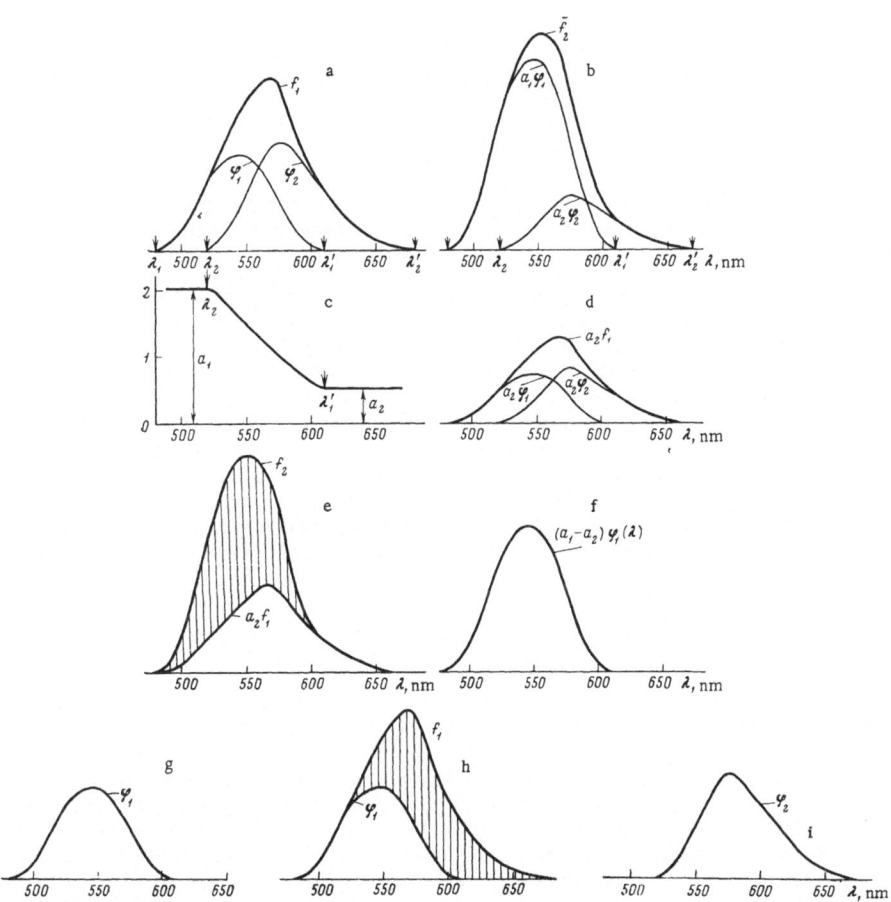

Fig. 4. Successive stages of separation of complex spectra into two components: a), b) original spectra and the component bands; c) ratio of the spectra $f_2(\lambda)/f_1(\lambda)$; d) spectrum a multiplied by a_2; e) elimination of band $\varphi_2(\lambda)$ [the shaded region represents the difference $a_2 f_1(\lambda) = (a_1 - a_2)\varphi_1(\lambda)$]; f) band $\varphi_1(\lambda)$ in the form obtained after elimination of $\varphi_2(\lambda)$ from f_2 and f_1; g) band $\varphi_1(\lambda)$ on correct scale; h) elimination of band $\varphi_1(\lambda)$ from $f_1(\lambda)$ [the shaded region represents the difference $f_1(\lambda) - \varphi_1(\lambda) = \varphi_2(\lambda)$]; i) band $\varphi_2(\lambda)$; λ_1 and λ_1' are the limits of band $\varphi_1(\lambda)$; λ_2 and λ_2' are the limits of band $\varphi_2(\lambda)$. Relative units, common to all the spectra, are used along the ordinate.

already mentioned that $\varphi_1(\lambda)$ and $\varphi_2(\lambda)$ differ in their spectral positions, i.e.,

$$\varphi_1(\lambda) \neq 0 \quad \text{when} \quad \lambda_1 < \lambda < \lambda_1',$$

$$\varphi_2(\lambda) \neq 0 \quad \text{when} \quad \lambda_2 < \lambda < \lambda_2', \tag{2}$$

where all four wavelength limits are different. We shall consider the specific case when the bands are shifted relative to one another in such a way that

$$\lambda_1 < \lambda_2 < \lambda_1' < \lambda_2'. \tag{3}$$

Figures 4a and 4b show examples of combinations of two bands of the type we are considering (the only difference between these two figures is that the bands are of different intensity – their profiles are the same).

We shall now plot the ratio $f_2(\lambda)/f_1(\lambda)$ as a function of the wavelength. It follows from our assumptions and from Fig. 4c that the plot of this ratio has two horizontal regions located at $\lambda < \lambda_2$ and $\lambda > \lambda_1'$. It is evident from the formulas in Eq. (1) and from the inequalities in Eqs. (2) and (3) that the ordinate of the first horizontal region is a_1, i.e., it is equal to the amplitude of the band $\varphi_1(\lambda)$ in the $f_2(\lambda)$ spectrum, whereas the ordinate of the second plateau a_2 is equal to the amplitude of the band $\varphi_2(\lambda)$ [following Eq. (1) we have assumed that the amplitudes of the two bands in the $f_1(\lambda)$ spectrum are equal to unity]. Thus, we can find the values of a_1 and a_2. Having multiplied $f_1(\lambda)$ by a_2 we obtain a spectrum in which the φ_2 band has exactly the same amplitude as in the f_2 spectrum. The only difference between the two spectra is that the $\varphi_1(\lambda)$ band has different amplitudes. Therefore, subtracting one spectrum from the other, we can eliminate completely the $\varphi_2(\lambda)$ band and obtain the $\varphi_1(\lambda)$ band with the amplitude equal to $a_1 - a_2$. In this way we find that

$$a_2 f_1(\lambda) = a_2 \varphi_1(\lambda) + a_2 \varphi_2(\lambda) \tag{4}$$

and, consequently,

$$f_2(\lambda) - a_2 f_1(\lambda) = (a_1 - a_2)\varphi_1(\lambda). \tag{5}$$

This proceeds of subtraction and its final result are illustrated in Fig. 4d-4i. We can now find the $\varphi_1(\lambda)$ profile because we know the values of a_1 and a_2. Subtracting the $\varphi_1(\lambda)$ band obtained in this way from the $f_1(\lambda)$ spectrum, we can find the $\varphi_2(\lambda)$ band. The true profiles of the components of the $f_2(\lambda)$ spectrum are obtained by multiplying $\varphi_1(\lambda)$ by a_1 and $\varphi_2(\lambda)$ by a_2. This completes the procedure of separation of the spectrum into its two components. Basically this is the way that Alentsev presented his method.

We can see that this method makes no assumptions about the nature of the functions $\varphi(\lambda)$, i.e., about the profiles of the infividual bands, apart from the assumption that there are spectral intervals in which the functions in question do not vanish. Therefore, the Alentsev method can be applied to the spectra f_1 and f_2 plotted in any coordinate system. The coordinate axes can even represent the actual distances taken from the chart of an automatic recorder without any corrections for the dispersion and sensitivity. An important advantage of the method is the existence of an experimental criterion of the validity of the method. This criterion is the presence of horizontal plateaus in the graph representing the ratio of the spectra.

We shall now consider more complex cases. We shall start by assuming that one of the bands encompasses completely the other band as shown in Fig. 5, i.e., we shall assume that only the first of the inequalities in Eq. (3) is satisfied. Although there are still two horizontal

Fig. 5. Relative positions of
two bands in the cases when the
band-separation procedure cannot be completed: a) graph of
the ratio of the spectra has only
one horizontal region; b) graph
of the ratio of the spectra has
two horizontal regions located
at the same level.

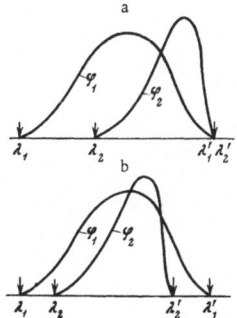

regions in the graph of the ratio of the spectra, the two regions are at the same level because
they are due to the same band (Fig. 5b). No difficulties are encountered in elimination of the
wider band from the combined spectrum and the consequent determination of the profile of the
$\varphi_2(\lambda)$ band (to within a constant factor). We shall denote the band obtained in this way by $\varphi_2'(\lambda)$.
However, we cannot calculate the normalization coefficient needed to eliminate the $\varphi_2'(\lambda)$ band
and to find the $\varphi_1(\lambda)$ band because the necessary horizontal plateau is absent. Therefore, the
problem cannot be solved if there is no additional information on the profile of the $\varphi_1(\lambda)$ band.*

We shall now assume that the $f_1(\lambda)$ and $f_2(\lambda)$ spectra consist of three individual bands,
i.e.,

$$f_1(\lambda) = \sum_{i=1}^{3} \varphi_i(\lambda), \tag{6}$$

$$f_2(\lambda) = \sum_{i=1}^{3} a_i \varphi_i(\lambda), \tag{7}$$

where

$$\varphi_i(\lambda) \neq 0 \quad \text{when } \lambda_i < \lambda < \lambda_i'. \tag{8}$$

We shall also assume that the bands are shifted relative to one another sufficiently so that at
any given point two (and not three) bands overlap, as shown in Fig. 6a, i.e., we shall assume that

$$\lambda_1 < \lambda_2 < \lambda_1' < \lambda_3 < \lambda_2' < \lambda_3'. \tag{9}$$

Then,

$$f_2(\lambda)/f_1(\lambda) = a_2 \quad \text{when } \lambda_1' < \lambda < \lambda_3 \tag{10}$$

and, consequently, we can eliminate $\varphi_2(\lambda)$ from the $f_1(\lambda)$ and $f_2(\lambda)$ spectra:

$$f_2(\lambda) - a_2 f_1(\lambda) = (a_1 - a_2)\varphi_1(\lambda) + (a_3 - a_2)\varphi_3(\lambda). \tag{11}$$

In view of the inequalities in Eq. (9) the $\varphi_1(\lambda)$ and $\varphi_3(\lambda)$ bands do not overlap at all and are even
separated by a spectral interval. Therefore, after the elimination of $\varphi_2(\lambda)$ there is no difficulty

* The author is grateful to A. Gailitis and Ya. Tsirulis for drawing attention to this point.

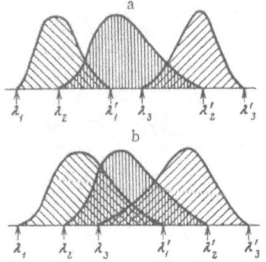

Fig. 6. Relative positions of three individual bands in complex spectra: a) only the neighboring bands overlap; b) all three bands overlap.

in establishing which part of the spectrum corresponds to $\varphi_1(\lambda)$ and which to $\varphi_3(\lambda)$. We can find $\varphi_2(\lambda)$ using the outer horizontal regions in the graph of the ratio $f_2(\lambda)/f_1(\lambda)$. We can see from Eqs. (6) and (7) and from the inequalities in Eq. (9) that these horizontal regions yield the values of the coefficients a_1 and a_2. Knowing all three coefficients (a_1-a_3) we can apply Eq. (11), find $\varphi_1(\lambda)$ and $\varphi_3(\lambda)$, and then eliminate them from $f_1(\lambda)$ to obtain $\varphi_2(\lambda)$. Thus, when the inequalities of Eq. (9) are satisfied in the case of three bands, the separation of a spectrum into its individual bands is only slightly more complicated than in the case of two bands. In particular, the problem can be solved completely if we know only the two spectra $f_1(\lambda)$ and $f_2(\lambda)$. The third horizontal region in the graph showing the ratio $f_2(\lambda)/f_1(\lambda)$ is experimental evidence that the inequalities in Eq. (9) are satisfied.

In general, when the $\varphi_1(\lambda)$ and $\varphi_3(\lambda)$ bands overlap (Fig. 6b), the third horizontal region is absent from the graph of $f_2(\lambda)/f_1(\lambda)$ and the situation is somewhat more complex but the problem is still soluble in many specific cases.

Let us assume that instead of Eq. (9), we can satisfy the inequalities

$$\lambda_1 < \lambda_2 < \lambda_3 < \lambda_1' < \lambda_2' < \lambda_3', \tag{12}$$

which differ from those in Eq. (9) because of the interchange of λ_1' and λ_3. This means that the two outer bands overlap each other. In this case the graph of the ratio $f_2(\lambda)/f_1(\lambda)$ gives only a_1 and a_3 and we can eliminate $\varphi_1(\lambda)$ or $\varphi_3(\lambda)$ from the spectrum but we cannot proceed any further without additional experimental data. However, if we have a third spectrum $f_3(\lambda)$, defined by

$$f_3(\lambda) = \sum_{i=1}^{3} b_i \varphi_i(\lambda), \tag{13}$$

we can eliminate $\varphi_1(\lambda)$ from this spectrum and from $f_1(\lambda)$. Then, we obtain two spectra each of which consists of just two bands:

$$\left.\begin{aligned} f_1^{(1)}(\lambda) &= f_2(\lambda) - a_1 f_1(\lambda) = (a_2 - a_1)\varphi_2(\lambda) + (a_3 - a_1)\varphi_3(\lambda), \\ f_2^{(1)}(\lambda) &= f_3(\lambda) - b_1 f_1(\lambda) = (b_2 - b_1)\varphi_2(\lambda) + (b_3 - b_1)\varphi_3(\lambda). \end{aligned}\right\} \tag{14}$$

In this way we have reduced the problem to the two-band case, which we know can be solved.

However, if the $\varphi_1(\lambda)$ band encloses completely the $\varphi_2(\lambda)$ band, i.e., if

$$\lambda_1 < \lambda_2 < \lambda_3 < \lambda_2' < \lambda_1' < \lambda_3' \tag{15}$$

[this system of inequalities differs from Eq. (12) by the interchange of λ_1' and λ_2'], the sequence of operations is initially the same as in the case when Eq. (12) applies. However, we cannot eliminate the $f_2(\lambda)$ band because the situation is exactly the same as in the case of two bands — one completely enclosed by the other. In the case of three bands we can find the amplitude and profile of the $\varphi_3(\lambda)$ band and the profile of the $\varphi_2(\lambda)$ but not its amplitude. Therefore, we cannot find the profile of the $\varphi_1(\lambda)$ band.

When we analyze a complex spectrum into its individual components we usually do not know the number of bands that it contains. We shall now consider what help we can get from the Alentsev method. Let us assume that the maximum possible number of bands is three but that we cannot exclude the possibility that there are just two bands. [The fact that the original band is not simple follows from the difference between the $f_1(\lambda)$ and $f_2(\lambda)$ spectra.] We already know of one case in which we can definitely say that the spectrum in question consists of three bands. This is the case when the inequalities of Eq. (9) are satisfied, i.e., when the graph of the appropriate ratio has a third horizontal region. If this region is absent, we are dealing with two possible cases: either the value of a_2 lies between a_1 and a_3 or outside this range. In the first case the graph of the ratio $f_2(\lambda)/f_1(\lambda)$ resembles the graph obtained for a spectrum consisting of two bands. The absence of the third band can be established only if we use an additional spectrum $f_3(\lambda)$. Then, eliminating one of the bands from the two pairs of spectra we must see whether the ratio $f_3^{(1)}(\lambda)/f_2^{(1)}(\lambda)$ obtained after such elimination remains constant. If it is not constant, there are three bands, whereas if it is constant there are probably only two bands. However, this does not exclude the possibility of three bands in the case when the ratio in question is constant because the conditions for deriving the third spectrum may have been chosen inappropriately so that the relevant determinant vanishes: $\| a_{ji} \| = 0$. Additional information is needed to eliminate this possibility.

In the second case, i.e., when a_2 lies outside the interval a_1-a_3, the graph of the ratio $f_2(\lambda)/f_1(\lambda)$ has an extremum which indicates that there are three bands. However, in order to separate these three bands we need a third spectrum. This applies only if the individual bands have a more or less bell-shaped profile, which need not be symmetric but must be free of secondary peaks. Otherwise, even two bands can give rise to a large number of extrema. Therefore, in practice it is best to check the absence or presence of an additional band by means of a third spectrum.

We shall now consider the general case when m spectra are available and each of them consists of n separate bands. In this situation we have:

$$\sum_{i=1}^{n} \varphi_i(\lambda) = f_1(\lambda), \tag{16}$$

$$\sum_{i=1}^{n} a_{ji}\varphi_i(\lambda) = f_j(\lambda), \tag{17}$$

where j varies from 2 to m. At each fixed value of λ Eqs. (16) and (17) represent simply a system of linear equations with unknowns φ_i and constant coefficients a_{ji}. The conditions governing the solubility of such a system are well known. However, the system we are considering has one special feature: we do not know the coefficients a_{ji} and, quite often, also the number n. We shall start by assuming that n is known. It is obvious that even then we cannot solve the system (17) in its general form. However, we can use the fact that $\varphi_i(\lambda)$ are functions which describe individual spectral bands associated with different species such as molecules, lattice defects, etc. Therefore, we shall assume that all $\varphi_i(\lambda)$ are different and that there is no part of the spectrum in which these functions are related linearly, i.e., we shall assume that none of the functions can be obtained from a linear combination of all the others. Moreover, we shall assume that all $\varphi_i(\lambda)$ have nonzero values in finite intervals and that they differ from one another at least on one side of the profile:

$$\varphi_i(\lambda) \neq 0 \quad \text{when} \quad \lambda_i < \lambda < \lambda_i'. \tag{18}$$

We shall label λ_i so that

$$\lambda_i < \lambda_{i+1}. \tag{19}$$

These assumptions are sufficient to begin successive elimination of individual bands in the same way as described for two or three bands. It follows from Eqs. (19), (16), and (17) that at the short-wavelength end of the graph of the ratio $f_j(\lambda)/f_1(\lambda)$ we have a horizontal region whose ordinate is a_{j1}. Having eliminated $\varphi_1(\lambda)$, we obtain a system of m−1 equations

$$\sum_{i=2}^{n} (a_{ji} - a_{j1}) \varphi_i(\lambda) = f_j^{(1)}(\lambda), \tag{20}$$

where j varies from 2 to m. In virtue of the inequalities stated in Eq. (19), we can eliminate $\varphi_2(\lambda)$ from the system and thus obtain m−2 equations with n−2 unknowns. For example, if we divide all the $f_j^{(1)}(\lambda)$ by $f_1^{(1)}(\lambda)$, we obtain the following system:*

$$\sum_{i=3}^{n} \frac{(a_{ji} - a_{j1})(a_{22} - a_{21}) - (a_{j2} - a_{j1})(a_{2i} - a_{21})}{a_{22} - a_{21}} \, \bar{\varphi}_i(\lambda) = f_j^{(2)}(\lambda), \tag{21}$$

where j varies from 3 to m. From this system we eliminate $\varphi_3(\lambda)$ and, so on. If m = n and the determinant of the system (16)-(17) does not vanish†

$$\| a_{ji} \| \neq 0, \tag{22}$$

we can apply the same method to obtain finally the $\varphi_n'(\lambda) = k_n \varphi_n(\lambda)$ band for which the coefficient k_n depends in a complex manner on a_{ji}. If, in addition to the inequalities in Eq. (18), similar inequalities apply to λ_i', i.e., if

$$\lambda_i' < \lambda_{i+1}', \tag{23}$$

the separation of the $f_j(\lambda)$ spectra into individual bands can be easily completed. In fact, the $\varphi_n(\lambda)$ band can be eliminated from the $f_n^{(n-2)}(\lambda)$ or $f_{n-1}^{(n-2)}(\lambda)$ spectra consisting only of the two bands $\varphi_{n-1}(\lambda)$ and $\varphi_n(\lambda)$. Since the functions $\varphi_n(\lambda)$ and $\varphi_n'(\lambda)$ differ only in respect of the constant coefficients, the $\varphi_n(\lambda)$ band can be eliminated by replacing it with the known band $\varphi_n'(\lambda)$.

---

*The spectrum $f_1(\lambda)$ or any other spectrum by which all the others are divided should be the widest spectrum, i.e., that which stretches as far as possible in the directions of the short and the long wavelengths. Then the ratio $f_j(\lambda)/f_1(\lambda)$ is not large at any wavelength. If it is not possible to select such a spectrum, the procedure of band elimination should start from the band $\varphi_1(\lambda)$ which is located at the shortest wavelengths and preference should be given to that spectrum which has the highest amplitude on the short-wavelength side. In this way we guarantee that this spectrum includes the short-wavelength band $\varphi_1(\lambda)$, i.e., the band which we are intending to eliminate. However, it may happen that the spectrum we select is the only one which includes $\varphi_1(\lambda)$. Then this spectrum should be excluded temporarily from consideration and the other spectra containing one band less should be analyzed. Having obtained all the $\varphi_i(\lambda)$ bands except $\varphi_1(\lambda)$, we can return to the spectrum omitted temporarily and eliminate from it all the bands found so far. This gives us the $\varphi_1(\lambda)$ band.

†We recall that according to Eqs. (16) and (17) all the coefficients a_{ii} are equal to unity.

In this way we find the band $\varphi'_{n-1}(\lambda)$ which differs from $\varphi_{n-1}(\lambda)$ by a constant coefficient k_{n-1}. Having eliminated these two bands from the $f_n^{(n-3)}(\lambda)$, $f_{n-1}^{(n-3)}(\lambda)$ or $f_{n-2}^{(n-3)}(\lambda)$ spectra, we obtain the $\varphi_{n-2}(\lambda)$ band again with an unknown coefficient. If this process is repeated $n-1$ times, we obtain all n bands. The coefficients k_i can be found, in general, if we know all the normalization factors used in the band elimination. However, the relevant formulas are far too cumbersome. It is simpler to find these coefficients directly by eliminating the $\varphi'_i(\lambda)$ bands from the $f_1(\lambda)$ spectrum. It is clear from Eqs. (16)-(17) that the normalization factors by which the functions $\varphi'_i(\lambda)$ must be multiplied in order to eliminate $\varphi_i(\lambda)$ from the $f_1(\lambda)$ spectrum are simply the reciprocals of the coefficients k_i. Having found all the $\varphi_i(\lambda)$ bands, we can now determine the unknown coefficients a_{ji} by eliminating successively the $\varphi_i(\lambda)$ bands from the $f_j(\lambda)$ spectra. In all, $2n^2 - 3n + 1$ separate operations are necessary in order to eliminate the individual bands.

Thus, the whole process of separation of a complex spectrum into its individual components can be divided into three stages: 1) determination of the profile of one band; 2) determination of the profiles of all the other bands; 3) calculation of the coefficients a_{ji}. If we are not interested in these coefficients, we can stop after the second stage and we then need to perform only $(n-1)^2$ operations. However, it is hardly sensible to omit the third stage (determination of a_{ji}) because the knowledge of these coefficients allows us to determine how changes in the conditions under which the various spectra are obtained affect the intensity of each band. This can be used to find the properties of the centers responsible for the various bands.

The sequence of operations in band elimination is shown schematically in Fig. 7 for five bands. The triangles in Fig. 7 symbolize the bands which are present in a spectrum at a par-

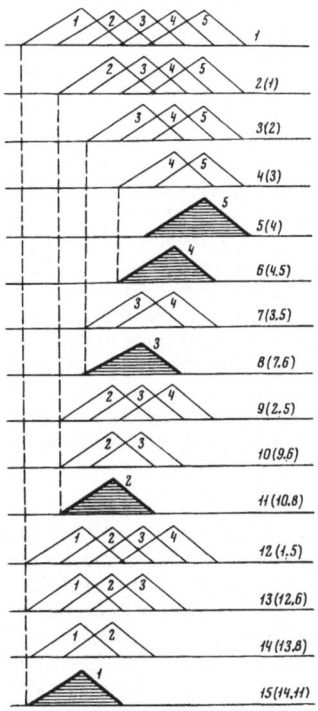

Fig. 7. Procedure employed in band elimination in the case when the inequalities of Eqs. (19) and (23) apply. The bands are represented by numbered triangles. The numbers in parentheses are the steps which are utilized in obtained a spectrum at any given stage. The shaded triangles represent the individual bands obtained as a result of the outlined analysis.

ticular step in the procedure. Steps 1-5 correspond to the first stage and steps 6-15 correspond to the second stage. Steps 12-15 represent also the last (third) stage in which the coefficients a_{ji} are determined. It should be remembered that the amplitudes of the bands shown in Fig. 7 are ignored. Therefore, all the $f_i(\lambda)$ spectra are identical.

If m < n, the elimination of individual bands cannot be completed (unless there are some special circumstances which will be discussed later), because the number of the $f_j(\lambda)$ spectra is insufficient. However, the situation is basically the same as in the m = n case: we are left with just one "band" $f_m^{(m-1)}(\lambda)$, which is not proportional to $\varphi_n(\lambda)$ but includes also n−m bands. If we do not know whether m < n, we can try to obtain the remaining bands and this procedure will be completed "successfully." If the inequalities of Eq. (19) are satisfied for at least the first m bands and the inequalities of Eq. (23) are obeyed by the last m bands, we find that the operations described above are performed easily. An analysis shows that even if we find "individual" bands in different ways [the $\varphi_1(\lambda)$ band can be obtained from any of the $f_j(\lambda)$ spectra, i.e., it can be found in m ways] we still fail to notice our error because we obtain results which differ only in respect of a constant factor, as in the m = n case. Nevertheless, each of the bands which we would then regard as "individual" still consists of n−m + 1 bands and the correct values of the coefficients a_{ji} are obtained only for i = 1 and i = n. [These coefficients can also be calculated from the graph of the ratio $f_j(\lambda)/f_1(\lambda)$ without elimination of any of the individual bands.]

It follows from our discussion that information on the number of individual bands in the $f_j(\lambda)$ spectra must be obtained from some other source such as the data on the chemical structure of the substance in question or the possible number of luminescence or absorption centers which can be present in that substance. The only thing that we can determine rigorously by the above method is the minimum number of bands. If the application of the method yields m different bands it follows that the spectra $f_j(\lambda)$ must contain at least m separate bands: if all these bands are associated with specific centers, the number of the bands is m; however, if some of these bands are complex, the number of individual bands is greater than m.

If m > n, the process of elimination of individual bands yields different results. When the operation is repeated n−1 times, we are left with m−n + 1 spectra consisting of just one band $\varphi_n(\lambda)$. Therefore, all the spectra obtained will be proportional to one another. This shows definitely that we have found all the individual bands in the original spectra or that we have chosen inappropriate conditions for recording the spectra. If the determinant is $\| a_{ji} \| = 0$, at least one of the equations in the (16)-(17) system can be obtained in the form of a linear combination of all the other equations. This linear relationship between the equations remains in force after the elimination of the $\varphi_1(\lambda)$, $\varphi_2(\lambda)$, etc., bands but the number of equations decreases. If we have r linearly dependent equations, we find that after elimination of the last band we are left with r + 1 spectra of which r are still linearly related, i.e., they are linear combinations of all the other spectra. However, over and above these r spectra we are left with one which cannot be combined with anything else. This means that all the other spectra are proportional to one another since a linear "combination" of one function is simply that function multiplied by a constant.

Thus, when we eliminate a certain number of individual bands and find that all the other spectra differ by a constant factor, we cannot draw the conclusion that all these spectra represent individual bands. However, the greater the variety of conditions under which the original spectra are recorded the greater is the likelihood of completion of the analysis in such a way that all the individual bands are obtained because the probability of accidental vanishing of the determinant $\| a_{ji} \|$ is reduced. In selecting the conditions for recording the original $f_j(\lambda)$ spectra we must try to achieve a situation in which changes in these conditions have different effects on the individual bands (this can usually be established in preliminary experiments). One should avoid any attempt to obtain the necessary number of the $f_j(\lambda)$ spectra by recording

Fig. 8. Special cases of
relative positions of five
bands. The bands are rep-
resented by triangles. A
thick line is used to identi-
fy band $\varphi_2(\lambda)$ whose profile
and position differ from one
case to another.

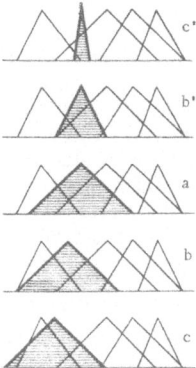

them under intermediate conditions which frequently differ only numerically from the extreme
cases already employed. For example, it is undesirable to use as one of the $f_j(\lambda)$ spectra a
luminescence spectrum recorded at some intermediate excitation intensity if we have already
recorded spectra at higher and lower intensities. This applies also to the concentration of a
mixture of substances if we are dealing with the absorption spectra of solutions. In such cases
there is a danger that the intensities of the individual bands characterized by the coefficients
a_{ji} are linear functions of the intensity, concentration, etc., so that the intermediate spectra
simply represent linear combinations of the spectra recorded under extreme conditions, i.e.,
there is a danger that $\| a_{ji} \| = 0$ with all the undesirable consequences.

Special attention should be paid to ensure that only the intensities of the individual bands
and not their profiles are altered by changing the conditions during recording of the spectra.
One should particularly avoid varying the temperature because although the temperature usually alters very strongly the ratio of the intensities, it often also affects the profiles of the individual bands. Therefore, the spectra $f_j(\lambda)$ measured at different temperatures are not necessarily comparable.

Let us consider how the complicating and simplifying factors, which we have identified in
the case of two and three bands, affect the analysis of spectra consisting of many bands. If one
of the inequalities of Eq. (23) is not satisfied, for example, if $\lambda_i' \geq \lambda_{i+1}'$, we do not obtain a horizontal region on the right-hand side in the graph representing the ratio of the spectra. We
have encountered this difficulty in the separation of a three-band spectrum into its components.
In the many-band case we again would fail to solve the problem completely. This reasoning applies also to the case when one of the inequalities of Eq. (19) is not satisfied, for example if
$\lambda_i = \lambda_{i+1}.$* In this case we should begin band elimination from the other end and reduce the
problem to that just described. However, the fact that one of the inequalities of Eq. (19) is not
satisfied will be discovered only after elimination of all the preceding bands, i.e., after the
computations have reached a fairly advanced stage. The results of these computations can be
used if we start eliminating the bands at the other end of the spectrum when we discover the
absence of the necessary horizontal region in the graph representing the ratio of the spectra.
In this case we obtain first not the $\varphi_n(\lambda)$ or $\varphi_1(\lambda)$ band but $\varphi_i(\lambda)$ at which we "got stuck" having
started the band elimination process from the short-wavelength end. In this case we again
cannot find the profile of the band which encompasses $\varphi_i(\lambda)$ but we can find all the other bands.
Thus, once more we obtain only a partial solution.

---

*We cannot have $\lambda_i > \lambda_{i+1}$ because we have labeled the bands in increasing order of λ_i.

In some cases the problem can be completely insoluble if the inequalities in Eqs. (19) and (23) are violated at the same time. The examples of such a situation are illustrated in Figs. 8b and 8b'. Generally speaking, the narrower the individual bands the smaller their overlap and, consequently, the easier the separation of a complex spectrum into its components. However, in some cases the problem is more difficult to solve or completely insoluble if only one band is narrower than the others or if the change in conditions reduces the width of one band much more than those of the others. This may also happen when one of the bands shifts with respect to the others. However, further shift or narrowing of such a band makes the problem partly soluble. All this can be seen in Fig. 8 which shows schematically the various possible cases of positions of five bands in which only one band differs from the rest in shift or width.

For the case represented in Fig. 8a the ratio of the spectra does not have a proper horizontal region but the spectrum can still be partly resolved into its individual components.

Figures 8b and 8b' show the situations in which the problem is insoluble since the $\varphi_2(\lambda)$ band has shifted or become narrower to the extent that the left-hand side horizontal region in the graph of the ratio of the spectra disappears (in the case represented by Fig. 8b this region is absent in the graph showing the ratio of the original spectra and in the case represented by Fig. 8b' the same region is absent from the graph showing the ratio of the spectra obtained after elimination of the first band). The problem is partly soluble in the cases represented by Figs. 8c and 8c' because of further shift or narrowing of the $\varphi_2(\lambda)$ band (in the case represented by Fig. 8c a horizontal region in the relevant graph appears on the left and in the case represented by Fig. 8c' this region appears in the middle of the graph showing the ratio of the original spectra).

Our discussion may have given an impression that various circumstances which complicate the difficult process of the separation of the spectra into their individual components are encountered so frequently that an analysis of the kind described is hardly ever successful. However, in fact one often encounters simplifying circumstances in which a horizontal region is found in the middle of the graph showing the ratio of the spectra. We have mentioned this point in discussing a spectrum consisting of three bands. A horizontal region in the middle of the graph representing the ratio of the spectra means that the individual bands do not overlap in the corresponding wavelength interval and that in this interval there is only one band. If we begin from this band, we can divide the problem into two independent stages. This immediately reduces strongly the number of the necessary band-elimination operations because the number of such operations depends on the square of the number of bands in a given spectrum. More-

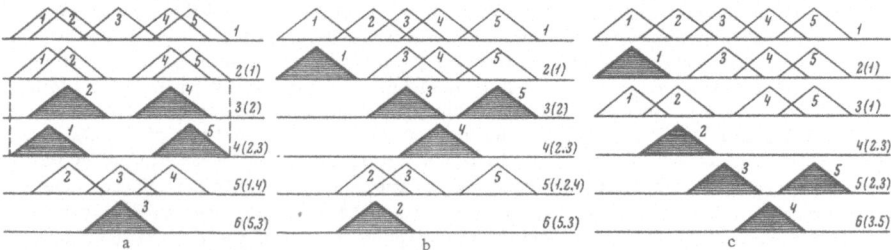

Fig. 9. Sequence of operations in the elimination of bands in those cases when additional horizontal regions are present in the graph of the ratio of the spectra. The notation is the same as in Fig. 7. a) One additional horizontal region in the middle of the graph representing the ratio of the spectra; b) two additional horizontal regions; c) three additional horizontal regions.

over, the necessary number of the original spectra is also reduced. The greater the number of horizontal regions in the graph of the ratio of the spectra the simpler is the solution of the problem. Thus, if a spectrum consists of five bands, a single additional horizontal region reduces the number of the necessary original spectra from five to four and sometimes even three. If two additional horizontal regions are present, we do not need more than three original spectra for five bands; in the case of three horizontal regions, only two spectra are needed. The number of necessary band-elimination operations decreases even more drastically. Figure 9 shows the sequence of operations in those cases when there are one, two, and three additional horizontal regions in the graph representing the ratio of the spectra. A comparison of Figs. 7 and 9 shows how far the problem can be simplified.

We have shown that there are many cases when the problem of separation of complex spectra into their individual bands is, in principle, soluble without any a priori assumptions as to the profiles and positions of these bands. However, we have not considered the influence of the scatter of the experimental results used to plot the $f_j(\lambda)$ spectra and of the associated random experimental error. Obviously, such a scatter not only makes it more difficult to find horizontal regions in the graph of the ratio of the spectra but it leads to inaccuracies in the determination of the ordinates of these regions so that a given band cannot be eliminated completely. If part of a band remains in a spectrum, we find that in the next graph of the ratio of the spectra there is no strictly horizontal region but one which is slightly inclined. The ordinate of such a region is even more difficult to determine with reasonable accuracy. This means that the next band will be eliminated even less accurately. Thus, the error in the successive band elimination operations is comulative. Hence, it follows that the original spectra must be determined with a high degree of precision if we wish to separate such spectra into their individual components.

It is very difficult to calculate the error permissible in various situations. Therefore, we shall consider some numerical experiments in which real conditions encountered in the separation of complex spectra are represented by idealized models. We shall consider spectra consisting of five bands because this number is sufficiently high to reveal the special features of the method and yet it is sufficiently small to enable us to complete calculations in a reasonable time.

We shall adopt the following procedure. We shall assume that the exact values of the five functions which represent individual bands $\varphi_i(\lambda)$ are known and tabulated. All these bands are asymmetric and strongly overlapping but the inequalities of Eqs. (19) and (23) are still satisfied. These bands are shown in Fig. 10. We can see that there is a central interval where all five bands overlap. We shall select a system of coefficients a_{ji} and use them to tabulate the exact values of $\sum_{i=1}^{5} a_{ji}\varphi_i(\lambda)$. The original spectra $f_j(\lambda)$ can be represented by adding a random error to each value of the sum. This error is found by random selection from the set of numbers 0,

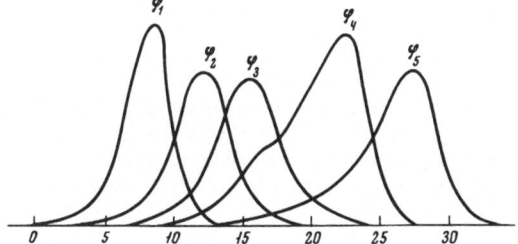

Fig. 10. Profiles and relative positions of five individual bands used in numerical experiments. The abscissa shows the numbers of the "experiment points" and the ordinate the "readings" of the spectroscopic instrument.

Fig. 11. Three original spectra in the first
numerical experiment. The spectrum $f_1(\lambda)$
is the sum of the amplitudes of the bands
shown in Fig. 10. The spectra $f_2(\lambda)$ and
$f_4(\lambda)$ are not shown because they differ less
from $f_1(\lambda)$ than the spectra $f_3(\lambda)$ and $f_5(\lambda)$.

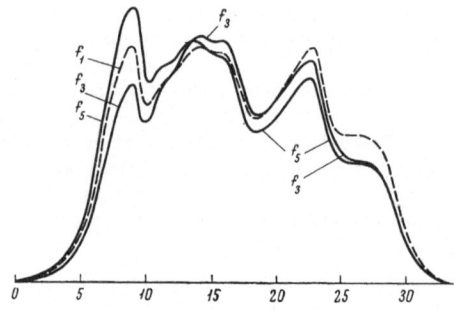

± 0.5 σ, ±σ, ±1.5 σ, ±2 σ, and ±2.5 σ, where σ is the rms scatter of the error being introduced.
This set is such that each value in the set is encountered in the proportion corresponding to the
ratio of the areas bounded by Gaussian curves with the rms scatter equal to σ and by the appropriate ordinates. In this way we obtain the five original spectra which have a random error
with an rms scatter σ. We then "forget" that these spectra are not "true" and we separate the
bands in accordance with the method described above. When the individual bands are obtained,
we "recall" the true band profiles specified earlier and we compare the results of our separation with these profiles. The values of σ will be different in the three cases we shall now consider.

First Case

We shall assume that the coefficients a_{ji} lie between 0.8 and 1.2, i.e., that they do not differ greatly from unity. The amplitudes, i.e., the maximum values of $\varphi_i(\lambda)$, of all five bands are
approximately equal. The value of σ is about 0.25% of the maximum intensity of the bands and
its absolute value is constant throughout the spectrum (this corresponds to a situation in which
the spectra would be measured without altering the sensitivity scale of the spectroscopic instrument). The spectra we are discussing are given in Fig. 11. We can see that the differences
between the three spectra shown in that figure are not very great. Figure 12 shows the bands
obtained by analysis of the spectra in Fig. 11. The thick curves in Fig. 12 represent the true
bands with the amplitudes which have been assumed in obtaining the original spectrum $f_1(\lambda)$.
We can see that, in spite of extremely small differences between the original spectra, the bands

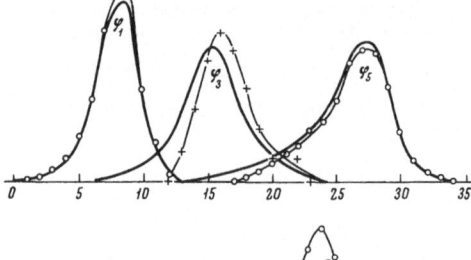

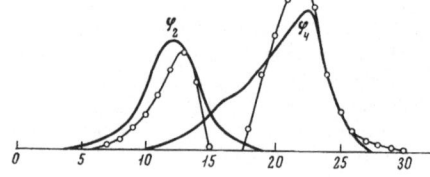

Fig. 12. Comparison of the bands obtained in the
first numerical experiment with the true band
profiles and amplitudes (thick curves). The amplitudes are the same as in the $f_1(\lambda)$ spectrum.

obtained are close to the true profiles: their maxima are shifted along the wavelength scale by not more than 10% of the width at half-maximum and the half-widths are within 15% of the true values. The results are particularly good for the two outer bands. For example, the asymmetry of the band on the extreme right can be seen quite clearly. The amplitudes are within 10% of the true values for all bands except the second (in this case the amplitude differs from the true value by a factor of 1.5). In view of the small differences between the three original spectra, the results obtained should be regarded as satisfactory.

We may occasionally find that the assumed experimental error is far too small and that the precision corresponding to such an error is practically unattainable. However, we must bear in mind that the assumed error simply represents the scatter of the experimental results and not the precision with which the true luminescence intensity (or the absorption coefficient) is known at a given wavelength.* The value of σ assumed in our calculations corresponds to 0.75 mm for a spectrogram whose amplitude in the chart is 30 cm. This is approximately equal to the random fluctuations observed in high-quality records. Naturally, this type of precision can be obtained if we ensure that the excitation source and the measuring apparatus are provided with stable power supplies. On the other hand, at the extreme ends of the spectrum, where the intensity is tens of times smaller than the maximum intensity, the relative experimental error may amount to a few percent and we have mentioned that the edges of the spectrum are extremely important in the correct elimination of individual bands. Therefore, high precision in recording of the spectra is essential, particularly when the differences between the spectra recorded under different conditions are not very large.

If there is a scatter in the experimental results, the horizontal part in the graph of the ratio of the spectra is not located in that region where only one band exists in the spectrum but it is shifted slightly toward the next (second) band. This occurs because at the very edge of the first band the error leads to such a scatter in the value of the ratio of the spectra that it is difficult to determine whether the ratio is constant. The importance of the error in the region between the first and the second bands is much reduced by the fact that the spectral intensity is now much higher than at the edge of the spectrum. Therefore, the ratio of the spectra is constant in the region where the second band is still unimportant compared with the first. However, by stating that the ratio of the spectra is constant we imply that the second band is absent and that the observed intensity is entirely due to the first band. This is why the "wings" of the bands obtained by analysis of the spectra are usually narrower than the true wings (Fig. 12). This is particularly clear in the region where the intensity of a band is less than 10% of the maximum value. The systematic error arising in this way must be born in mind in studies of profiles of individual bands.

The sequence of operations in the band-elimination process affects the precision in which they can be isolated. The first band is determined least accurately because it is affected by all the errors committed in the elimination of the other bands. If we know the true values of the coefficients a_{ji}, we can check at each step the correctness of the normalization coefficients. It is found that the first normalization coefficients are determined with an error slightly below 2%. The error in the subsequent normalization coefficients exceeds 10%. The normalization coefficients needed in the elimination of the $\varphi_3(\lambda)$ band cannot be determined because of incomplete elimination of the $\varphi_2(\lambda)$ band (the appropriate horizontal region is absent in the graph representing the ratio of the spectra). Therefore, we first have to eliminate $\varphi_5(\lambda)$ and then $\varphi_4(\lambda)$.

It is evident from Fig. 12 that the maximum of the $\varphi_3(\lambda)$ band is shifted most. In the case of the $\varphi_2(\lambda)$ and $\varphi_4(\lambda)$ bands the edges overlapping the $\varphi_3(\lambda)$ band are distorted most.

---

*The error associated with the inaccuracies in the calibration of the apparatus in respect of the wavelength or the energy does not affect the feasibility of the analysis of a spectrum since such an analysis can be carried out for any calibration and when the bands are finally obtained the scale can be altered in accordance with the calibration.

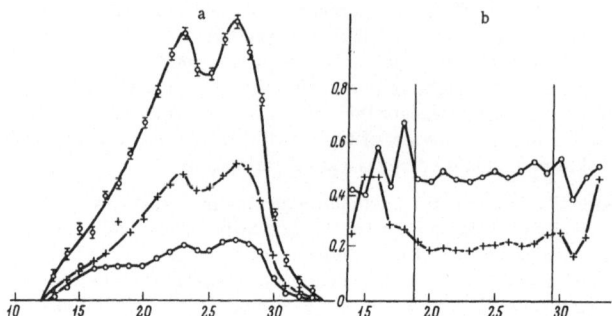

Fig. 13. Case corresponding to $\|a_{ji}\| = 0$: a) spectra obtained after elimination of two bands in the second numerical experiment; b) graphs of the ratio of the ordinates of the spectra plotted in Fig. 13a. The vertical lines are the precise limits of the horizontal region.

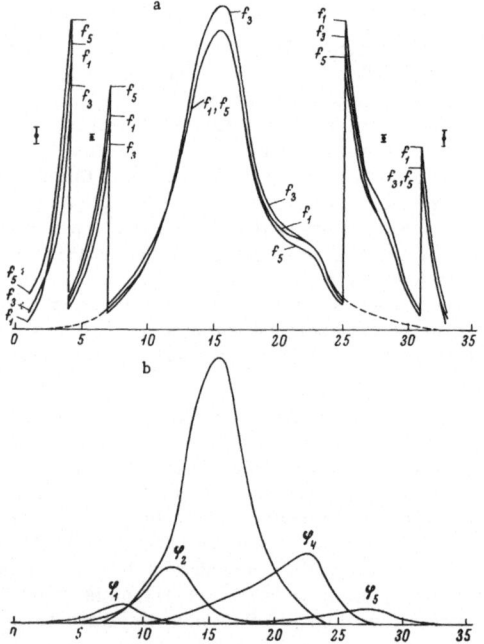

Fig. 14. a) Three initial spectra in the third numerical experiment (the centra band is much stronger than the other four bands); b) amplitudes and profiles of individual bands corresponding to one of the original spectra in Fig. 14a.

Second Case

In this case the conditions are the same as in the first case but $\|a_{ji}\| = 0$ so that two rows in this determinant are linear combinations of the other three rows. After the elimination of the $\varphi_1(\lambda)$ and $\varphi_2(\lambda)$ bands we obtain three spectra which are practically identical (apart from the amplitude). These spectra are shown in Fig. 13a. The ratios of the first two spectra to the third are plotted in Fig. 13b. We can see that, within the limits of the scatter, the two graphs in Fig. 13b have horizontal regions. The first case considered above differs from the present case only in respect of two coefficients: in the former case $a_{23} = a_{35} = 1.1$ whereas in the latter case these coefficients are equal to 1.2. This is sufficient to alter the situation to the extent that in the first case the problem is soluble but in the second it is insoluble.

Third Case

The coefficients a_{ji} and the band profiles, i.e., the functions $\varphi_i(\lambda)$, are the same as in the first case but the amplitudes differ quite strongly: the central band is approximately four times as strong as its two neighbors and these neighbors are three and five times as strong as the outer bands. The value of σ is assumed to be variable. In the region where the intensity is more than 10% of the maximum we have $\sigma = 0.25\%$ of the maximum intensity. Outside this region the absolute value of σ decreases by a factor of 3 and when the intensity is less than 1% of the maximum value, σ decreases again (this time by a factor of 3). The meaning of this reduction in the error is this: when the intensity is reduced by a factor of 10, we are using a different sensitivity scale of the spectroscopic apparatus. This reduces the absolute value of the error but not by the same factor by which the sensitivity increases because strays, noise, etc., become more important. The three original spectra obtained for this case are shown in Fig. 14. They differ so little that the scale along the ordinate has to be altered in order to exhibit the differences clearly. The bands obtained as a result of analysis of these spectra are shown in Fig. 15, where they are compared with the true band profiles. We can see that the agreement be-

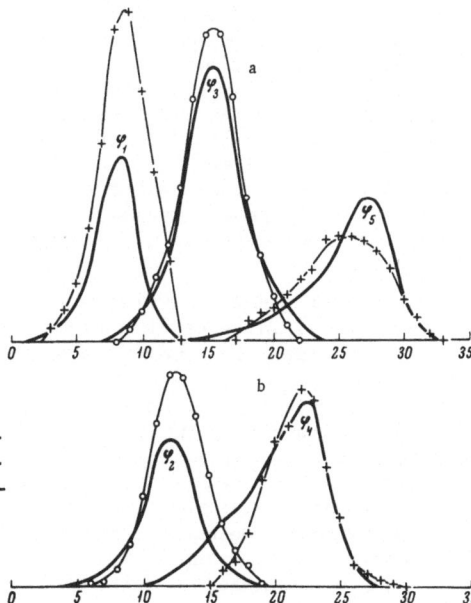

Fig. 15. Comparison of the profiles and amplitudes of the bands obtained in the third numerical experiment with the true profiles. For convenience the bands are shown in two parts and on different scales so as to ensure that their amplitudes are comparable.

tween the calculated and the true profiles is now poorer but the situation is more difficult: the amplitudes of the outer two bands are only 5-7% of the central band and these outer bands are practically hidden in the background of the other bands. The predominance of the central band is responsible for the horizontal region in the graph representing the ratio of the spectra. At the outer edges the horizontal regions practically disappear because of the large scatter of the points. Therefore, it is necessary to follow the band elimination sequence shown in Fig. 9a. This affects the precision with which the band profiles can be determined. The position of the central band and its half-width are practically identical with the true values but the amplitude of this band differs by 12% from the correct value. The second and the fourth bands have "lost" the parts located within the region corresponding to the maximum of the central band. This has occurred because the constancy of the ratio of the spectra in this region suggests misleadingly that the combined intensity is entirely due to the central band. The poorest results are obtained for the weakest band $\varphi_5(\lambda)$ whose maximum is shifted by a quarter of its width. The "wings" of the second, third, and fourth bands are much narrower than the true wings (Fig. 15). This systematic error has been mentioned in considering the first case above.

Discussion of the Results of Numerical Experiments

One of the aims of our numerical experiments was to find the conditions under which a spectrum could be still separated into its individual bands. Therefore, we deliberately selected difficult cases. In practice one often encounters easier situations in which the coefficients a_{ji} of the original spectra differ more than in our three cases. Moreover, the bands may overlap to a lesser extent. In either case it would be possible to separate the spectra into individual bands even if the precision of the measurements were lower.

However, we must bear in mind that the errors in the separation of spectra into their components arises not only because of the lack of precision in the measurements. There are at least two additional sources of error. The first of them is the approximate nature of the calculations. In each arithmetical operation we introduce an error related to the number of significant places to which this operation is carried out. The number of operations which have to be performed on a given spectrum runs into tens and some of them involve subtraction of two comparable quantities. Therefore, one must stress the great importance of the need to work with a sufficient number of significant figures. The experimental errors usually amount to a few units in the third or even fourth significant place. Therefore, in order not to lose the experimental accuracy all the calculations must be performed to at least four significant figures and it is best to use five such figures because repeated successive operations may result in an accumulation of large rounding-off errors. In practice, this means that when the original spectra do not differ greatly from one another, i.e., when the differences between the coefficients a_{ji} are small, and the spectra consist of more than two bands, it is not permissible to use a slide rule.

An additional source of error is related to the determination of the ordinate of the horizontal part of the graph representing the ratio of the spectra. This error results from the scatter of the points in the horizintal regions and the consequent lack of definition of these regions. The reliability can be increased by making horizontal regions as long as possible so that the ordinates of large numbers of points can be averaged out. However, this automatically results in inclusion of the points that are systematically different because of the presence of a second band (this systematic difference is difficult to detect because of the large scatter). These points are useful in the sense that we can determine whether a horizontal region does exist but inclusion of these points may give rise to a systematic error particularly when the horizontal region is in the middle of a spectrum and is in the form of a flat-topped extremum. If the interval in which the ratio is assumed to be constant is extended too much, we find that the average ordinate of the horizontal region is understimated in the case of a spectral maxi-

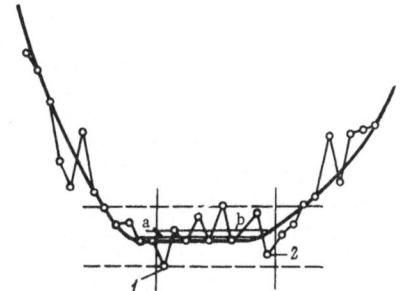

Fig. 16. Recommended method for finding
the limits of a horizontal region in the graph
of the ratio of the spectra. The thick continuous curve is the ratio of the true spectra, which cannot be found experimentally.
The points represent the "experimental"
values of the ratio of the spectra. Numbers
1 and 2 are used to denote the limits of the
horizontal region found by the recommended
method. The dashed horizontal lines are the
boundaries of the interval on the ordinate in
which the points corresponding to the horizontal region are located. The continuous
horizontal lines are the levels of the horizontal region found by averaging the ordinates
of a) all the points from 1 to 2, inclusive,
and b) all the points lying between the dashed
horizontal lines.

mum and overestimated in the case of a spectral minimum. Therefore, it is best to use the
smallest number of points in the determination of the average ordinate of the horizontal part:
if this is done the random scatter introduces smaller errors than the systematic deviation because in subsequent operations the systematic errors are cumulated and the random scatter is
averaged out. In our calculations we usually employed the following empirical rule: "the horizontal region ends where the experimental points deviate systematically, even though this deviation lies within the limits of the scatter." In other words, the horizontal region is considered
as bounded by horizontal lines and not vertical ones.

The continuous curve in Fig. 16 represents the ratio of the spectra obtained ignoring the
scatter of the points, and the points plotted in this figure deviate at random from the curve with
an rms scatter of $\sigma = 0.5$ relative units. To the left of the point denoted by 1 we have six other
points which show clearly a systematic deviation. This is also true of the region to the right
of point 2. Therefore, points 1 and 2 should be regarded as the ends of the interval which corresponds to the horizontal part of the graph representing the ratio of two spectra. We can see
that the horizontal region found in this way is shifted slightly to the right relative to the true
region. However, there is no objective criterion for a more accurate determination of the
limits of this region. If we simply use the scatter of the points in the vertical direction we
must include the whole of that region which is enclosed between horizontal lines in Fig. 16. We
can see that this region includes parts which are definitely not horizontal. If we average out the
ordinates of all the points in the narrower region in accordance with the rule formulated above,

we obtain a value which is only 0.35σ different from the true value. A better agreement cannot be expected because the arithmetic mean of ten measurements of the same quantity differs, according to the probability theory, from the true value by an amount whose rms scatter is 0.32σ, where σ is the rms deviation of a single measurement. However, if we average the ordinates of all the points lying between the horizontal lines, we find that the level obtained in this way differs from the true value by 0.56σ. This demonstrates that the rule we have adopted in finding the ordinate of the horizontal part gives more accurate results.

Thus, our numerical experiments show that the Alentsev method for the separation of complex spectra into individual bands is valid even when the bands overlap strongly and the spectra differ little from one another.

Literature Cited

1. V. G. Tyazhelova, Zh. Prikl. Spektrosk., 10:22 (1969).
2. Y. Uehara, J. Chem. Phys., 50:961 (1969).
3. I. I. Antipova-Karataeva, S. F. Arkhipova, and B. N. Grechushnikov, Zh. Prikl. Spektrosk., 10:480 (1969).
4. M. B. Fok, Zh. Prikl. Spektrosk., 11:926 (1969).

APPLICATION OF THE GENERALIZED ALENTSEV METHOD IN ANALYSIS OF THE BLUE LUMINESCENCE SPECTRUM OF ZnS

E. E. Bukke, T. I. Voznesenskaya, N. P. Golubeva,
N. A. Gorbacheva, Z. P. Ilyukhina,
E. I. Panasyuk, and M. V. Fok

An investigation was made of the blue luminescence spectra of a large number of ZnS and ZnS:Cl single crystals, films, and powders. An analysis of these spectra by the generalized Alentsev method indicated that they consist of four bands differing only in the positions of the maxima. This analysis led the authors to reject the hypothesis that the luminescence of ZnS is due to a random distribution of donor—acceptor pairs.

The Alentsev method for the separation of complex spectra into component bands was used in a study of the blue liminescence of zinc sulfide samples containing chlorine and those in which no chlorine was introduced deliberately [1].

The blue luminescence band of zinc sulfide is complex because its profile depends on the excitation conditions [2]. There is no agreement about the nature of the luminescence centers responsible for the blue luminescence band. It has been suggested that these centers may include vacancies and interstitial atoms. According to some workers, the luminescence centers include chlorine and the blue luminescence of "pure" zinc sulfide is due to accidental traces of chlorine. Other workers are of the opinion that chlorine atoms do not participate in the luminescence. These contradictions may arise from the existence of several types of center responsible for the blue luminescence of zinc sulfide, which may give rise to closely spaced bands that are not normally separated.

The complex nature of the blue luminescence of zinc sulfide is supported also by our preliminary experiments. A comparison of the luminescence spectra of films prepared under different conditions indicated that these spectra can differ quite considerably even within the blue part of the spectrum. This suggested the presence of several types of center responsible for the dark and light blue luminescence of zinc sulfide. On the other hand, Era, Shionoya, Washizawa, and Ohmatsu [3] are of the opinion that this luminescence is due to recombination between impurities in a random distribution of donor—acceptor pairs and that each of these pairs gives rise to its own luminescence spectrum which depends on the distance between the donor and acceptor and is therefore slightly shifted compared with the spectra of the other pairs. Such spectra have been encountered in some $A^{III}B^V$ compounds and they consist of series of lines which merge in the long-wavelength part of the spectrum to form a wide band.

According to Era et al., the wide blue luminescence band of zinc sulfide is complex although there are no lines at its short-wavelength edge. Era et al. report observations according to which the maximum of the blue band shifts in the direction of longer wavelengths when the excitation intensity is reduced or when the luminescence decays after the end of excitation. This is in agreement with the assumption of a random quasicontinuous (in respect of the donor—acceptor distance) of impurity pairs because the pairs with large donor—acceptor distances emit longer wavelengths and are characterized by lower probabilities of interimpurity recombination than the pairs in which the donor and acceptor are close to one another. Therefore, when the excitation density is reduced, the relative importance of the former pairs increases and this results in a shift of the whole liminescence spectrum. However, the shift of the maximum cannot be regarded as a proof that a given band is due to donor—acceptor pairs. If a band consists of two strongly overlapping components, which are due to luminescence centers of different nature and properties, the position of the maximum in the combined luminescence spectrum is a function of the relative intensities of the individual bands and it varies with the excitation density and during decay. In particular, if the band located at shorter wavelengths has a faster response, the maximum of the combined band will shift in the same direction as that observed in the case of bands due to donor-acceptor pairs. An elementary numerical example shows that, in this case, the width of the band changes very little. Figure 1 shows the shift of a spectral band consisting of two Gaussian components. The shift is due to a change in the contributions of the components to the total band. The shift and the associated change in the half-width are of the same order of magnitude as those observed by Era et al. This discussion demonstrates that the attribution of the blue luminescence of zinc sulfide to a random set of donor—acceptor pairs cannot be regarded as experimentally established.

According to the hypothesis of a random distribution of donor—acceptor pairs, a given crystal contains tens if not hundreds of types of pair which have different donor—acceptor distances and, consequently, have luminescence bands located at different wavelengths. If the blue luminescence of zinc sulfide were indeed due to such a set of pairs, it would be practically impossible to analyze the corresponding spectrum into individual bands because one would have to record spectra under as many sets of excitation conditions as there are individual bands, i.e., at least under tens of different conditions. Moreover, all the spectra obtained in this way should differ significantly. The precision which would be needed in recording the spectra needed in such an analysis would be truly fantastic.

In view of this situation, we approached the problem in a different way. We used several different forms of zinc sulfide, which included single crystals prepared under different conditions, films produced by condensation from the paper phase, and powder phosphors. We measured the spectra emitted by each type of sample under different excitation conditions and we separated the spectra into the individual components. Since the conditions of preparation were very different, the samples should have contained different sets of donor—acceptor pairs. These sets should have differed in respect of the donor—acceptor distance in the pairs whose concentration was highest. Therefore, the "individual bands" obtained by analyzing the spectrum should be different for each sample. On the other hand, if the luminescence were due to not a set of donor—acceptor pairs but to luminescence centers of just a few types, the bands ob-

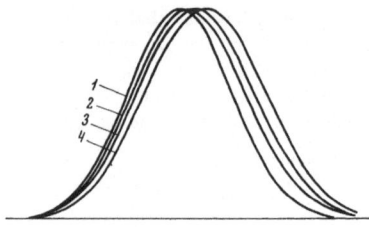

Fig. 1. Positions and profiles of a band, which is a combination of two Gaussian curves, plotted for different relative intensities of the component bands. Curves 1-4 correspond to intensity ratios 1:0, 1:0.25; 1:0.5, 1:1.

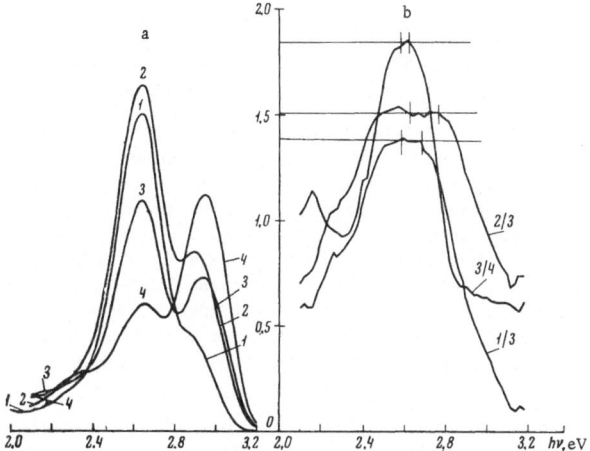

Fig. 2. a) Luminescence spectra of crystal No. 1 (without correction for the spectral sensitivity of the instrument) and b) ratio of the ordinates of the spectra of crystal No. 1. (a): 1) Excitation with λ = 366 nm; 2) excitation with λ = 366 nm + λ = 313 nm (passed through a UFS-6 filter); 3) excitation with λ = 313 nm (relative intensity 1); 4) excitation with λ = 313 nm (relative intensity 5). b): The numbers alongside the curves represent the spectra whose ordinates were divided to obtain the ratios. The vertical lines indicate the ends of the horizontal regions over which the ratio was averaged to obtain the normalization coefficients. The positions of the horizontal lines correspond to the coefficients found in this way.

Fig. 3. Examples of separation of the spectra. a) Results of elimination of a band located at 2.6 eV. The numbers alongside the curves represent the spectra which were subtracted from one another (the curves are plotted on different vertical scales in order to fit them in the same figure). b) Ratio of band 1–3 to band 4–3. The continuous straight line represents the average value of the ratio in the horizontal region; the dashed lines are shifted relative to this line by 2%.

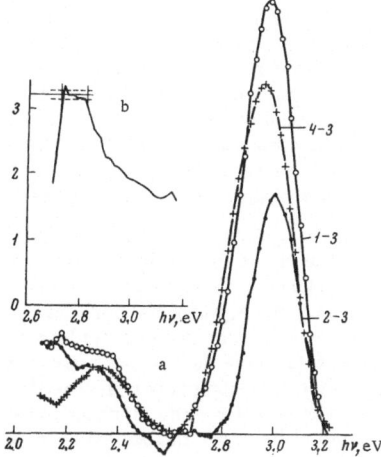

tained in all cases should be the same. In this case, the luminescence spectra of all this great variety of samples should be resolvable into a small number of different bands and the bands could then be regarded as truly individual, i.e., they could each be associated with a definite type of luminescence center differing strongly in its structure from the other centers.

The overlap of the bands was reduced by carrying out all the measurements at liquid nitrogen temperature since at this temperature the bands were normally much narrower. The samples were bonded with Wood's alloy to the lateral surface of a copper can attached to one end of a Textolite tube placed inside a Dewar flask. The "stray" luminescence was reduced in intensity by using a flask made of high-quality vitreous silica. Liquid nitrogen was poured only into the copper can. In such a thermostat the temperature of the sample differed by not more than 3-5 deg from that of liquid nitrogen. This arrangement avoided errors resulting from the presence of boiling nitrogen bubbles which would have disturbed measurements if the sample were immersed in liquid nitrogen. The luminescence was excited and observed directly through the walls of the flask because no special windows were needed in view of the high quality of vitreous silica.

The luminescence was excited with a PRK-4 mercury lamp whose emission was stabilized by an electronic unit. The unit was controlled by a P-209A high-power triode which shunted part of the ballast resistance in the lamp circuit. The stabilizer maintained the light flux constant to within 0.2%. The λ = 313, 366, and 405 nm mercury lines were selected by optical filters specially designed for these lines. The luminescence was passed through a UM-2 monochromator to an FÉU-51 photomultiplier which was supplied from a stabilized VS-16 rectifier. The photocurrent was amplified with a balanced dc amplifier in which 6Zh1Zh tubes were used. The amplifier was operated in the electrometric mode and its output was matched to the input of an ÉPP-09 automatic recorder with a balanced cathode follower based on a 6N15P tube. The filament and the anode circuits of the amplifier and cathode follower were supplied with power from well-stabilized rectifiers.

The null-point drift at the output of the apparatus did not exceed 0.2 mV/h after heating for one hour (the full-scale deflection of the automatic recorder corresponded to 10 mV). The maximum noise amplitude was about 0.05 mV. Photocurrents from 1×10^{-11} A upward could be recorded for input resistances of 10^9 Ω. The influence of noise and of the drift of the null point was reduced by recording the spectra at points separated by intervals of 0.02 eV and at each point the photocurrent was recorded for at least 30 sec. In the intervals between the points the photomultiplier was screened from the luminescence light flux and the dark current was recorded. During this time the monochromator was tuned to the next wavelength. This procedure was equivalent to continuous recording with a time constant of 30 sec but much less time was needed because there was no need to wait for the end to transient processes. Moreover, this procedure enabled us to check regularly the dark current. All this increased the precision with which the spectra were recorded. The relative rms deviation of each measurement did not exceed 0.3%, with the exception of the edges of the spectra, where the error could be considerably higher because the absolute values of the readings were small.

All the calculations necessary for the separation of the spectra into individual bands were carried out directly on the readings taken from the recorder chart without allowance for the spectral dependence of the sensitivity of the apparatus. The correction for the sensivity was made in the final stage, when the individual bands were already separated. Consequently, the random scatter which was necessarily introduced by the sensitivity correction did not reduce the precision with which the spectra were determined and this raised the accuracy of the computations. In the preceding paper [4] Fok showed that the error resulting from the random scatter of the experimental results increased at each stage in the elimination of an individual band. Therefore, the same error introduced at the end of the band-elimination procedure had much less influence on the final result than the error applied in the data used in the calculations. In the description given below we shall always understand the "spectra" and "bands" to be the results obtained without correction for the spectral dependence of the sensitivity of the apparatus, unless stated to the contrary. Our calculations were carried out to the fifth significant place (four places represented the precision of our measurements and the fifth was a margin of safety necessary to ensure that the error due to rounding-off did not distort the results).

All the spectra obtained in the present investigation covered the whole visible region and sometimes extended even to the ultraviolet or infrared wavelengths. However, the luminescence maxima of all the crystals subjected to a detailed study were located in the blue region. The existence of luminescence bands overlapping those which we investigated made the band-elimination process very difficult because we could never be sure that there was only one band at the edge of the spectrum. Nevertheless, in some cases, we were able to complete the analysis of the spectrum into its individual components. In most cases, we were unable to vary the excitation conditions sufficiently to obtain the necessary number of significantly different luminescence spectra of the same sample and it was therefore necessary to use the spectra of groups of two or three samples prepared under different conditions.

We shall now consider in detail the procedure used in the separation of the spectra into their components. Figure 2a shows the luminescence spectra of a single crystal No. 1 (Appendix) whose spectrum we were able to separate completely. It is evident from Fig. 2 that the spectra obtained depended strongly on the excitation conditions. They consisted of two easily distinguishable bands in the blue part of the spectrum but the luminescence intensity did not fall to zero in the ultraviolet and yellow parts of the spectrum. Figure 2b shows the graphs of the ratios of these spectra. The random scatter of the results is stressed by plotting these graphs in the form of broken lines passing exactly through all the experimental points. It is clear from Fig. 2b that all the three graphs have a wider or narrower horizontal region near $h\nu = 2.6$ eV. In all three cases, this region is located at practically the same photon energy and, therefore, its existence can be assumed to be established reliably even in the case when it extends over just two points.* Apart from this horizontal region, all three graphs have an additional but not very pronounced horizontal region at the short-wavelength end. However, the scatter of the experimental points in the second region is so large that the value of the normalization coefficient can be calculated only very approximately. A careful examination of this figure shows also that at least one other band must exist between the two horizontal regions. This follows from the observation that the ratio of the spectra varies in a complex manner in the central region and that the variation is different for different pairs of spectra (the graph showing the ratio of the spectra 3 and 4 intersects twice the corresponding graph of the ratio of the spectra of 1 and 3).

Figure 3a shows the "spectra" obtained as a result of elimination of a band located near 2.6 eV. This figure supports strongly our hypothesis of the complex nature of the spectrum in the region where the photon energy is higher than 2.6 eV. We can see that the bands located in this region vary in their widths and positions. This means that there are at least two separate bands in this region. An even more complex spectrum is observed in the $h\nu < 2.6$ eV region. Figure 3b shows the graph of the ratio of the two wider bands of Fig. 3a. This ratio has a horizontal ledge near 2.8 eV and a tendency to reach a constant level at the short-wavelength end (the points corresponding to the long-wavelength end are omitted simply because the ordinates used in the ratio are small and known only with a large error).

---

*The width of the horizontal region and its exact position naturally depend on the relative intensities of the neighboring elementary bands because, in practically all cases, the region extends to parts of the spectrum where the neighboring bands make a contribution although this contribution may be small compared with the random scatter.

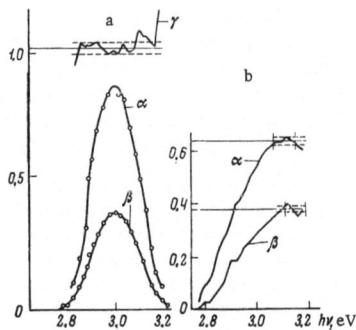

Fig. 4. Examples of separation
of the spectra. a) Comparison of
the difference between bands 1-3
and 4-3 and band 2-3 in Fig. 3a:
α is the band obtained by subtracting band 4-3 from band 1-3; β is
band 2-3 of Fig. 3a; γ is the ratio
of bands α and β. b) Ratio of
band I to band 4-3 (curve α) and
to band 1-3 (curve β). The continuous straight lines give the
average values of the ratio in the
horizontal region and the dashed
lines are displaced by 2%.

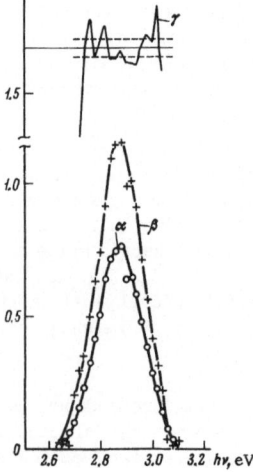

Fig. 5. Comparison of the bands
obtained by elimination of band I
from bands 4-3 and 1-3; α) individual band deduced from band
4-3; β) individual band deduced
from band 1-3; ratio of bands α
and β.

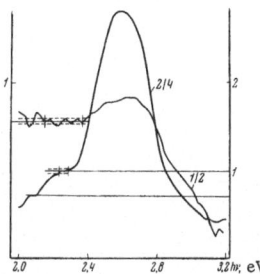

Fig. 6. Graphs showin the ratios
of the spectra with definite horizontal regions on the long-wavelength side. The numbers alongside the curves identify the spectra whose ordinates are divided
by one another. The vertical lines
denote the regions in which averaging was carried out in order to
determine the normalization coefficients. The positions of the continuous horizontal lines represent
the coefficients found in this way
and the dashed lines are displaced
by 2%.

We used the more pronounced horizontal region and eliminated the band at longer wavelengths. In this way, we obtained the band shown in Fig. 4a. This figure also includes the narrower of the short-wavelength bands of Fig. 3a as well as a graph of their ratio. We can see
that the ratio is constant (within the limits of relatively small scatter) everywhere with the exception of the band edges where the precision is very low. Since this band was deduced by two
methods, we can regard it as a simple component, i.e., a band which represents luminescence
centers of just one type. We shall call it band I. Figure 4b shows the graphs of the ratio of this
band to the wider bands of Fig. 3b. At the short-wavelength end we again observe horizontal
regions but, in this case, the scatter is considerably greater because of the low precision of the

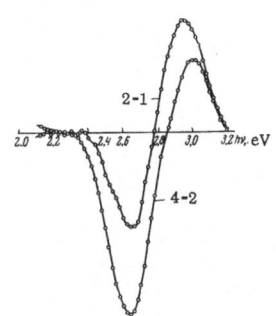

Fig. 7. Results of elimination
of a band located near 2.3 eV.
The numbers alongside the
bands indicate the spectra subtracted from one another.

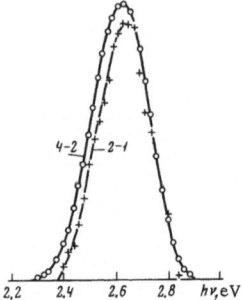

Fig. 8. Comparison of two bands deduced from the "spectra" in Fig. 7 after elimination of bands I and II. The curves are designated in the same way as in Fig. 7.

initial measurements (in this region the luminescence intensity was low). Figure 5 shows the bands obtained by elimination of band I from the two wide short-wavelength bands shown in Fig. 3a. We can see that these bands have the same profiles. Therefore, we may assume that they belong to the same type of center. We shall use the designation II for the band corresponding to these centers. The large scatter in the value of the ratio is due to the fact that the bands are obtained as a result of double and triple subtraction of the spectra.

In order to separate the bands with longer wavelengths it is best to start again from the initial spectra but use them in a different combination. Figure 6 shows the graphs of the ratio of the spectra 2 and 4 and the spectra 1 and 2. These graphs have horizontal regions at long wavelengths and, in one case, there are even two such regions. This means that there are at least two separate bands in the $h\nu < 2.3$ eV range. However, we ignored them initially and simply eliminated them from the spectra using the horizontal regions bounded by vertical lines. Such elimination produced the spectra shown in Fig. 7. Next, we subtracted bands I and II from these spectra and obtained the bands shown in Fig. 8. For convenience, we altered the sign and selected the ratio of the scales in such a way that the right-hand edges of the bands coincided. We found that once again the spectra differed considerably so that they must have consisted of at least two individual bands. These bands are shown in Fig. 9. The band located at the longest wavelengths was obtained least accurately because the bands from which it was deduced differed only a little from one another. Therefore, it was necessary to average out the positions of three neighboring points and to draw curves through these average points. The curve obtained in this way was subtracted from the band 4-2 in Fig. 8 in order to isolate the second band. Therefore, the long-wavelength edge of this band was not determined very accurately.

Thus, the spectrum of the blue luminescence of the single crystal under investigation was resolved into four bands and only one showed a considerable scatter. There were strong reasons for believing that bands I and II were elementary. However, this could not be said of the

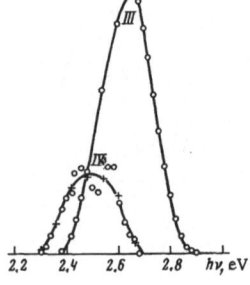

Fig. 9. Postulated individual bands into which the bands of Fig. 8 can be resolved. The circles represent results of subtraction and the crosses represent the result of averaging of three neighboring points.

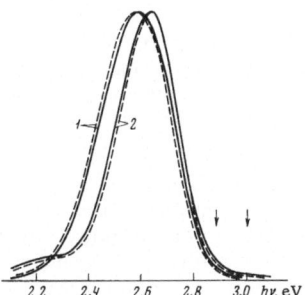

Fig. 10. Luminescence spectra
of one pair of single crystals
(Nos. 3 and 4) used to isolate
two long-wavelength individual
bands: 1) crystal containing
chlorine; 2) crystal free of
chlorine. The continuous curves
were obtained by excitation with
λ = 313 nm and the dashed curves
by excitation with λ = 366 nm.
The vertical arrows indicate the
positions of maxima of bands I
and II.

other two bands (III and IV) because these were obtained in just one way. Bands III and IV were
observed also in other cases when we had to use spectra of several samples. We investigated
four pairs of combinations of five single crystals and in each pair one crystal contained chlorine introduced deliberately, whereas the other crystals was not doped with chlorine. The
spectra of these crystals differed strongly in the relative intensity in the region where band IV
was earlier found to have a large scatter (Fig. 10). In this way, we were able to separate this
band with a greater accuracy. Moreover, we also obtained band III more precisely. It was found
that all four combination of single crystals yielded the same pair of bands and these differed
little from the corresponding bands deduced from the spectra of single crystal No. 1. Fig. 11
shows the bands deduced from the spectra of single crystal No. 1 as well as the bands deduced
from the spectra of one of the four pairs of single crystals. We can see that the positions of
the bands coincide although the profiles differ somewhat. The differences cannot be significant
because of the poor accuracy with which band IV was obtained. Therefore, we concluded that
the two bands in Fig. 11 were due to specific luminescence centers and we called them band III

Fig. 11. Comparison of the
long-wavelength bands deduced
from the spectra of single crystal No. 1 (dashed curves) with
the bands deduced from the
spectra of single crystals Nos.
2 and 4 (continuous curves).

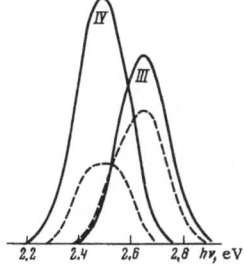

and band IV. Band II in the spectra of pairs of single crystals was weak and could be deduced only with low precision. However, within the limits of this precision it coincided with band II deduced from the spectra of single crystal No. 1. Band I was almost absent from the spectra of the pairs of single crystals and could not be isolated reliably. All that we could say was that some weak band was located in the region where band I was found in the spectrum of single crystal No. 1.

Thus, in all the cases under consideration, we were dealing with the same set of four individual bands which differed only in their intensity. The spectra under consideration also included bands at longer wavelengths. These were manifested strongly in the afterglow because the short-wavelength bands decayed faster. However, we did not study these bands because it would have been very difficult to separate them from the spectrum in view of their low intensity. In some cases, we did use the afterglow spectra in order to eliminate a green band from the initial spectra. The error due to our inaccurate knowledge of the green band profile was slight because the amplitude of this band was small compared with the bands at shorter wavelengths.

We also studied the spectra of powder phosphors as well as the spectra of films condensed on hot substrates. We found that, in both cases, the spectra had stronger green bands and, therefore, it was difficult to separate them out into individual bands. In the case of powders, we used the results of our investigation of single crystals and we reduced the random error by employing the average profiles of bands I-IV found by combining the results of several methods. This made it easier to resolve the spectrum into known bands and the problem could be solved completely. The results of such a resolution are presented in Fig. 12 which also shows the difference between the initial spectrum and the sum of the bands into which it was decomposed. For the sake of clarity, this difference is plotted on a larger scale. In fact, the difference does not amount to more than 2%. In order to determine whether our resolution of the spectrum into individual bands was reliable, we expanded the same spectrum into a system of bands in which bands I, II, IV, and the short-wavelength edge of the green band were the same as before, whereas band III was shifted by 0.01 eV, i.e., by about 4% of its half-width. We found that, in this case, the difference between the sum of the bands and the initial spectrum was much larger

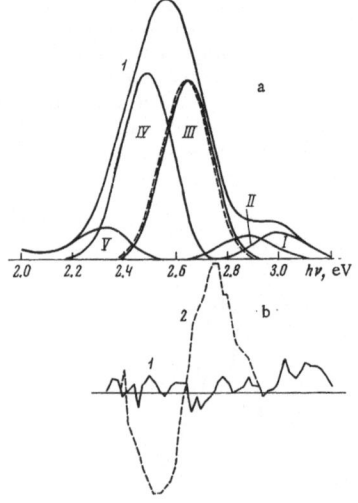

Fig. 12. Separation of the spectra. a) Separation of the spectrum of powder phosphor No. 7 into individual bands: 1) initial spectrum; I)-IV) individual bands; V) green band known inaccurately; the dashed curve represents band III shifted by 0.01 eV. b) Difference between the initial spectrum and the sum of individual bands: 1) individual bands obtained from the spectra of single crystals; 2) band III shifted by 0.01 eV. The scale in Fig. 12b is ten times as large as the scale in Fig. 12a.

and it was systematic. The error of 2% represented the precision of our measurements: 0.3% was contributed by the initial spectrum and the rest by the individual bands because they were obtained as a result of repeated subtraction and, consequently, the errors associated with the individual bands were several times larger. We concluded that the blue luminiscence spectra of powder phosphors consisted also of four bands.

The spectra of films were also studied by a different method: the spectrum of one of the films was resolved into the known bands and the spectra of three other films (recorded under different excitation conditions) were analyzed by the Alentsev method. All these spectra were weak at short wavelengths. We were able to isolate only the bands which we revealed earlier as III and IV. We could not separate bands I and II: instead, we obtained one wider band with a maximum shifted somewhat toward longer wavelengths relative to band II. It was not possible to determine whether this was a single band and whether the observed shift was real. This was because the precision with which the spectra of the films were measured was considerably less than that attained in other measurements and the intensity of the luminescence in the relevant part of the spectrum was low. Ignoring for the moment the usually weak short-wavelength bands, we concluded that the strongest bands (III and IV) were always the same. This was in conflict with the hypothesis that the blue luminescence of zinc sulfide was due to a random set of donor—acceptor pairs. On the other hand, it did not follow from our experiments that the blue lumi-nescence centers could not be donor—acceptor pairs. It was quite likely that they were pairs in which donors and acceptors were separated by very short or the shortest possible distances. The difference between the energies of the Coulomb interaction between donors and acceptors in the case of these two types of pair was of the order of a few tenths of an electron volt and, consequently, they could be separated from one another by the Alentsev method because they gave rise to individual bands. The absence of the luminescence of the pairs with large donor—acceptor distances could be due to two reasons.

First, according to the well-known principle of charge compensation on a microscopic scale, the donors and acceptors are frequently associated with one another because, during the preparation of crystals at high temperatures they are almost all ionized and can therefore attract each other. The degree of association of donors and acceptors increases during cooling because an appreciable mobility of the ions is also retained at lower temperatures.

Secondly, the theory of recombination within impurity pairs predicts an exponential fall of the rate of such recombination with reduction in the overlap of the wave functions of the donor and acceptor. At any given distance between the donor and acceptor the overlap is governed by the radii of the Bohr orbits of an electron and a hole which increases as the donor and acceptor levels become shallower. Therefore, carrier recombination between shallow donors and ac-ceptors occurs quite easily but if these impurities form deep levels the recombination in pairs with longer donor—acceptor distances becomes highly unlikely because the wave functions of the donor and acceptor fall off rapidly even at distances of the order of the atomic spacing. Both these factors reduce the rate of recombination in pairs of impurities in which the donor—accep-tor distances are large and, when the two factors act together, luminescence of this kind may be completely negligible.

Our conclusion that bands I and II are also simple bands is less convincing because, in one case, instead of these two bands, we obtain a single band. However, since bands I and II are ob-tained for single crystals and for powders, we may conclude that they belong to specific centers and not to a random set of donor—acceptor pairs. This is also supported by the profiles of all the bands we obtained, which are shown in Fig. 13. In this figure we plotted not the readings taken from the recorder chart but the number of quanta per unit frequency interval. We can see that all the bands have the same half-width and occur in pairs. Within each pair the distance be-tween the maxima is 0.16 eV, whereas the two pairs are separated by 0.41 eV. Such a regular distribution of bands can hardly be accidental. It is more likely that the bands are of the same origin.

Fig. 13. Profiles of individual bands
in the spectrum of blue luminescence
of zinc sulfice plotted in the coordinates (N_ν, hν). The dashed curves
represent the region obtained with
lower precision.

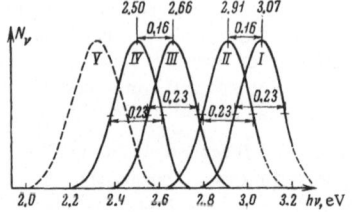

However, it is worth pointing out that the behavior of these bands differs somewhat. For example, bands I and II appear more easily when the crystals are prepared in a reducing atmosphere, whereas band IV is observed more easily in crystals containing chlorine. In view of the absence of sufficiently reliable and convincing data, we shall not make any hypotheses on the origin of the centers responsible for the various bands. This could be done only by comparing the spectra of crystals containing not only chlorine but also other coactivators. Moreover, it would be necessary to determine whether there were any such pairs of bands in the long-wavelength part of the spectrum and how the structure of the crystal of the crystal lattice affects these bands. It would be reasonable to carry out all these experiments only if the spectra were determined to at least 0.3% (as in our experiments), and an even higher precision would be desirable.

In conclusion, we must mention that since all our investigations were carried out at liquid nitrogen temperature, the conclusions apply only to this temperature. However, since the probability of recombination in donor—acceptor pairs is a weaker function of the temperature than the probability of thermal liberation of electrons and holes, it follows that at higher temperatures the likelihood of recombination in pairs with larger donor—acceptor separations can only be smaller since the charges localized at these impurities will be liberated thermally before charges of the opposite sign can reach them by tunneling. On the other hand, if the temperature is lowered we are likely to reach a situation in which the probability of thermal liberation of charges will become less than the probability of recombination in pairs characterized by large donor—acceptor distances. Then, if the hopping conduction is weak, the contribution of such pairs to the luminescence spectrum will increase. However, under all conditions the relative contribution of the pairs with large donor—acceptor distances cannot be greater than the value obtained by multiplying the ratio of the concentration of such pairs to the concentration of pairs with short donor—acceptor distances by the ratio of the corresponding recombination probabilities. If our conclusions relating to the association of donors and acceptors and the small radii of the Bohr orbits are correct, these two ratios must be much smaller than unity. Therefore, even under the most favorable excitation conditions we can hardly expect strong blue luminescence of zinc sulfide due to carrier recombination in a random set of donor—acceptor pairs.

The question now arises as to under what conditions the luminescence associated with the recombination in pairs with large donor—acceptor distances can appear. We have mentioned that such recombination may occur if electrons and holes have large-radius hydrogen-like orbits and this requires that at least the following two conditions be satisfied simultaneously: a) the energy levels of donors and acceptors should be shallow and b) the electrons and holes should be in an unpolarized state. We can easily see that neither of these two conditions is satisfied by the blue-luminescence centers. In fact, the short-wavelength edge of the band located at the shortest wavelengths lies at about 3.3 eV, whereas the optical width of the forbidden band of zinc sulfide at liquid nitrogen temperature is at least 3.7 eV. Thus, at least 0.4 eV is converted into heat in each recombination event in such centers. If this energy is evolved at the moment of charge localization, the levels at which such localization occurs cannot be regarded as shallow, i.e., the first condition is not satisfied. However, if the losses just mentioned arise on polarizationof the crystal lattice by the recombining charges, it means thatthese charges are in the polaron state, i.e., the second condition is not satisfied. If both conditions are satisfied, the energy lost as a result of recombination cannot be large. This means that the energy of the emitted quanta must be close to the forbidden band width, i.e., the luminescence bands in question must lie in the ultraviolet part of the spectrum.

APPENDIX

List of Samples Investigated

a) Single Crystals Grown from the Vapor Phase in Closed (But Not Sealed)

Ampoules by the Grillot Method

No. 1, a "pure" ZnS single crystal grown at T = 1360°C in 27 h;* a slow stream of NH_3 was used in the outer tube;

No. 2, a "pure" ZnS crystal grown at T = 1380°C in 25 h; an H_2S stream was used in the outer tube.

b) Single Crystals Grown by the Sublimation Method in a Stream of Gas

No. 3, a "pure" ZnS crystal grown at T = 1320-1340°C in 4 h in a stream of H_2S;

No. 4, a ZnS:Cl single crystal grown at T = 1270°C in 18 h in a stream of H_2S with traces of HCl;

No. 5, a ZnS:Cl single crystal, grown at T = 1270°C in 17 h in a stream consisting of a mixture of 25% H_2S and 75% HCl;

No. 12, a "pure" ZnS single crystal grown at T = 1340°C in 6 h in a stream of H_2S.

c) Powder Samples

No. 6, a ZnS:Cl powder prepared by heating with an NH_4Cl flux (3%) at T = 900°C applied for 30 min in a stream of argon;

No. 7, a powder of ZnS:Cl into which chlorine was introduced by electrolysis [5] after the original material was heated at T = 900°C for 30 min in a stream of argon; the electrolysis was carried out at T = 900°C applied for 12 min (the current was 5 mA).

d) Films Prepared by Condensation on Substrates in a System

in Which the Residual Pressure was 10^{-5} mm Hg

ZnS, coactivators, and Zn were evaporated from different sources with independently controlled temperatures. The temperature in the condensation zone was about 500°C. The films prepared with an excess of zinc (ZnS−Zn) were deposited on substrates which were first coated with a zinc layer.

No. 8, a ZnS:Cl film (ZnS evaporator temperature 900°C, NH_4Cl evaporator temperature 0°C);

No. 9, a ZnS film (ZnS evaporator temperature 1100°C);

No. 10, ZnS-Zn film (ZnS evaporator temperature 900°C; Zn evaporator temperature 740°C);

*The values of the temperature always refer to the crystal itself.

No. 11, a ZnS:Cl film (ZnS evaporator temperature 900°C, BaCl$_2$ evaporator temperature 320°C).

The single crystals and the films were prepared from ZnS of the phosphor grade made at the "Red Chemist" factory. This material was first deoxidized by heating in a stream of H$_2$S. Powder phosphors were prepared from "specially pure" ZnS made in our laboratory, also deoxidized in a stream of H$_2$S.

The individual bands I and II were separated in their final form from four spectra of sample No. 1. These bands were weak in the other samples but, in some cases, the existence of two bands in the appropriate part of the spectrum could be seen quite clearly. Bands III and IV were established reliably for all the spectra. They were separated in the final form by averaging the bands obtained from the spectra of the following combinations of single-crystal samples: Nos. 3 and 4; Nos. 3 and 5; Nos. 2 and 4; Nos. 12 and 4; Nos. 12 and 5.

Literature Cited

1.   E. E. Bukke, T. I. Voznesenskaya, N. P. Golubeva, N. A. Gorbacheva, Z. P. Kaleeva, E. I. Panasyuk, and M. V. Fok, Zh. Prikl. Spektrosk., 12:1047 (1970).
2.   V. F. Tunitskaya, Zh. Prikl. Spektrosk., 10:1004 (1969).
3.   K. Era, S. Shionoya, and Y. Washizawa, J. Phys. Chem. Solids, 29:1827 (1968); K. Era, S. Shionoya, Y. Washizawa, and H. Ohmatsu, J. Phys. Chem. Solids, 29:1843 (1968).
4.   M. V. Fok, This volume, p. 1.
5.   T. I. Voznesenskaya and M. V. Fok, Opt. Spektrosk., 15:249 (1963).

PREPARATION OF ZINC SULFIDE CRYSTALS AND NATURE OF BLUE LUMINESCENCE CENTERS IN SELF-ACTIVATED ZnS

Z. P. Ilyukhina, E. I. Panasyuk, V. F. Tunitskaya, and T. F. Filina

A description is given of the method used in growing zinc sulfide single crystals of various "types." These high-purity crystals were grown by a sublimation method in a gas stream using two temperature gradients. The luminescence of self-activated ZnS prepared in this way was investigated. It was found that the blue luminescence band was complex. The nature of the luminescence centers responsible for the four individual bands in the blue region was determined.

CHAPTER I

PREPARATION OF CRYSTALS

Introduction

Zinc sulfide is one of the most promising crystalline phosphor materials and, therefore, much work has been done on its luminescence.

The basic task of current investigations is to determine the nature of the centers responsible for any given luminescence band, especially in vies of the disagreements on this subject. The disagreements arise because, for a long time, experimenters have not paid sufficient attention to the control of the preparation conditions and have found it difficult to reproduce these conditions exactly.

When crystalline phosphors are prepared at high temperatures it is essential to make suitable allowances for all the processes occurring in the material being prepared. The nature of gaseous, liquid, and solid substances which come into contact with this material is of great importance. These substances may react with the material under study or they may act as sources of impurities. The growth of crystals at high temperatures in various gaseous media followed by cooling at different rates may produce a great variety of defects due to departures from stoichiometry and these may alter the luminescence emitted by zinc sulfide independently of any changes in the charge used to prepare such crystals.

For any given composition of the initial charge, the activator impurities can be introduced into the ZnS lattice in different ways and they can manifest themselves differently in the luminescence, depending on the ambient medium, heating temperature, cooling rate, presence of defects associated with deviations from stoichiometry, etc. For example, copper-activated crystalline phosphors prepared from the same intial charge may exhibit either predominantly green or predominantly blue luminescence, depending on the crystal-growth temperature and the rate of cooling [1].

The luminescence of any given sample is also affected by the accidental impurities which are usually present in the initial charge. Thus, the luminescence emitted by ZnS will depend also on the degree of contamination of the raw materials used in the preparation of phosphors.

This brief discussion explains the contradictions between the published results and the reason why there is as yet no agreed view on the nature of luminescence of zinc sulfide.

We must obviously start by preparing ZnS samples with a wide range of physical properties and they must be prepared under the same rigidly controlled conditions (ensuring a high degree of purity). They must then be investigated in various ways. It is necessary to consider simultaneously not only the optical and electrical properties of such samples but also to take account of their crystal structure. In this task one must use single crystals of ZnS which ought to satisfy the various requirements in respect of their physical properties as well as in respect of their shape and external appearance. In studies of the optical properties one must have pure and very thin undoped plate-like crystals. Crystals of the same shape but activated with large amounts of impurities are needed in investigations of electroluminescence concentrated in the form of streaks. In studies of photoluminescence, cathodoluminescence, electroluminescence, and electrical conductivity in strong pulsed fields one requires both plate-like and "bulk" crystals, some of which must be "pure" and the others doped with various impurities in different concentrations.

The purpose of our investigation was to develop a method by which we could grow ZnS crystals of different types with highly reproducible optical and electrical characteristics. It was also our aim to determine the nature of the luminescence centers in zinc sulfide.

§1. Selection of the Method

We considered sublimation and hydrothermal methods as well as synthesis from the vapor phase and growth from the melt. We found that, in our case, the most suitable method was the sublimation in a gas stream. The purity of crystals would be ensured by high-temperature recrystallization of zinc sulfide in the gas stream and it would be possible to grow crystals which were not in contact with the walls of the enclosure. In this way, we grew crystals of dimensions sufficient for the investigation of various luminescence properties of zinc sulfide.

Crystals of ZnS and CdS have been grown from the vapor phase by recrystallization of the sulfide in a stream of gas [2, 3], using apparatus in which quartz tubes are placed in a furnace with a definite temperature gradient along its length. A stream of gas or a mixture of gases is supplied to the reaction zone inside a quartz tube. In this method it is possible to prevent oxidation of the ZnS and to employ various atmospheres in which samples exhibiting different types of luminescence can be grown. This method is used primarily to grow pure (undoped) crystals.

Doped crystals are usually grown by a different sublimation method in which ZnS is recrystallized in closed ampoules (the special case of this method is the gas-transport reaction technique). This method is adopted because the growth of doped crystals by recrystallization in a moving stream of gas suffers from very poor reproducibility [3].

However, recrystallization of ZnS in closed ampoules sometimes creates difficulties in respect of the required purity. The "phosphor" grade of zinc sulfide usually contains a relatively large number of impurities which are incorporated into the lattice of ZnS during crystal growth. This is why crystals grown in closed ampoules practically always exhibit green or blue luminescence at room temperature without deliberate doping.

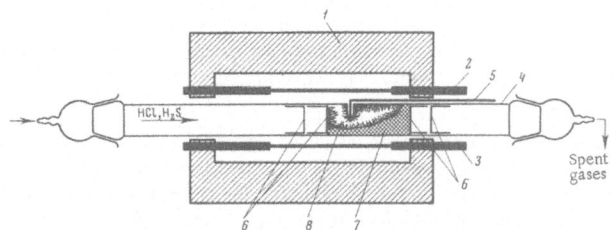

HCl, H₂S

Spent gases

Fig. 1. Schematic diagram of the apparatus used to grow ZnS crystals by sublimation in a stream of gas moving along a tube subjected to two temperature gradients: 1) furnace with Silit heaters; 2), 3) separate heating sections; 4) quartz tube; 5) thermocouple; 6) quartz screens; 7) initial zinc sulfide charge; 8) newly grown crystals.

The variant of the sublimation method used in our investigation has the advantage of a gas stream which removes some of the impurities from the reaction zone and helps to transport the matter into the growth zone so that the growth operation can be performed much more speedily.

§ 2. Methods Used in Preparation of ZnS Crystals of Different "Types"

We grew zinc sulfide crystals in the apparatus shown schematically in Fig. 1.

We carried out preliminary experiments in which we grew crystals at various temperatures, in different temperature gradients, and in various gaseous media (H_2S, H_2, H_2S + HCl, H_2S + H_2, Ar, NH_3) without special additions and with various activators (Cu, MN, In, etc.).

We established that

a) crystals of different shapes (thin plates, "bulk" prisms) were formed at definite temperatures: the "bulk" prisms grew at temperatures above 1300°C in a weak temperature gradient (3-5 deg/cm), whereas plate-like crystals grew at 1000-1100°C in a strong temperature gradient (35-50 deg/cm);

b) the crystallization temperature was found to be independent of the presence of dopants in the crystals being prepared;

c) the rate of transport of zinc sulfide from the sublimation zone to the crystallization region depended on the sublimation temperature (T_{subl}), the temperature gradient, the dimensions of the particles in the initial charge, the weight and composition of this charge, and the rate of flow of the gaseous mixture.

The method of recrystallization in a stream of moving gas enabled us to grow zinc sulfide crystals of different "types" by suitable selection of the technological conditions which ensured sufficient supersaturation with zinc sulfide vapor in the crystallization zone.

The optimal conditions for the growth of "pure" (not specially doped) plate-like ZnS crystals were as follows.

The temperature in the sublimation zone was 1300-1340°C and in the crystallization zone it was 1000-1100°C; the temperature gradient was 30-50 deg/cm; the rate of flow of the gases (H_2S, H_2) was 3-5 cm³/min (p ~ 760 mm Hg).

The starting material was zinc sulfide of the "phosphor" grade obtained from the "Red Chemist" factory. This material was first deoxidized by slow heating in a stream of H_2S from room temperature to 500°C. The same conditions were employed in all cases in order to ensure that the ZnS grains obtained in this way were approximately of the same size (the weight of the charge was 100 g).

When special activating admixtures were introduced into the charge the temperature in the sublimation zone was altered in such a way as to maintain the original rate of transport of zinc sulfide into the crystallization zone. For example, in the growth of crystals doped with copper and manganese, whose sulfides had lower vapor pressures than that of zinc sulfide, we increased T_{subl} by 10-30 deg in the case of Cu and by 20-50 deg in the case of Mn (depending on the impurity concentration), compared with T_{subl} for undoped crystals. The crystallization temperature T_{cr} was kept constant.

The same considerations were applied when the gas medium was changed. For example, when we introduced chlorine into the gas stream ($H_2S + HCl$; $H_2 + HCl$), which produced volatile zinc chlorides by interaction with ZnS, we reduced T_{subl} by 30-50 deg, depending on the concentration of the chlorine.

In spite of the fact that the growth conditions for each "type" of crystal were carefully selected and rigorously maintained, the reproducbility of the properties of the crystals was poor. This was due to the smallness of the growth zones of the "bulk" and plate-like crystals so that the slightest fluctuation in the conditions resulted in the crystallization of zinc sulfide outside the zone where the temperature was maintained rigorously. The reproducibility of the results could be improved by increasing the dimensions of the crystallization zone. This was done by applying not only a temperature gradient along the reaction zone but also a temperature gradient along its height.

The furnace was used (Fig. 1) had two separately controlled sections (lower and upper). This feature and the changes in the position of the reaction zone within the furnace enabled us to vary the horizontal and vertical temperature gradients from 3 to 50 deg/cm.

Figure 2 shows typical isotherms in the reaction zone during the growth of plate-like crystals. The isotherms are shown at intervals of 40 deg. The vertical temperature gradient is 25-30 deg/cm. The crystallization zone is shown shaded. In this case, the reaction zone was half-filled with the initial charge. The vertical temperature gradient altered the shape of the 1100°C crystallization zone in such a way that it became wider along the length of the tube. The plate-like crystals grew best in this crystallization zone.

This change in shape of the crystallization zone improved greatly the reproducibility of the properties of our crystals. This was particularly important when one activator was re-

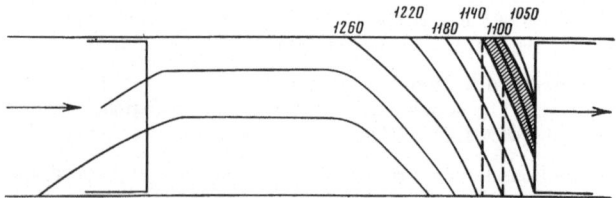

Fig. 2. Isotherms in the reaction zone during growth of plate-like crystals. The shaded region is the crystallization zone. The dashed lines represent the crystallization zone before application of a vertical temperature gradient.

Fig. 3. Typical isotherms in the
reaction zone during growth of
"bulk" crystals.

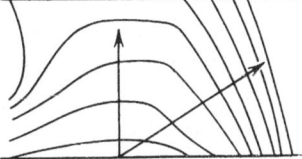

placed with another. When the composition of the charge or the gaseous mixture was altered,
it was necessary to change T_{subl} in such a way as to maintain the previous carefully selected
rate of transport of zinc sulfide. In the presence of a vertical temperature gradient the temperature conditions did not have to be controlled as carefully as before. Crystals were obtained in practically all cases.

The introduction of the second (vertical) temperature gradient had an even greater effect
on the crystallization zone of the "bulk" crystals, which grew at higher temperatures (1300-
1350°C) and in weaker horizontal temperature gradients (1-3 deg/cm). Figure 3 shows typical
isotherms in the reaction zone which were obtained during the growth of the "bulk" crystals.
These isotherms are plotted at intervals of 5 deg/cm (the height of the reaction zone was 5 cm).
The arrows show the directions along which the temperature decreased.

In this case, the reaction zone was filled completely with the initial charge in order to
prevent the transport of zinc sulfide to the colder part of this zone and to ensure crystallization
only in the high-temperature zone. During heating the charge settled down and small prisms
of zinc sulfide grew on its surface without making contact with the walls of the quartz tube.
Some crystals were also formed on the upper part of the tube. In this way we were able to produce "pure" and doped crystals with highly reproducible properties. Moreover, we were able
to vary the dopant concentration within a wide range.

The precision with which the concentration of an activator in a crystal could be controlled
was much poorer than that attainable in the case of powder phosphors. This was due to the
fact that the activator concentration in crystals of the same batch could vary for a number of
reasons. This concentration was governed not only by the amount of the activator in the initial
charge but also by the position of the crystal in the reaction zone, the morphology (shape) and
the number of defects in the crystal in question, and the concentration of the activator vapor
which varied from one point to another (this concentration was governed by the volatility of the
activator in a given gaseous atmosphere).

The time needed to grow crystals ranged from 5 to 6 h in the case of plates 10-100 μ thick
and from 15 to 20 h in the case of plates 0.5 mm thick or prisms of 10-30 mm³ volume. The
areas of the plates ranged from 5-7 to 50 mm². The yield of optical-quality crystals was 10-30
per batch. The luminescence properties of these crystals were completely identical in the
qualitative and quantitative sense. Small quantitative differences were found between different
batches.

§3. Preparation of Self-Activated ZnS Crystals

Most of the published investigations of the luminescence centers have been carried out on
powders and crystals which were either doped with various impurities and, therefore, represented complex systems, or on "pure" samples which were not specially doped but emitted
luminescence even at room temperature. This indicated that such "pure" samples had some
accidental impurities. In view of this, it seemed desirable to prepare first crystals of self-activated ZnS under conditions ensuring a sufficiently high degree of purity and to investigate the
properties of these crystals. This should be followed by studies of the more complex systems
in the form of activated crystals.

We used the designation "self-activated ZnS" for crystalline phosphors prepared without deliberately introduced additions as well as samples to which monovalent anions (Cl, I, etc.) or trivalent cations (Al, etc.) were added. At low temperatures (77°K) all these crystalline phosphors emitted blue luminescence with a maximum in the 465-475 nm region. The blue luminescence observed at room or lower temperatures has been attributed so far to the same luminescence centers. However, the nature of these centers is still a controversial subject. According to some workers [4-6], the blue-luminescence centers are complexes consisting of Cl_S and V_{Zn} defects. Others [7, 8] are of the opinion that the blue luminescence of ZnS is due to an excess of Zn or due to sulfur vacancies (V_S).

We studied the role of various defects (Cl_S, V_S, V_{Zn}, Zn_i) in the formation of the blue-luminescence centers by preparing pure crystals under conditions which minimized as far as possible the possibility of contamination with chlorine or other impurities. We also prepared crystals which were doped with large amounts of chlorine. We determined the spectral characteristics of both types of sample as well as the temperature dependence of the luminescence intensity. We also studied thermoluminescence, luminescence decay, and infrared properties as a function of deviation from the stoichiometric composition.

§ 4. Preparation of "Pure" (Undoped) Crystals

It is known that zinc sulfide absorbs chlorine quite readily and that small traces are usually present in the initial charge used in the preparation of zinc sulfide phosphors. In view of this, we prepared undoped crystals in such a way as to reduce as far as possible the amount of chlorine (as well as oxygen and its compounds) from the initial charge. This was done by multiple recrystallization in a stream of H_2S. Crystals were grown in a room in which no chlorides had been used for a long time and in which no reagents containing chlorine were stored.

The chlorine-removal operation was carried out in two stages. First, we heated slowly (7-8 h) a zinc sulfide powder from room temperature to 500°C in a hydrogen sulfide atmosphere. The partly dechlorinated and deoxidized reagent was subjected to a further heat treatment (at 500, 700, 900, 1100°C) in H_2S during the crystal-growth stage. The moisture and chlorides evolved during recrystallization were removed from the reaction zone by a stream of H_2S:

$$ZnCl_2 + H_2S \rightleftarrows 2HCl + ZnS,$$
$$ZnO + H_2S \rightleftarrows ZnS + H_2O.$$

The crystallization temperature in the case of "bulk" crystals was 1320-1335°C and in the case of plates it was 1100°C. In both cases, the sublimation temperature was 1330-1340°C. The rate of flow of H_2S during the heat treatment was 20-30 cm^3/min and during the crystal-growth stage it was 1-3 cm^3/min. The crystals were cooled at a rate of 5 deg/min.

The crystals grown without coming into contact with the walls of the quartz tube were optically transparent, did not emit visible luminescence at room temperature, and exhibited ultraviolet (327 and 340 nm) and blue ($\lambda_{max} \approx 465$ nm) luminescence at liquid nitrogen temperature. Chemical analysis indicated that these crystals did not contain chlorine (the analysis was carried out by the nephelometric method, whose sensitivity was $1 \times 10^{-3}\%$ Cl in ZnS).

In this way we prepared thin (30 μ) plates of ZnS which were needed in the investigation of the structure of the lower edge of the conduction band by the thermoreflection method [9] and "bulk" prisms which were used in a study of the nature of the defects responsible for the blue band at $\lambda_{max} \sim 465$ nm and for the ultraviolet luminescence bands of ZnS.

§ 5. Preparation of Chlorine-Doped ZnS Crystals

The main difficulty in the preparation of ZnS:Cl crystals was their purity. Zinc chloride formed at high temperatures (1300°C) interacted with quartz, producing zinc oxide:

$$SiO_2 + 2ZnCl_2 \rightleftarrows 2ZnO + SiCl_4.$$

This oxide was incorporated in the ZnS lattice and gave rise to a green luminescence band. To aboid such contamination we always added a reducing agent (H_2S) to the chlorinating gas (HCl). This led to the following reaction with zinc oxide during crystal growth:

$$ZnO + H_2S \rightleftarrows ZnS + H_2O.$$

(The moisture evolved in this reaction was removed by the gas stream.)

The crystallization temperature T_{cr} during the growth of the "bulk" crystals ranged from 1200 to 1320°C, whereas in the case of plate-like crystals it was $T_{cr} = 1100$°C. In both cases, the sublimation temperature was $T_{subl} = 1270\text{-}1320$°C. The rate of flow of the gas during crystal growth was 3-5 cm³/min. The crystals were cooled at a rate of ~5 deg/min.

Samples containing various amounts of chlorine were grown by altering the composition of the gaseous mixture (from 5% HCl + 95% H_2S to 90% HCl + 10% H_2S). However, we were unable to prepare samples in which the concentration of chlorine increased in a regular manner: within one batch this concentration fluctuated considerably due to different vapor pressures of zinc sulfide and zinc chloride: those crystals which grew in colder regions always contained more chlorine.

The ZnS:Cl crystals grown on the surface of the charge (without contact with the quartz tube) were well-faceted optically transparent six-faced prisms emitting blue luminescence with $\lambda_{max} \approx 470$ nm (the green luminescence band was either absent or very weak). The "bulk" ZnS:Cl crystals prepared in this way had a relatively high electrical conductivity: the crystals with ~2×10^{-2}% Cl had a dark resistivity $\rho_d \approx 10^7$ Ω·cm. These were the crystals in which p–n junctions were fabricated and investigated [10-13].

Thin ZnS:Cl plates of different thicknesses (~100 μ, 30-50 μ) and with different chlorine concentrations were grown in order to investigate the energy band structure by the thermoreflection method [9] and to study the nature of the electroluminescence centers responsible for luminous points and streaks (these crystals were doped with copper and manganese by thermal diffusion) [14, 11].

The nature of the defects responsible for the ultraviolet luminescence of zinc sulfide was studied in "bulk" ZnS:Cl crystals with different chlorine concentrations; these crystals were excited by electron bombardment [16].

Investigations of the crystal structure indicated that our samples were sphalerite microtwins containing 10-30% of stacking faults, depending on the concentration of chlorine.

§6. Treatment of Grown Crystals

We assumed that the crystals grown as described above had stoichiometric compositions. However, at high temperatures some deviations from the periodicity of the crystal lattice were possible because of the presence of excess zinc or sulfur.

The presence of such intrinsic láttice defects and their participation in the formation of the blue-luminescence centers was investigated by subjecting the grown crystals to heat treatment in molten sulfur or zinc for periods ranging from 2 to 54 h. Such immersion in sulfur or zinc took place in evacuated (p $\approx 10^{-5}$ atm) ampoules kept at 750°C. In the case of heat treatment in sulfur, an ampoule was placed in a vertical furnace with a definite temperature gradient because of the danger of explosion. The lower part of such an ampoule, where the crystals were located, was kept at 750°C and the upper one at 250°C. The sublimated sulfur condensed in the upper part of the ampoule and flowed back to the lower part. In this way, we ensured that the zinc sulfide crystals were immersed in molten sulfur throughout the heat treatment.

We found that the crystals treated in molten sulfur and zinc retained their optical transparency. The "pure" crystals did not emit luminescence at room temperature before or after such treatment. The blue luminescence was still emitted at liquid nitrogen temperature. The ZnS:Cl crystals retained the blue luminescence at room and liquid nitrogen temperatures.

CHAPTER II

LUMINESCENCE OF SELF-ACTIVATED ZnS CRYSTALS

§ 1. Luminescence Spectra

We investigated the luminescence of self-activated crystals of ZnS, ZnS:Zn, ZnS:S, ZnS:Cl, ZnS:Cl:Zn, and ZnS:Cl:S. The luminescence spectra of these crystals included ultraviolet and blue bands. At this stage we did not investigate the infrared part of the spectrum.

The ultraviolet luminescence band was studied by Blazhevich, Lavrov, and Panasyuk [16]. They found that electron bombardment of chlorine-free crystals at 77°K produced luminescence bands at 327.5, 340, and 360 nm, whose intensities were two orders of magnitude higher than the intensity of the blue band. The ultraviolet bands were observed most clearly in the ZnS crystals which were closest to the stoichiometric composition. Treatment in molten zinc and sulfur and the introduction of chlorine reduced the intensity of these bands.

We investigated in detail the blue part of the spectrum. Up to now most of the workers have regarded the blue luminescence of self-activated ZnS ($\lambda_{max} \sim 470$ nm) as a single band due to the same centers in chlorine-doped and in "pure" crystals. There is no agreement as to whether these centers always include chlorine atoms (assumed to be present in "pure" crystals as an accidental impurity) [4, 5] or whether chlorine simply acts as a catalyst and is not one of the components of these centers [7, 8].

Some workers (see, for example, [5]) have drawn attention to the presence of a second blue band ($\lambda_{max} \approx 415$ nm) and have also attributed the corresponding luminescence centers to the presence of chlorine.

However, it has been demonstrated for powder phosphors [17] and will be shown below that the long- and short-wavelength ends of the blue band differ strongly in their properties. This confirms the conclusion reached in [18] that the blue band of self-activated ZnS crystals is complex.

Moreover, the experiments which will be described below not only confirm the differences mentioned above but also indicate that the properties of the long-wavelength end of the blue band are different in chlorine-doped and chlorine-free crystals. When the temperature was lowered the chlorine-doped samples exhibited a shift of the luminescence band in the direction of longer wavelengths, whereas no such shift was observed in the samples free of chlorine. Moreover, the long-wavelength luminescence emitted by the ZnS and ZnS:Cl crystals was characterized by different temperature dependences of the intensity. This led us to the conclusion that the long-wavelength part of the blue band of self-activated ZnS was due to the presence of several types of center of different origin and that the main contribution to the luminescence of the chlorine-free crystals was made by centers different from those responsible for the predominant band exhibited by the chlorine-doped samples. This made it necessary to study the nature of the

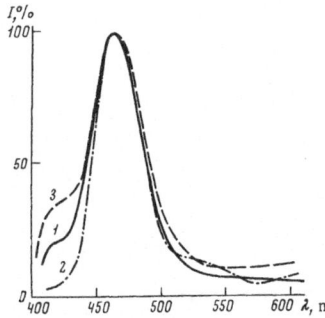

Fig. 4. Luminescence spectra of ZnS crystals with different deviations from stoichiometry: 1) ZnS; 2) ZnS:S; 3) ZnS:Zn.

luminance centers responsible for each of the individual bands in the blue luminescence spectrum.

The nature of the centers responsible for any given band could best be studied by isolating the properties of each of these bands. However, we were unable to prepare crystals which would exhibit only one of these elementary bands and we had to study their behavior only by making an allowance for the interaction with the other bands.

The visible luminescence spectra of the chlorine-free ZnS crystals with different deviations from stoichiometry are plotted in Fig. 4 (77°K, λ_{exc} = 312 nm). At room temperature these crystals did not luminesce. It is evident from Fig. 4 that these crystals had a luminescence band with a maximum at 465 nm and, in the case of crystals with excess zinc, they had a peak at about 415 nm. Figure 5 shows the spectra of ZnS obtained for λ_{exc} = 356 nm at two temperatures: 15 and 77°K. In this case, the peak at 415 nm was practically absent but because of the higher intensity of the exciting radiation we were able to study shorter wavelengths and to observe a sharp rise in the ultraviolet region. The 465 nm maximum was practically unaffected by variations of the temperature and of the excitation wavelength λ_{exc}.

The spectra of the ZnS:Cl crystals (Figs. 6a and 6b) exhibited a definite peak in the region of 415 nm when they were excited with light of λ_{exc} = 312 nm but not when they were excited with λ_{exc} = 365 nm. The shift of the blue band maximum in the direction of longer wavelengths, reported in [18, 20] was observed clearly when the temperature was lowered. This shift displaced the 465 nm peak to 475 nm.

An examination of the luminescence spectra indicated that, in the case of the chlorine-free crystals, the bands at 415 and 465 nm were of different origin, whereas in the case of

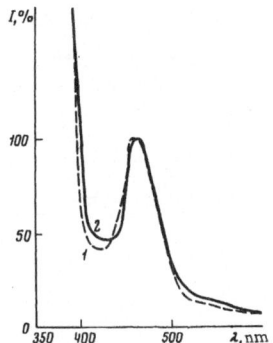

Fig. 5. Luminescence spectra of the same ZnS crystal excited with λ_{exc} = 365 nm line at T = 77°K (1) and 15°K (2).

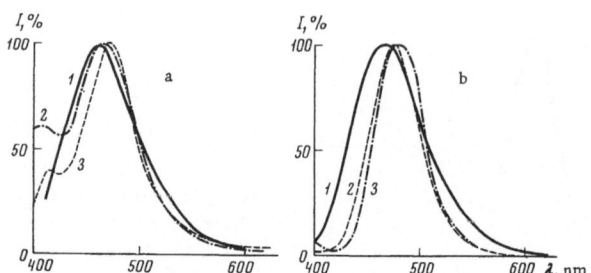

Fig. 6. Luminescence spectra of a ZnS:Cl crystal excited
with λ_{exc} = 312 nm (a) and λ_{exc} = 365 nm (b) at T = 293°K
(1), 77°K (2), and 15°K (3).

chlorine-doped crystals these bands were supplemented by other bands associated with chlorine
centers. Since the blue luminescence band of the chlorine-doped samples was located at longer
wavelengths, we assumed that the additional band resulting from the presence of chlorine had a
peak at wavelengths longer than 465 nm. It is evident from curve 3 in Fig. 6b that there were
some indications of a peak in the region of 490 nm.

The properties of each of the postulated bands could be investigated only if the complex
band were separated into its components. We used differences between the behavior of different
parts of the spectrum and separated the bands in the following manner. We applied the method
used in [17] and separated the regions at 405, 425, and 490 nm representing the luminescence
due to the short- and long-wavelength centers. We found that, as in [17], the width of the elementary band at 490 nm was much greater (32 nm) than the width of the bands at shorter wavelengths (5 nm). We assumed that, in the case of the chlorine-free crystals, the long-wavelength
luminescence was solely due to the 465 nm band (the "chlorine" band did not appear in the
spectrum of these crystals). In the case of the chlorine-doped crystals, the long-wavelength
region was dominated by the contributions of the luminescence centers responsible for the band
at λ_{max} = 490 nm. The centers responsible for the 465 nm band made a smaller contribution.
Naturally, the ratio of the intensities of these two bands should vary with the temperature and
with λ_{exc} if the centers responsible for these bands had different properties.

A comparison with the results reported elsewhere in the present volume [21] indicated
that no significant errors were made in our separation procedure. When the blue luminescence
of ZnS was analyzed by the method developed by Fok on the basis of the ideas put forward by
Alentsev [22], it was found that the blue luminescence of self-activated zinc sulfide consisted of
at least four bands with maxima at 496, 466, 426, and 405 nm.

Therefore, we shall use the results given in [21] and assume that the blue luminescence
observed in our experiments consisted of two long-wavelength bands 466 and/or 496 nm and two
short-wavelength bands at 426 and 405 nm.

§ 2. Temperature Dependence of the Intensity of Blue

Luminescence, Thermoluminescence,   Decay

of Afterglow

A. Experimental Method

Our crystals were ground to a depth of 0.3-0.5 mm on each side in order to remove the
surface defects and to obtain plane-parallel samples. These samples were attached with Wood's

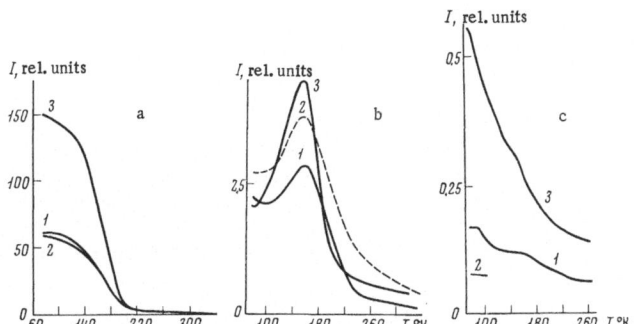

Fig. 7. Temperature dependences of the intensities of the 466 nm (a), 426 nm (b), and 405 nm (c) luminescence bands of various chlorine-free crystals: 1) ZnS; 2) ZnS:S (4-h treatment); 3) ZnS:Zn (2-h treatment).

alloy to a massive copper block placed in a Dewar flask or helium cryostat fitted with quartz windows. The excitation was provided by a PRK-4 lamp with a filter for the separation of the 365 nm line.* We also used a cuvette filled with a saturated aqueous solution of $CuSO_4$. The luminescence bands were selected with a UM-2 monochromator.

The temperature dependence of the luminescence brightness was recorded under steady-state conditions during slow cooling. This procedure was adopted in order to avoid the complicating influence of thermoluminescence. The intensity of the exciting radiation was reduced with the aid of attenuating grids (this was necessary in a study of the kinetics of thermal quenching). In studies of thermoluminescence an excited crystal was heated at a rate of 10.6 deg/min. The 415 nm thermoluminescence of the chlorine-free crystals was too weak to be recorded even by the most sensitive photomultiplier with multialkali cathodes. Therefore, special measures were taken to reduce the photomultiplier noise, which consisted of cooling the photocathode to 260°K with dry ice, in accordance with the Vasil'ev method [23]. This raised the signal-to-noise ratio by a factor of 2.

B. Temperature Dependence of Luminescence Intensity

The temperature dependences of the luminescence intensity of the individual blue bands are plotted in Figs. 7a-7c and in Figs. 8a and 8b.

1. The 466 nm band of the "pure" ZnS exhibited thermal quenching at temperatures as low as 100°K (Fig. 7a). In the case of those crystals which were treated in molten sulfur or zinc (the duration of such treatment is given in the caption of Fig. 7), the thermal quenching started at even lower temperatures. The sulfur treatment had practically no influence on the intensity of the 466 nm band (curve 2). On the other hand, a crystal treated in zinc for 2 h emitted, at 77°K, luminescence which was 2.5 times stronger than that emitted by "pure" ZnS.

In contrast to the 466 nm band, which was quenched even at low temperatures, the 426 nm band increased in intensity when the temperature was raised (Fig. 7b), reaching its maximum intensity in the region of 160°K. However, at room temperature the 426 nm band was strongly quenched.

*The short-wavelength bands were excited more effectively by the $\lambda_{exc} = 312$ nm line. However, in this case, the behavior of the separate bands was less differentiated. Therefore, although measurements with the $\lambda_{exc} = 312$ nm line were made, the results will not be given here.

Z. P. ILYUKHINA ET AL.

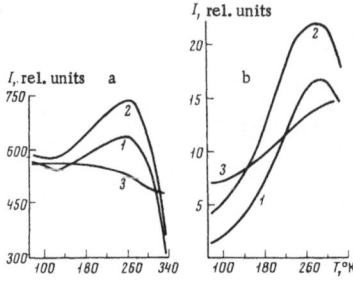

Fig. 8. Temperature dependences
of the intensities of 496 +466 nm (a)
and 426 nm (b) luminescence bands
of various chlorine-doped crystals:
1) ZnS:Cl; 2) ZnS:Cl:S; 3) ZnS:Cl:Zn.

The zinc treatment reduced somewhat the intensity of the low-temperature luminescence
in the region of 426 nm but it enhanced the high-temperature luminescence in the same region
(curve 3 in Fig. 7b).

The 405 nm band of the same crystals was very weak (as indicated by the values of the
ordinate in Fig. 7c), particularly in the case of crystals treated in molten sulfur. Therefore,
we were unable to study this band without some admixture of the 426 nm band. The influence of
the latter band was clearly visible in the 150-160°K range (Fig. 7c). The rest of the temperature dependence of the 405 nm band was not affected by the 426 nm contribution. We found that
the 405 nm band experienced strong thermal quenching even at low temperatures. Moreover,
this band was enhanced by the zinc treatment and reduced by the sulfur treatment (only a part
of the curve for the sulfur-treated samples is shown in Fig. 7c because it was strongly distorted
by the 426 nm band).

Since the blue luminescence bands of the "pure" crystals were quenched at low temperatures, these crystals did not exhibit luminescence at room temperature.

2. The temperature dependences of the luminescence intensity of the various bands exhibited by the chlorine-doped crystals differed from the corresponding temperature dependences
of the "pure" ZnS. This was true of the long- and short-wavelength bands (Figs. 8a and 8b).*
When the temperature was raised, the intensities of the 496 and 426 nm bands of the ZnS:Cl crystals, which were not treated in sulfur or zinc, increased right up to 260 and 280°K,respectively.

The sulfur treatment increased somewhat the intensities of both bands. The zinc treatment altered the temperature dependences in such a way that the maxima disappeared and the
curve for the 496 nm band approached the curve obtained for the 466 nm band of the "pure"
crystals.

Since the luminescence intensities of the long-wavelength bands were proportional to the
excitation density (this was checked specially), we were able to calculate the thermal activation
energy of the luminescence centers from the thermal quenching curves. These calculations
were performed using the formula

$$\ln\left(\frac{I_0}{I} - 1\right) \propto f\left(\frac{10^3}{T}\right).$$

We found that the activation energy for the 466 nm band was ~0.15 eV and that for the 496 nm
band was ~0.57 eV. The latter value was in good agreement with the results obtained by Era,
Shionoya, Washizawa, and Ohmatsu [24, 25] for chlorine-doped crystals similar to those used

---

*The 405 nm band of the chlorine-doped crystals could not be measured because of its low in-
tensity or absence.

in the present study: Era et al. reported a similar temperature dependence of the luminescence intensity and found that the activation energy was ~0.64 eV.

C. Thermoluminescence

Kaleeva et al. [18] determined the "complete" thermoluminescence curves of our crystals without any division into individual bands. They found that the "pure" crystals exhibited two thermoluminescence peaks: one was located at 110°K (0.093 eV) and the other at 160°K. They established that the first peak was associated with excess zinc in interstices and the second peak was due to sulfur vacancies.

The introduction of chlorine produced a wide thermoluminescence peak at 160°K, whose origin was different from that of the corresponding peak exhibited by the "pure" crystals (we attributed the former peak to chlorine). We also found that the chlorine-doped crystals exhibited low-temperature luminescence due to a group of shallow traps adjoining the conduction band (the origin of these traps was not determined). The depth of the chlorine-peak level was ~0.25 eV (calculated in [26]).

We determined separately the thermoluminescence associated with the short- and long-wavelength bands.

1. The thermoluminescence curves of the main (long-wavelength) bands of the "pure" (466 nm) and chlorine-doped (496 nm) crystals were, as expected, completely identical with the thermoluminescence curves of the same crystals obtained without division of the luminescence into bands: the "pure" crystals exhibited two thermoluminescence peaks at 110 and 157°K. The zinc treatment enhanced strongly the 110°K peak but did not alter the 157°K peak; the sulfur treatment enhanced slightly the former and destroyed the latter peak (Fig. 9).

The chlorine-doped crystals had a strong peak at 160°K which was practically unaffected by the sulfur treatment (up to 60 h) but was destroyed completely by a 4-h treatment in molten zinc (Fig. 10).

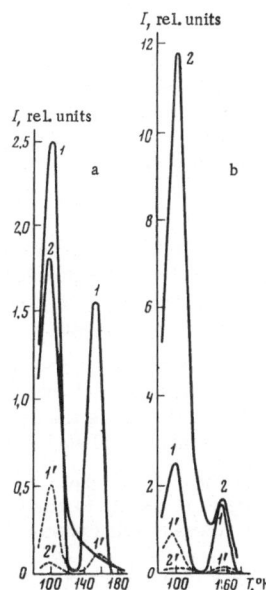

Fig. 9. Thermoluminescence spectra of chlorine-free crystals. The continuous curves represent the 466 nm luminescence band and the dashed curves represent the 426 nm band. a) Influence of sulfur treatment: 1), 1') ZnS; 2), 2') ZnS:S (4 h). b) Influence of zinc treatment: 1), 1') ZnS; 2), 2') ZnS:Zn (2 h).

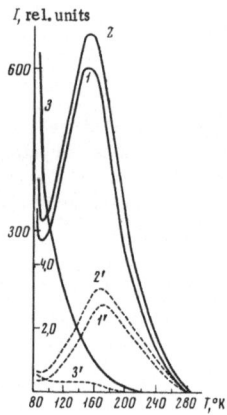

Fig. 10. Thermoluminescence
of chlorine-doped crystals. The
continuous curves represent the
496 nm luminescence band (the
ordinate scale on the left) and
the dashed curves represent the
426 nm band (the ordinate scale
on the right). 1), 1') ZnS:Cl; 2),
2') ZnS:Cl:S (60 h); 3), 3') ZnS:
Cl:Zn (4 h).

TABLE 1

Decay time, sec	I_{496}	I_{415}	I_{496}/I_{415}	Decay time, sec	I_{496}	I_{415}	I_{496}/I_{415}
0	704643	550	1281	20	2123	0.5	4246
5	9650	3.1	3113	30	1370.3	0.3	4568
10	4728	1.2	3940	45	805.3	0.15	5369

2. The thermoluminescence curves obtained in the 426 nm range differed from those found for the long-wavelength bands. The "pure" ZnS crystals (Fig. 9) exhibited the 110 and 157°K peaks but the ratio of their intensities was different from that for the long-wavelength bands. This ratio was affected by the sulfur and zinc treatments.

The ZnS:Cl crystals did not have a thermoluminescence maximum at 160°K but at a higher temperature (~172°K). These crystals did not exhibit the thermoluminescence associated with shallow levels. An excess of sulfur did not affect the 172°K peak but excess zinc destroyed it completely (Fig. 10). This observation indicated that the excess zinc destroyed the trapping levels responsible for the 160°K peak of the chlorine-doped crystals because the effect of zinc was the same in the case of peaks associated with the long- and short-wavelength bands.

D. Decay of Afterglow

The decay of the afterglow confirmed that the centers responsible for the long- and short-wavelength luminescence bands were different. We found that, like the temperature dependences of the luminescence intensity and the thermoluminescence curves, the decay was specific to each wavelength range and each type of crystal.

In the case of the chlorine-doped crystals, we observed very rapid decays of the after-glow in the 415 nm range and a much slower decay of the 496 nm band. It is evident from Table 1 that the ratio I_{496}/I_{415} changed during the decay of the afterglow emitted by the ZnS:Cl crystals in such a way that the I_{496} component became stronger.*

Since the luminescence of the "pure" crystals was weak, the afterglow of these crystals was even weaker and we were unable to study the decay of the afterglow of these crystals with sufficient accuracy, even when cooled photomultipliers were employed.

*The measurements were carried out at 77°K.

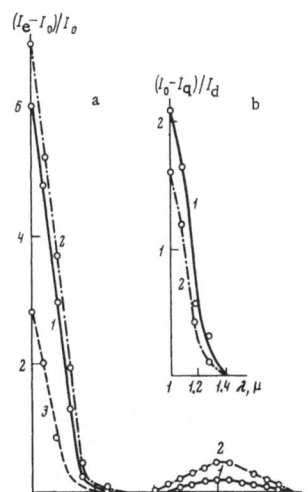

Fig. 11. Infrared sensitivity spectra
of chlorine-doped crystals: a) enhancement; b) quenching. 1) ZnS:Cl;
2) ZnS:Cl:S (60 h); 3) ZnS:Cl.Zn
(4 h).

§3. Sensitivity to Infrared Radiation

The sensitivity of our crystals to infrared radiation was described in [27]. The study reported in that paper was concerned with the enhancement and quenching of the blue luminescence
which resulted from infrared illumination at liquid nitrogen temperature. The enhancement
was calculated from the formula $(I_e - I_0)/I_0$, where I_0 is the luminescence intensity in the absence of infrared radiation and I_e is the infrared-enhanced intensity. The quenching was determined at the moment of excitation and deduced from the formula $(I_0 - I_q)/I_d$, which was deduced by Fok [28]. Here, I_q is the steady-state luminescence intensity under the simultaneous
action of ultraviolet and infrared radiation, and I_d is the discontinuous change in the luminescence intensity which decays rapidly after the end of the infrared illumination. In the case of
the chlorine-doped crystals, this type of quenching was observed also at room temperature but
the effect was weaker. No enhancement was observed at room temperature. The enhancement
and quenching effects were determined for the whole blue band without resolving it into the long-
and short-wavelength components. The measurements were thus basically applicable to the
466 nm band in the case of the "pure" crystals and to the 496 nm band (or the 496 + 466 nm band)
in the case of crystals doped with chlorine. The main results obtained in [27] can be summarized as follows.

A. Chlorine-Doped Crystals

a) Enhancement (Fig. 11a). The ZnS:Cl crystals did not exhibit the 1.2-1.3 μ sensitivity
peak typical of activated crystals. Instead, at the short-wavelength end of the spectrum these
crystals exhibited a strong peak with a maximum at $\lambda < 1$ μ.

The crystal closest to the stoichiometric composition as well as that treated in sulfur exhibited also a sensitivity peak with a maximum at 2.5 μ. A short (4-h) treatment in molten zinc
destroyed completely the long-wavelength sensitivity and weakened the enhancement at short
wavelengths. These results indicated that the infrared sensitivity was associated with the presence of zinc vacancies.

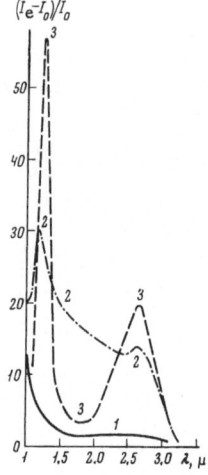

Fig. 12. Spectra of the
infrared-induced en-
hancement of the lumi-
nescence of chlorine-
free crystals: 1) ZnS; 2)
ZnS:S (4 h); 3) ZnS:Zn
(2 h).

b) Quenching (Fig. 11b). The nature of the quenching spectrum resembled the short-wave-length part of the enhancement spectrum. Infrared radiation of longer wavelengths had no quenching influence.

When a crystal was treated in molten zinc, the quenching was restricted only to certain wavelengths. Thus, the occurrence of quenching in the chlorine-doped crystals should also be attributed to zinc vacancies.

B. Chlorine-Free Crystals

No quenching was observed for the chlorine-free crystals in the investigated range of in-frared wavelengths [27]. However, these crystals exhibited a quite strong enhancement and the enhancement-sensitivity spectra differed strongly from the corresponding spectra of the chlorine-doped crystals (Fig. 12). The spectrum of "pure" stoichiometric ZnS is represented by curve 1 in Fig. 12. After a 2-h treatment in molten zinc (curve 3) the spectrum assumed the shape typical of phosphors and activated crystals consisting of two maxima at 1.3 and 2.6 μ. In this case, the excess zinc acted in a different way from that in the chlorine-doped crystals, be-cause it behaved as an ordinary activator. It was interesting to note that the zinc treatment produced also a sharp thermoluminescence peak at 110°K and an infrared sensitivity peak at 1.3 μ; these peaks were probably of the same origin.

In contrast to the chlorine-doped crystals, an excess of sulfur in the chlorine-free sam-ples altered considerably the infrared sensitivity spectrum, which changed to a broad band ex-tending over the whole infrared region.

The different infrared sensitivities of the chlorine-free and chlorine-doped crystals showed that the amount of chlorine in the former must have been small and that the properties of the blue luminescence were different for the chlorine-free and chlorine-doped samples. This provided further evidence in support of our hypothesis that the principal blue luminescence cen-ters in these two groups of crystals were of different origin.

CHAPTER III

DISCUSSION OF RESULTS

§ 1. Chlorine-Doped Crystals

Our chlorine-doped crystals were grown in a stream of H_2S (30%) and HCl (70%). Consequently, the conditions were such as to produce $ZnCl_2$, which — according to [29] — dissolve in ZnS and produce not only Cl_S defects but also V'_{Zn} and V''_{Zn} defects which are needed to satisfy the valence compensation principle. The energy levels of the defects in ZnS:Cl crystals (Fig. 13) were deduced in [29]. It was pointed out in that paper that (a) the valence compensation is achieved in accordance with the scheme $[Cl_S^{\cdot}] = 2[V''_{Zn}]$; and (b) the energy positions of the levels of V''_{Zn} and V'_{Zn} are related by the simple equation $E_{V''_{Zn}} = 5E_{V'_{Zn}}$, which was deduced from theoretical calculations and from a comparison of these calculations with the experimental data.

The reported ESR results indicate that the V''_{Zn} and $Cl_S^{\cdot}$ defects in chlorine-doped crystals have a tendency to form complexes of the $(V_{Zn}Cl_S)'$ type. This tendency increases with increasing defect concentration so that when this concentration reaches $C_{Cl} = 10^{-4}$ g-atom per mole of ZnS at T = 1373°K, almost all the zinc vacancies are associated with Cl ions [29]. In this way, we obtain a system in which there are two main types of defect and concentrations of these defects are approximately equal:

$$[(V_{Zn}Cl_S)'] = [Cl_S].$$

Moreover, the chlorine-doped crystals may contain defects such as V'_{Zn} and V_S as well as interstitial zinc, which we shall denote by Zn_i (sulfur is unlikely to occur in the interstitial form because of its size). In the presence of chlorine in the atmosphere surrounding the crystals during their growth, the sulfur vacancies are most likely to be filled with chlorine and, therefore, the number of such vacancies in the chlorine-doped crystals should be very small.

The ESR methods show that excited self-activated zinc sulfide crystals contain three types of defect, one of which (that producing a strong signal A) can be attributed with assurance to the $(V_{Zn}Cl_S)'$ complexes. There is no agreement about the nature of the other two types of defect: they are variously attributed to sulfur vacancies [30], interstitial zinc atoms [21], or chlorine occupying sulfur sites [32]. In our opinion, these defects may be of different origin in different samples, particularly since the g factors can vary considerably (in one case, it is reported that g = 2.000, i.e., a donor band is formed and this band is most likely to be due to chlorine defects).

It is logical to assume that the strongest signal (A) is due to the centers responsible for the strongest luminescence band in chlorine-doped crystals (496 nm). Hence, we can attribute the 496 nm band to the $(V_{Zn}Cl_S)'$ complexes.

Era, Shionoya, Washizawa, and Ohmatsu [24, 25] suggested that centers of this kind are responsible not only for the long-wavelength band but the whole of the blue luminescence spec-

Fig. 13. Positions of trapping levels in the energy scheme of ZnS:Cl phosphors, taken from [29].

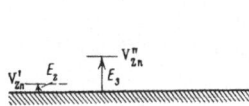

trum, including the short-wavelength components. The occurrence of the j and t shifts* in ZnS:Cl crystals is attributed by Era et al. to the presence of a system of donor−acceptor pairs with different donor−acceptor distances. According to these workers, the blue luminescence of self-activated ZnS:Cl is due to the acceptor states $V_{Zn}^{\prime\prime}$ in the $(V_{Zn} Cl_S)'$ complexes.

We cannot agree with this hypothesis because it implies a continuous distribution of the donor−acceptor distances in pairs. Our experiments and the results reported in [21] show that the blue luminescence spectrum consists of several definite bands, each of which appears independently of the other bands and has its own special properties. Some of these bands may be due to donor−acceptor pairs and, therefore, a continuous distribution of donor−acceptor distances may be a characteristic of one of the bands (it would be responsible for the considerable width of such a band). However, none of the bands found in the blue luminescence spectrum fits this description.

In view of this, we shall assume that the centers responsible for the short-wavelength band at 426 nm have a structure different from that of the centers responsible for the 496 nm band. The structure of these centers can be understood by considering the ESR data.

Studies of the ESR of the ZnS:Cl crystals have established that these crystals contain centers of two additional types. Since the number of sulfur vacancies in the chlorine-doped crystals is small, they can hardly be detected by the ESR method and we must assume that one of the signals is due to interstitial zinc (Zn_i) and the other to Cl_S centers. We shall show later that the interstitial zinc can be attributed, with a high degree of probability, to the 466 nm band, which is also exhibited by the chlorine-doped crystals. The properties of this band will be considered in the next section which will deal with the chlorine-free crystals. Here, we shall discuss the nature of the 426 nm band which should be attributed to the Cl_S defects. When a crystal is excited, free electrons may be captured by the $Cl_S^{\cdot}$ traps which are thus converted to paramagnetic Cl_S defects. We may assume that such defects are responsible for the ESR and the short-wavelength luminescence. Chemical considerations show that the probability of the existence of such defects is quite high.

A $Cl_S^{\cdot}$ trap can act as a luminescence center only if it can be reached by electrons and holes. Electrons transferred to the conduction band as a result of excitation can be captured quite easily by such traps but the situation is more complicated in the case holes. When the excitation is in the form of λ_{exc} = 312 nm radiation, holes are formed in the valence band because of the band−band absorption and then they can be transferred to acceptor levels of the centers existing in a crystal. However, in the case of excitation with the λ_{exc} = 365 nm line, it is necessary to ensure either direct excitation of the relevant centers or the transfer of holes from other absorption centers. Our experiments show that the luminescence centers responsible for the 426 nm band of the chlorine-doped crystals are hardly excited by the λ_{exc} = 365 nm line (the intensity of the 426 nm band at 77°K is negligible). Therefore, the holes needed for the 426 nm luminescence are supplied by the centers responsible for the other bands such as the 466 and 496 nm bands. However, our calculations indicate that the $V_{Zn}^{\prime\prime}$ level of the complexes is separated by 0.57 eV from the edge of the valence band, i.e., E_3 in the scheme of Fig. 13 is equal to 0.57 eV. Therefore, the probability of the liberation of holes from this level at 77°K is very low. The energy gap separating the $V_{Zn}^{\prime}$ level, denoted by E_2, from the valence band is 0.12 eV because $E_3 = 5E_2$. It follows that the liberation of holes from these levels at 77°K is also unlikely. The depth of the level responsible for the quenching of the 466 nm band is also of the order of 0.15 eV (see §2 in Chap. II).

It follows from our discussion that the intensity of the 426 nm luminescence band excited by the λ_{exc} = 365 nm line at 77°K should be very low but the intensity of the same band should

---

*These shifts are labeled in the same way as in [24, 25].

be quite high if it is excited by the λ_{exc} = 312 nm line. When the temperature is raised, the intensity of this band should increase even when it is excited with the λ_{exc} = 365 nm line.

In view of this, we shall assume that the $(V_{Zn}Cl_S)'$ complexes are responsible for the 496 nm luminescence band and the Cl_S centers are responsible for one of the short-wavelength bands which predominates in the spectrum of the ZnS:Cl crystals and which exhibits its characteristic properties in the 426 nm region.

Let us now see whether these hypotheses are in agreement with the experimental data obtained in our investigation.

1. We shall start by considering the temperature dependences of the intensities of the long- and short-wavelength bands (Figs. 8a and 8b). The complexes responsible for the long-wavelength band have relatively deep levels and are excited efficiently by the λ_{exc} = 365 nm line at liquid nitrogen temperature. This is in agreement with the strong luminescence observed in the region of this band. In contrast to this strong (at 77°K) long-wavelength band, the short-wavelength component is very weak, in agreement with the proposed explanation. When the temperature is raised, we find that the intensity of the short-wavelength band rises rapidly because of the thermal quenching of the other centers (Fig. 8b). The 496 nm band also increases somewhat in intensity with increasing temperature. This can be explained by the transfer of charge from the luminescence centers which are present in our chlorine-doped crystals and which are responsible for the luminescence band at 466 nm. This band is quenched at very low temperatures, as indicated clearly by the temperature dependences plotted in Fig. 4 for the chlorine-free crystals whose luminescence spectrum is dominated by this band.*

The temperature dependences show also that the quenching of the long-wavelength band begins somewhat earlier than that of the short-wavelength component and that at high temperatures the short-wavelength band increases continuously in importance.

This result was confirmed by the experiments of Koda and Shionoya [33]: they also reported that the luminescence peak of the ZnS:Cl crystals shifts toward higher energies when the temperature is raised. This shift can be explained by assuming that the centers responsible for the long- and short-wavelength luminescence bands of the ZnS:Cl crystals are separated by different gaps from the appropriate energy band and that the Cl_S centers have deeper levels than the complexes. This is in agreement with the lower activation energy of the complexes than that of free defects. When the excitation wavelength is λ_{exc} = 312 nm, the 426 nm band is quite strong also at 77°K, in agreement with our results (Fig. 6a) and with our interpretation.

2. The experiments involving the treatment of crystals in molten sulfur and zinc provide additional support for the proposed interpretation.

According to the published data, sulfur is not readily absorbed by chlorine-doped crystals. Therefore, the sulfur treatment should have little influence on the characteristics of both luminescence bands, which is in agreement with the experimental results. Some increase in the intensity of these bands may be due to an increase in the number of zinc vacancies.

---

*The intensity of the long-wavelength luminescence band decreases somewhat in the 77-130°K range. This is in good agreement with the hypothesis that at low temperatures the 466 nm band represents a large proportion of the long-wavelength luminescence. Therefore, in this temperature range the temperature dependence of the intensity of the 466 nm band, which is quenched, governs the behavior of the long-wavelength part of the spectrum. When this band is strongly quenched (above ~130°K) the intensity is governed by the temperature dependence of the 496 nm band which is affected by the transfer of carriers from the quenched centers responsible for the 466 nm band.

The intensity of the 426 nm band at 77°K increases by a factor of over 5 after the treatment in molten zinc. This can be explained as follows. The heat treatment of the ZnS:Cl crystals in molten zinc shifts to the right the equilibrium of the reaction

$$[(V_{Zn}Cl_S)'] \rightleftarrows [(Cl_S)'].$$

Consequently, the number of the $(V_{Zn}Cl_S)'$ complexes, responsible for the long-wavelength luminescence, decreases, whereas the number of the $Cl_S^{\cdot}$ centers, which are not associated with $Z_{Zn}^{\shortmid\shortmid}$ defects, increases. Moreover, the zinc treatment raises the number of the centers responsible for the 466 nm band, which acts as the source of holes for the centers associated with the 426 nm band. A comparison of the enhancement of the short-wavelength luminescence with the increase in the number of the $Cl_S^{\cdot}$ centers in ZnS:Cl:Zn provides further evidence in support of our hypothesis that free Cl_S donors are responsible for the short-wavelength luminescence.

Similar comments can be made about the luminescence centers responsible for the 496 nm band in view of the observed reduction in the number of the $(V_{Zn}Cl_S)'$ complexes and weakening of the long-wavelength band. However, in this case, the experimental results (Fig. 8) indicate that the 77°K luminescence does not decrease in intensity. This observation can be explained quite easily bearing in mind the temperature dependences plotted in Fig. 7. A comparison of Figs. 7 and 8 shows that, in this case, the luminescence is not due to the 496 nm band but is mostly due to the 466 nm band which increases strongly in intensity as a result of the treatment in molten zinc. Since the thermal quenching of the 466 nm band begins at very low temperatures and the contribution of the 496 nm band to the luminescence spectrum becomes small, the nature of the temperature dependences of the luminescence intensity changes. Thus, maxima are not observed in the 496 and 426 nm range (such maxima are exhibited by crystals which have not been treated in molten zinc) and the dependences approach that recorded for the 466 nm band.

3. Thermoluminescence corresponding to the various bands should be considered in conjunction with the temperature dependences. The features of these dependences are responsible for the fact that the short-wavelength band, which is hardly noticeable at low temperatures, does not have a characteristic thermoluminescence peak which is observed on the liberation of electrons from the shallowest levels forming a continuum and which is exhibited so clearly by the long-wavelength band. When the temperature is raised, the transfer of holes to the short-wavelength luminescence centers is accelerated and this results in a redistribution of the intensity.

The depth of the chlorine traps responsible for the 160°K thermoluminescence peak has been calculated in [26] without separation of the blue luminescence spectrum into its component bands: this calculation yielded 0.25 eV. This depth applies to the long-wavelength band, i.e., to the Cl_S components of the complexes. The depth of the free Cl_S donors is greater.

4. The t shift [24, 25] is exhibited also by our chlorine-doped crystals (Table 1) and the shift is in the direction of longer wavelengths because the short-wavelength band has a faster response and the long-wavelength band increases steadily in importance during decay.

However, these experimental results, which are in agreement with those reported in [24, 25], can be explained in a way different from that adopted by Era et al., who attributed the slow decay to the direct radiative recombination of electrons within donor—acceptor pairs without transfer of these electrons to the conduction band.

Our investigations of the short-wavelength thermoluminescence covered the range from 77°K (and from 15°K in the case of ZnS:Cl phosphors [34]) and they indicated that a very large thermal light-sum is stored in crystals at extremely shallow trapping levels whose depth is 0.06 eV or less. These levels form a continuum stretching to the edge of the conduction band. They are formed mainly in the presence of chlorine but their nature is not yet fully established. When an excited sample is stored at 15°K in darkness, this light-sum decreases considerably, whereas the light-sum stored in the 160°K peak remains practically constant, i.e., the phosphorescence during decay at 77°K is solely due to the de-excitation of these shallow levels. The presence of such a large light-sum, which is released at low temperatures, proves that the probability of the liberation of electrons from shallow traps is high. It is assumed in [25] that at low temperatures the thermal transfer of electrons from these 0.05 eV levels to the conduction band is impossible because the probability of such transfer is 10^{-29} sec^{-1} if the frequency factor is 10^{12}-10^{13} and the temperature is T = 6°K. However, in calculation of the same probability under our conditions, we have assumed an average depth of 0.03 eV in view of the presence of a level continuun extending right up to the conduction band. The temperature in our experiments was 15°K. For these parameters the probability of thermal transfer is 10^2 sec^{-1} and at 77°K for 0.06 eV the probability is 10^8 sec^{-1} (for a frequency factor of $10^{12.5}$ assumed in [35]). Consequently, thermal transfer of electrons from the trapping levels in the continuun to the conduction band is quite likely. This is supported also by the hyperbolic nature of the decay curves.

At 77°K the luminescence centers associated with the short-wavelength luminescence are practically inactive (see points 1 and 3 above) and it follows that the afterglow spectrum should be dominated by the 496 and 466 nm bands.

5. The proposed interpretation of the energy levels in the ZnS:Cl crystals should explain also their infrared properties. However, since the available results apply to the unresolved spectrum, i.e., primarily to the 496 nm band, they cannot be regarded as providing direct support for our hypotheses but they should not be in conflict with our explanation.

The existence of two sensitivity peaks in the case of enhancement by infrared radiation of λ = 1 and 2.5 μ wavelengths (Fig. 11) is in full agreement with the hypothesis that the crystals under investigation have two groups of trapping levels: a continuum of levels which are joined to the conduction band and are responsible for the 2.5 μ sensitivity peak [34], and a chlorine level at 0.25 eV, which is responsible for the thermoluminescence peak at 160°K and the infrared sensitivity peak at λ = 1 μ.

The enhancement of the blue luminescence is due to the liberation of electrons from these levels by infrared radiation, followed by the transfer of these electrons to the conduction band. Next, the electrons drop to the luminescence centers and give rise to the radiation observed at 77°K. At room temperature there is no enhancement because these levels cannot be filled.

The quenching spectrum has only one peak which is due to the transfer of electrons by infrared radiation from the valence band to a hole level of the V_{Zn}^{n} centers, which are responsible for the 496 nm luminescence band.

After treatment in liquid zinc the infrared properties of the ZnS:Cl crystals change drastically (Fig. 11). The enhancement at λ = 2.5 μ disappears and that at 1 μ becomes much weaker. The quenching effect and the thermoluminescence peak at 160°K are both destroyed.

The results can be explained if we adopt the proposed model. The zinc treatment fills zinc vacancies. Consequently, the hole levels associated with the V_{Zn}^{n} centers are destroyed so that electron transitions to these levels from the valence band become impossible (since these transitions are responsible for the quenching in the untreated crystals, the quenching effect disappears). Moreover; the recombination of electrons liberated by infrared radiation from the electron levels is no longer possible, i.e., the enhancement becomes weaker.* The charge state of the Cl_{S}^{-} centers, which are the components of the $(V_{Zn} Cl_{S})'$ complexes, changes because the V_{Zn}^{n} components of these complexes become filled with zinc. Therefore, these levels no longer act as traps. Consequently, the thermoluminescence peak at 160°K disappears. Thus, the infrared properties of the ZnS:Cl crystals are not in conflict with our basic assumptions relating

---

*The luminescence intensity does not decrease because the 466 nm band increases in intensity.

to the nature of the luminescence centers responsible for the 496, 466, and 426 nm luminescence bands.

§ 2. Chlorine-Free Crystals

The chlorine-free crystals have luminescence bands with maxima at 466, 426, and 405 nm, i.e., they must have at least three types of luminescence center.

The properties of the 466 nm band differ sharply from the properties of the 496 nm band of the chlorine-doped crystals. The properties we have in mind are thermal quenching which occurs at very low temperatures (Fig. 7) and which is responsible for the absence of the blue luminescence of the "pure" crystals at room temperature. The other important property is the strong enhancement of the intensity of the 466 nm band which results from the introduction of excess zinc. The zinc treatment also enhances the thermoluminescence peak at 110°K which is due to interstitial zinc [18]. Hence, it is likely that the centers associated with the 466 nm band are due to excess zinc.*

Chemical considerations show that the formation of interstitial zinc is very likely: zinc ions have very small radii and, therefore, can enter easily the interstitial voids without disturbing zinc vacancies. This is confirmed by the fact that relatively short (~2-h) treatments in molten zinc are sufficient to reach the desired effects: zinc vacancies begin to fill only after prolonged immersion in zinc. The low activation energy of the interstitial zinc centers (~0.15 eV) is in good agreement with the ease of transfer of electrons and holes from such centers to the deeper centers associated with the 496 and 426 nm bands, as observed experimentally (the last two bands increase in intensity when the temperature is raised).

The chlorine-free crystals have a strong thermoluminescence peak at 157°K which cannot be due to chlorine [18] but must be attributed to sulfur vacancies. The depth of the levels responsible for this peak was not determined in [18] because the peak was not sufficiently well defined. However, in view of the relatively high temperature of this peak, the thermal depth of the V_S levels should be greater than the thermal depth of the Zn_i levels. The centers responsible for the 405 nm band of ZnS could be associated with these vacancies.

Let us now consider how other results of our investigation fit the proposed interpretation.

1. The temperature dependences indicate that the 426 nm band increases in intensity (as in the case of the chlorine-doped crystals) when the temperature is raised. This increase can be explained quite easily by the arrival of carriers from other centers present in the chlorine-free crystals. These carriers may be provided by the centers responsible for the 466 nm band which is quenched at lower temperatures. This is naturally accompanied by a fall in the intensity of the 466 nm band (Fig. 7). Thus, once again the short-wavelength part of the spectrum predominates at higher temperatures (this is also observed in the chlorine-doped crystal). However, since there are practically no centers responsible for the 496 nm band, the rise in the intensity of the short-wavelength luminescence of the "pure" samples occur only in the range up to T = 160°K and to 260°K, as observed in the chlorine-doped crystals (this is due to the fact that carriers are supplied only by the 466 nm centers).

2. The results of treatment in molten sulfur and zinc can also be explained satisfactorily. The experiments show that the sulfur treatment destroys almost completely the 405 nm band which is not observed even when the $\lambda_{exc} = 312$ nm line is used (Fig. 7). This is to be expected because the sulfur treatment should fill the V_S vacancies with sulfur, which destroys the V_S

---

*Note added in proof. In a recent paper Gutan et al. [37] demonstrated that the 466 nm band and the thermoluminescence peak at 110°K were both associated with interstitial zinc. This was deduced from mass spectroscopy and from luminescence of ZnS single crystals.

traps, as indicated by the thermoluminescence curves. The sulfur treatment does not affect the 466 nm band (Fig. 7), which is again in agreement with the proposed interpretation.

Different results are obtained after treatment in molten zinc: short treatment results in the penetration of the zinc into interstices and the formation of a large number of Zn_i levels. This increases strongly the intensity of the 466 nm band. The 77°K intensity of the 426 nm band decreases slightly, probably because of the transfer of excited electrons from traps to the Zn_i capture levels whose number is increased strongly by the zinc treatment. However, when the temperature is raised, the transfer of electrons from the de-excited Zn_i traps is very large and this enhances strongly the 426 nm band.

The 405 nm band is also enhanced by the zinc treatment: this can be explained by the formation of sulfur vacancies during this treatment.

3. The strong effect of infrared radiation on the "pure" crystals is observed only for the samples treated in molten sulfur or molten zinc (Fig. 12).* The appearance of an enhancement-sensitivity peak can logically be linked to the 110°K thermoluminescence peak resulting from the presence of the Zn_i levels. The depth of these levels is less than the depth of the chlorine traps (160°K) and, therefore, the sensitivity peak is located at longer wavelengths (1.3 μ instead of 1 μ). The sensitivity peak at 2.6 μ is probably due to the liberation of electrons from the continuum of the very shallow levels and the recombination of these electrons at the centers responsible for the 466 nm band. Naturally, the intensity of this peak increases strongly when the number of such centers becomes larger.

Little information can be deduced on the influence of sulfur from the strongly broadened sensitivity spectrum represented by curve 2 in Fig. 12. However, it is clear that the introduction of sulfur destroys the V_S levels so that the luminescence is dominated by the 466 nm band. The strong broadening of the spectrum is probably due to the fact that the very large sulfur ions displace the Zn_i ions from their earlier positions. In view of the different displacements, the level structure becomes diffuse, which is manifested as a broad infrared sensitivity spectrum.

CONCLUSIONS

It follows from the above discussion that practically all the results obtained in an experimental investigation of the properties of ZnS and ZnS:Cl crystals are in agreement with the proposed interpretation of the nature of the luminescence centers responsible for the blue luminescence bands located at 496, 466, 426, and 405 nm.

This allows us to propose a scheme of the energy levels of the centers found in our crystals. Since each of the blue luminescence bands is fairly wide even at liquid helium temperature, it is possible that these bands can be decomposed into simpler subbands. In particular, the postulated complexes may consist of identical partners separated by different distances, i.e., they can be donor—acceptor pairs.

We shall assume that the optical width of the forbidden band of cubic ZnS is the average of the values deduced from measurements of the absorption in nonluminescent crystals [36], which is ~3.9 eV, and the value reported in [9] for luminescent crystals, which is 3.69 eV. Thus, we shall start with the average value of 3.79 eV. The two values reported in [36, 9] were obtained at room temperature. We can calculate the forbidden band width at 77°K using the generally adopted temperature coefficient of 5×10^{-4} eV/deg. It follows that the forbidden band at

---

*Once again we shall consider only the 466 nm band because the study of the influence of infrared radiation was carried out without resolution of the spectrum in its component bands.

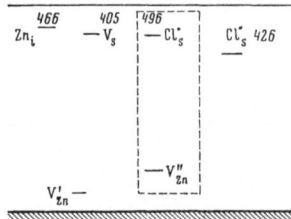

Fig. 14. Proposed energy level scheme of self-activated
ZnS:Cl crystals.

77°K should be 0.11 eV wider. Thus, we can assume that the forbidden band at 77°K is about 3.9 eV wide, as shown in the energy level scheme in Fig. 14.

The nature of the levels responsible for the various luminescence bands is indicated in the energy scheme. However, the transitions responsible for the luminescence bands are omitted deliberately because there is as yet no direct evidence in support of the proposed interpretation of these transitions. Moreover, the energy level scheme of Fig. 14 ignores the nonradiative transitions and the fact that there may be some additional luminescence bands. We have also ignored the influence of an admixture of hexagonal structure which may be present in the basically cubic lattice of ZnS crystals.

We regard the nature of the individual centers responsible for the four blue luminescence bands of ZnS crystals as reliably established. This interpretation can be summarized as follows:

the 2.5 eV (496 nm) band is due to the $(V_{Zn}Cl_S)'$ complexes;
the 2.66 eV (466 nm) band is due to the interstitial zinc Zn_i;
the 2.91 eV (426 nm) band is probably due to free Cl_S donors;
the 3.07 eV (405 nm) band is probably due to sulfur vacancies V_S.

The activation energies of these centers can be arranged in the following sequence:

$$E_{Zn_i} \approx E_{V_S} < E_{(V_{Zn}Cl_S)'} < E_{Cl_S}.$$

We must bear in mind that although the resolution of the blue luminescence spectrum of the "pure" ZnS crystals at 77°K yields a weak band with a maximum at 496 nm, the properties of this band are quite different from the 496 nm band observed in the chlorine-doped samples. For example, the intensity of the former band does not increase with rising temperature and the band does not exist at room temperature; carriers are not transferred from the centers responsible for this band to the centers associated with the short-wavelength luminescence; and so on. It is probable that the 496 nm band of the "pure" single crystals is of a different nature, as expected because of the absence of chlorine in such crystals. This band was not studied in detail in our investigation.

Our results show that the contradictions between the earlier hypotheses relating to the nature of the luminescence centers in self-activated crystals can be attributed to the fact that the spectrum consists of several bands. The samples used by different workers were prepared under different conditions. Therefore, one particular type of center predominated in these samples. The blue luminescence was then ascribed to that particular center and the whole blue spectrum was attributed to such centers. Although the conclusions drawn in these various cases were correct in the strictly local sense, they conflicted with the conclusions reported by other workers.

We are grateful to M. V. Fok and A. N. Georgobiani for their valuable advice.

LITERATURE CITED

1. H. Ortmann and H. Hartmann, Z. Naturforsch., 16a:903 (1961).
2. T. B. Tomlinson, J. Electron., 2:166 (1956).
3. H. Gobrecht, H. Nelkowski, and P. Albrecht, Z. Naturforsch., 16a:857 (1961).
4. A. M. Gurvich, Usp. Khim., 35:1495 (1966).
5. H. Samelson and A. Lempicki, Phys. Rev., 125:901 (1962).
6. F. A. Kröger and H. J. Vink, J. Chem. Phys., 22:250 (1954).
7. A. A. Bundel' and A. I. Rusanova, Izv. Akad. Nauk SSSR, Ser. Fiz., 13:173 (1949).
8. A. A. Bundel', Izv. Akad. Nauk SSSR, Ser. Fiz., 26:527 (1962).
9. A. N. Georgobiani and H. Friedrich, Abstracts of Papers presented at All-Union Conf. on Chemistry and Physics of $A^{II}B^{VI}$ Compounds, Uzhgorod, 1969 [in Russian].
10. A. N. Georgobiani and V. I. Steblin, Electroluminescent p−n Junctions in Zinc Sulfide, Preprint from Lebedev Physics Institute, Academy of Sciences of the USSR [in Russian] (1966).
11. A. N. Georgobiani and V. I. Steblin, Optical and Electrical Properties of p−n Junctions in ZnS, Preprint from Lebedev Physics Institute, Academy of Sciences of the USSR [in Russian] (1967).
12. A. N. Georgobiani and V. I. Steblin, Fiz. Tekh. Poluprov., 1:934 (1967).
13. A. N. Georgobiani and V. I. Steblin, Fiz. Tekh. Poluprov., 1:931 (1967).
14. V. A. Ryzhkov and B. T. Fedyushin, Opt. Spektrosk., 13:721 (1962).
15. V. E. Oranovskii and M. P. Volovei, ZhETF Pis. Red., 5:256 (1967).
16. A. I. Blazhevich, A. V. Lavrov, and E. I. Panasyuk, Izv. Akad. Nauk SSSR, Ser. Fiz., 33:980 (1969).
17. V. F. Tunitskaya, Zh. Prikl. Spektrosk., 10:1004 (1969).
18. Z. P. Kaleeva, E. I. Panasyuk, V. F. Tunitskaya, and T. F. Filina, Zh. Prikl. Spektrosk., 10:819 (1969).
19. V. L. Levshin and V. F. Tunitskaya, Opt. Spektrosk., 18:328 (1965).
20. V. L. Levshin and V. F. Tunitskaya, Opt. Spektrosk., 9:223 (1960).
21. E. E. Bukke, T. I. Voznesenskaya, N. P. Golubeva, N. A. Gorbacheva, Z. P. Ilyukhina, E. I. Panasyuk, and M. V. Fok, This volume, p.23.
22. M. V. Fok, this volume, p. 1.
23. R. V. Vasil'ev, O. N. Karpukhin, and V. L. Shlyapintokh, Zh. Fiz. Khim., 35:461 (1961).
24. K. Era, S. Shionoya, and Y. Washizawa, J. Phys. Chem. Solids, 29:1827 (1968).
25. K. Era, S. Shionoya, Y. Washizawa, and H. Ohmatsu, J. Phys. Chem. Solids, 29:1843 (1968).
26. T. C. Reshetina and V. F. Tunitskaya, Zh. Prikl. Spektrosk., 12:295 (1970).
27. V. F. Tunitskaya, Zh. Prikl. Spektrosk., 12:722 (1970).
28. M. V. Fok, Fiz. Tekh. Poluprov., 4:1009 (1970).
29. A. M. Gurvich, Lectures on Physical Chemistry of Crystal Phosphors [in Russian], Izd. MIÉM, Moscow (1967).
30. K. A. Müller and J. Schneider, Phys. Lett., 4:288 (1963).
31. V. S. Gavrilov and V. A. Shutilov, Fiz. Tverd. Tela, 8:621 (1966).
32. R. H. Kasai and Y. Otomo, J. Chem. Phys., 37:1263 (1962).
33. T. Koda and S. Shionoya, Phys. Rev., 136:A541 (1964).
34. É. Ya. Arapova, V. L. Levshin, N. V. Mitrofanova, T. S. Reshetina, V. F. Tunitskaya, V. V. Shchaenko, and S. A. Fridman, Izv. Akad. Nauk SSSR, Ser. Fiz., 30:573 (1966).
35. G. Baur, N. Riehl, and P. Thoma, Z. Phys., 206:229 (1967).
36. M. N. Alentsev and E. I. Panasyuk, Opt. Spektrosk., 5:207 (1958).
37. V. B. Gutan, V. S. Kupev, A. V. Lavrov, and E. I. Smagina, Opt. Spektrosk. (in press).

USE OF CRYSTAL PHOSPHORS IN DETECTION
OF ELECTROMAGNETIC RADIATIONS

V. L. Levshin, N. V. Mitrofanova, Yu. P. Timofeev,
S. A. Fridman, and V. V. Shchaenko

Experimental and approximate theoretical investigations were made of the optical and thermal effects of infrared radiation on the photoluminescence of activated crystal phosphors, mainly those based on zinc sulfide. The relative efficiencies of the optical and thermal effects on the luminescence of screens were estimated and compared. It was found that these effects could be used in practice to study the spatial distributions of the coherent radiation fields of lasers operating in a wide range of wavelengths from near infrared to the millimeter waves. Copper- and cobalt-doped ZnS phosphors were developed for the near infrared region. The quantum efficiency of the optical quenching of the afterglow of these phosphors by infrared radiation ($\lambda_{IR} \approx 1.3~\mu$) could be as high as 10%. Temperature differences of less than 1 deg C (in an interval of 100 deg above room temperature) and corresponding infrared radiation powers down to 1 mW/cm^2 were detected with specially developed temperature-sensitive crystal phosphors in which the thermal effect of infrared rays was utilized in a wide range of wavelengths. The highest rate of the thermal quenching of the luminescence of these phosphors was 25% per degree Celsius, which was close to the theoretical limit deduced from an analysis of the kinetics of recombination radiation based on the band model of crystal phosphors. These phosphors were found to be suitable for determination of temperature fields in various practical applications.

INTRODUCTION

The appearance of high-power lasers capable of generating directional coherent beams has made it necessary to develop methods for investigating the spatial and temporal characteristics of electromagnetic radiations of wavelengths ranging from the near infrared to short radiowaves. Hence, it has become necessary to find various means for detection of electromagnetic quanta outside the range of sensitivity of the photoelectric effect and of photographic emulsions.

Many infrared detectors are already available and most of them operate on the basis of the thermal effect. The threshold sensitivity of these instruments is quite satisfactory for the detection and investigation of the fields of most of the laser sources. However, such thermal detectors cannot be used directly to determine the spatial distribution of the radiation field except as movable probes: it is not possible to record the instantaneous pattern of the whole radiation field.

Other methods for investigating infrared radiation include the use of luminescent screens coated with films of powder phosphors sensitive to such radiation.

The development of the luminescent methods of visualization of electromagnetic fields is the subject of the present paper. The paper will describe the conditions which must be maintained during synthesis of new phosphors and the determinations of the principal parameters that govern the action of these phosphors in infrared detection.

Infrared radiation can be detected and investigated by means of luminescent screens on the basis of the optical and thermal effects of infrared radiation.

The optical effect is the stimulation or quenching, by infrared rays, of the phosphorescence of a screen excited first with ultraviolet rays. Under the action of infrared rays the light sum stored by the screen phosphor is partly liberated (this enhances the phosphorescence intensity) or this light sum is reduced because of nonradiative transitions induced by infrared rays. The presence of infrared radiation and its distribution across the surface of a screen are determined from the contrast between the stimulated and quenched parts of the screen.

Much work on the stimulating effect of infrared radiation has been done in the Soviet Union and elsewhere. The theoretical and practical aspects of the stimulation effect have been studied quite thoroughly. The stimulation effect is more sensitive to infrared radiation than the quenching but the duration of a stimulated flash is relatively short and the flash amplitude decreases as the light sum of the phosphor in question is gradually depleted. Therefore, in those cases when it is necessary to record the time-average rather than the instantaneous distribution of the em field density, the optical quenching method has definite advantages. This approach makes it possible to observe the results of the action of infrared radiation over an interval of time and to record photographically the quenched distribution of the luminescence emitted by a screen for a considerable time after infrared illumination.

The optical methods can be used in investigations of infrared rays of wavelengths up to $3-4\ \mu$ but most of the luminescence screens are sensitive only up to $1.5-1.7\ \mu$. Therefore, it is necessary to use the thermal quenching of the luminescence of screens in investigations of infrared radiation of longer wavelengths.

In the thermal method one uses the thermal quenching effect which reduces the intensity of luminescence at the points where a screen in heated by the radiation being investigated. Much higher incident radiation powers are needed in the thermal method but satisfactory results have been obtained for many lasers and other modern high-power infrared and radiofrequency sources.

We shall consider in detail the possibility of detection of infrared radiation and of radio waves as a result of optical and thermal interaction with luminescent screens. In the optical case we shall consider a ZnS:Cu:Co phosphors whose sensitivity maximum is located at $\lambda = 1.3\ \mu$. The thermal quenching method will be illustrated by considering ZnS:CdS:Ag:Ni phosphors. These phosphors were studied first in the laboratory and then used in some practical applications.

The change in the temperature as a result of the interaction between infrared radiation and a phosphor can be deduced not only from the quenching of the luminescence of the phosphor but also from changes in the color of that luminescence. In the case of two-activator phosphors such as ZnS:Ag:Sm a change in the temperature by a few degrees can alter drastically the color of the luminescence which is transformed from blue into red or conversely.

The thermal quenching and the change in the color of luminescence can be used conveniently to determine the temperature distributions on the surfaces of complex objects. Such distributions are very important in some problems encountered in modern technology and medicine. Preliminary investigations of practical applications have given encouraging results.

In later chapters we shall consider the optical and thermal properties of the aforementioned phosphors (ZnS:Cu:Co, ZnS·CdS:Ag:Ni, ZnS:Ag:Sm) and of other materials. We shall also calculate the optimal conditions for the optical quenching, thermal quenching, and change of color of the luminescence of crystal phosphors used in investigations of the structure of electromagnetic fields and of temperature distributions over various complex bodies.

CHAPTER I

MECHANISM OF OPTICAL AND THERMAL INTERACTION BETWEEN ELECTROMAGNETIC RADIATIONS AND ACTIVATED CRYSTAL PHOSPHORS. POSSIBLE APPLICATIONS OF SUCH INTERACTIONS

§ 1. Historical Review and Potential Applications of Crystal Phosphors in Studies of Electromagnetic Radiations

The quenching and stimulating effects of infrared radiation on the afterglow of phosphors were first observed by Becquerel [1]. A detailed investigation of these phenomena was made by Lenard [2], who gave a qualitative description of the infrared stimulation of excited phosphors. The experimental investigations were made in parallel with theoretical applications of stimulation and quenching in investigations of the Fraunhofer lines in the solar spectrum and in photography [3-7]. However, these investigations were not continued.

In the Soviet Union the first investigations of the effect of infrared radiation on excited phosphors were carried out by Levshin, Antonov-Romanovskii, and Tumerman [8]. They studied the possibility of utilization of the quenching of excited phosphors in infrared photography. The sensitivity of zinc and cadmium sulfide crystal phosphors to infrared rays of different wavelengths was determined quantitatively. The first infrared photographs were recorded.

During the Second World War these investigations were started again and greatly expanded. New very efficient flash phosphors based on CaS, SrS, and CaS·SrS, activated with two rare-earth elements (Ce and Sm or Eu and Sm), were synthesized [9-14].

After the end of the War it was found that similar investigations were carried out in the USA by Urbach, Schulman, and others under the leadership of O'Brien [15-16]. The results obtained by the American investigators were very close to those reported by the Soviet authors.

Schulman, and others under the Leadership of O'Brien [15-16]. The results obtained by the American investigators were very close to those reported by the Soviet authors.

Fridman and Cherepnev [18] were the first to develop ZnS:Cu:Co phosphors exhibiting long-lived afterglow and, as demonstrated in later investigations, sensitive to the quenching effect of infrared rays. Urbach, Nail, and Pearlman [17] prepared ZnS·CdS:Ag:Ni phosphors exhibiting strong thermal quenching.

Since the War numerous experimental investigations have been made of the infrared sensitivity of alkaline-earth phosphors and of the ZnS and CdS group. Detailed studies have been made of the kinetics of stimulation flash and de-excitation of phosphors. Some practical applications, including phosphor dosimeters for hard radiations [19], were developed.

The quenching effect in phosphors has been used in photocopying processes and in investigations of damage to art objects, which were illuminated with infrared radiation and photographed with the aid of these phosphors [20, 21].

The stimulation and quenching of the luninescence of crystal phosphors by infrared radiation have not found a wide range of applications because of the relatively low sensitivity, narrow range of the sensitivity extending only to 1.5 μ (there are other efficient detectors which can work in this range), and relatively poor resolving power of luminescent screens which fail to reproduce finer details.

In the last decade the situation has changed drastically because of the development of high-power coherent and directional light sources in the form of lasers [22]. The relatively low sensitivity of crystal phosphors to infrared radiation became less important because of the high powers available. On the other hand, the extreme simplicity and convenience of luminescent screens and the possibility of fast and complete recording of the structure and density of electromagnetic fields made these screens extremely effective in studies of the nature and intensity of the output radiation of lasers.

Luminescent screens are currently employed in studies of the mechanism of laser action and the space—time characteristics of the radiation fields of lasers are being recorded with these screens (these characteristics are very important in quantum electronics) [23, 24]. Visualization and quantitative studies of the spatial distribution of the intensity of the output radiation of a laser are essential in the determination of some of the most important laser characteristics such as the angular divergence of a beam, which determines possible applications of lasers in communications, the mode structure, the spiking operation, and the diffraction and interference between various coherent beams. Investigations of these characteristics of laser radiation are also essential in many practical applications.

The range of the wavelengths which can be generated in lasers has become extremely wide, because these wavelengths can be varied by altering the active substance and the excitation conditions. In this way monochromatic radiation can be generated at wavelengths ranging from near infrared to short radiowaves.

The high and continuously growing output powers of lasers, which exceed considerably the output powers of thermal sources, make it possible to record infrared fields by means of less sensitive but more convenient methods. One of these methods utilizes the effect of infrared radiation on luminescent screens, which makes it possible to visualize infrared radiation fields and to study them quantitatively. In this case a screen acts as a nonselective infrared detector, which makes it possible to visualize radiation of $\lambda \sim 10 \mu$ (these wavelengths cannot be made visible with the aid of the optical effect of infrared radiation on phosphors).

In view of the importance of the studies of radiation fields of lasers and the possibility of visualizing these fields with the aid of phosphors, a decision was made at the P. N. Lebedev Physics Institute to carry out detailed investigations of the optical and thermal effects of electromagnetic radiations on various classes of crystal phosphors, concentrating mainly on those based on ZnS.

It was established that in the near infrared range one can use successfully the phosphors whose luminescence is either quenched (ZnS:Cu:Co) or stimulated (SrS·CaS:Eu:Sm, CaS·SrS: Ce:Sm) to solve completely the problems encountered in investigations of laser radiation and in other practical applications. It was found that luminescent screens could be used to study the radiation emitted by various gas lasers and by solid-state lasers with rare-earth ions (CaF_2: Tm^{2+}, $CaWO_4:Nd^{3+}$, etc.).

Our investigations demonstrated that the best results in detection of laser radiation with crystal phosphors were achieved for laser output radiation of $\sim 1.3 \mu$ wavelength. This wavelength lies at the limit of sensitivity of other optical radiation detectors (image converters, photocells, photographic plates) and, therefore, the luminescent method has considerable advantages. The use of ZnS:Fe phosphors developed earlier with the participation of the present authors [25] makes it possible to extend the low-temperature optical sensitivity of the luminescent screen method to $\lambda \approx 3 \mu$.

In the laser radiation range $\lambda > 3\ \mu$ one can use satisfactorily the thermal effect of infrared radiation on ZnS · CdS:Ag phosphors doped with Ni (a quenching agent) or with Sm (a second activator) whose thermal sensitivity maxima can be made to lie near room temperature (this done by suitable selection of the composition and the excitation intensity) [26]. We found that the ZnS · CdS:Ag:Ni phosphors exhibited a strong thermal quenching, whereas in the case of the ZnS:Ag:Sm phosphors we observed a change in the color of luminescence, which allowed us to distinguish easily changes in the screen temperature amounting to 1-3 deg C. This sensitivity was quite sufficient for studies of the field of the output radiation of carbon dioxide lasers, which emit in the region of 10 μ. The intensity of the radiation generated by such lasers was so high that it was necessary to apply water cooling to the substrates of luminescent screens [27].

It was also found that the ZnS · CdS:Ag:Ni and the ZnS:Ag:Sm phosphors could be used in the visualization of the diffraction and interference patterns in the millimiter ($\lambda = 2.37$ mm) range of waves emitted by sources of moderate intensity [28]. In such cases a phosphor was deposited as a thin film on a special absorbing substrate and it was used in vacuum (these measures improved considerably the thermal sensitivity).

The high thermal sensitivity of these phosphors can be used in other ways. For example, it has been suggested in the USA that ZnS · CdS:Ag:Ni phosphors can be used in medicine [29] in the diagnostics of malignant tumors, which raise the temperature locally by several degrees and emit thermal radiation of $\lambda \sim 10\ \mu$ wavelength, which cannot be detected by other means.

Direct visualization and determination of temperature fields are also needed in various branches of modern technology, such as studies of the thermal conditions in and close to electronic and semiconductor devices.

The purpose of the present investigation was to study the parameters governing the relevant processes in the phosphors mentioned above as well as the more important characteristics of these phosphors (duration of afterglow, depth of trapping levels, values of light sums). The aim was to determine the optimal working conditions for these phosphors, their sensitivity thresholds, and the range of practical applications.

§2. Estimates of Principal Parameters Governing Recombination Processes in II − VI Crystal Phosphors

The luminescence mechanism and the kinetics of the recombination process which extends, in contrast to the discrete-center luminescence, over the whole of a crystal have been analyzed quite exhaustively for some simple cases [13, 30-33]. These analyses have been based on the band model of solids with a suitable allowance for the influence of local impurity and defect levels, acting as radiative and nonradiative recombination centers as well as nonequilibrium-carrier trapping centers. Theoretical expressions are available for the luminescence efficiency, the duration and light sum of the afterglow, the sensitivity to infrared radiation and temperature, etc. These expressions include the nonequilibrium carrier density and the probabilities of various elementary processes. However, quantitative calculations of these luminescence characteristics which are necessary in comparisons of the theory with the experimental data and in the evaluation of various practical applications of phosphors cannot be carried out without the knowledge of the various parameters representing the crystal lattices of the phosphors in question and of the impurities present in them. In the case of real phosphors these parameters cover a very wide range. Therefore, it is important to estimate the limits of variation of these parameters, which is usually done empirically.

In the experimental part of the present paper (Chaps. II and III) we shall consider mainly the optical and thermal quenching in II–VI crystal phosphors activated with specially selected pairs of impurities [34].

The host lattices of these phosphors are of mixed ionic–covalent type. In view of this and because of the presence of various lattice defects and residual impurities, the energy band structure of such phosphors is quite complex. This complexity makes it possible to prepare phosphors with very different properties by altering their composition. However, this aspect also makes it difficult to carry out theoretical calculations because of the need to allow for many factors. In a preliminary analysis of the capabilities of II–VI crystal phosphors as detectors of electromagnetic radiation based on the optical and thermal quenching effects one can apply an approximate treatment. In this treatment we can use a simplified band model of an extrinsic semiconductor with two types of recombination center, one radiative and the other nonradiative.

Luminescence centers are formed by heavy metals (Ag, Cu, Au, etc.) or by rare-earth ions (Sn, Tm, Eu, etc.) which usually give rise to acceptor levels. Quenching centers are formed by Co, Ni, and Fe, which give rise to deep donor levels. Shallow electron traps, which are easily emptied at temperatures < 293°K are formed by the introduction of Cl, Al, etc. Some levels are formed also by intrinsic lattice defects (V_{Zn}, V_S, Zn_i, etc.) for their complexes.

Figure 1 shows the energy level scheme and the transitions occurring during luminescence in the simplest phosphor.

In the excitation of phosphors we must distinguish two basically different cases: $h\nu_e < E_G$ and $h\nu_e > E_G$, where ν_e is the frequency of the exciting light and E_G is the forbidden band width. In the first case the internal photoelectric effect can occur only as a result of detachment of an electron from an activator (transition 1' in Fig. 1) and the probability of such detachment is governed by the concentration and the nature of the activator. In the second case an electron is detached from a host ion (transition 1).

Experiments show that the absorption coefficient $\varkappa$ of an activator can reach 100–200 cm^{-1} for activator concentrations $C_a = 10^{17}$–10^{18} cm^{-3}, i.e., the absorption cross section of the exciting light is $\sigma_e = \varkappa/C_a \lesssim 10^{-15}$ cm^2. On the other hand, the cross section at the maximum of the activator absorption band is related to the oscillator strength f of an optical transition by [31]

$$\sigma_{e\,max} = \frac{\varkappa_{e\,max}}{C_a} = 1.3 \cdot 10^{-17} \frac{1}{\gamma} \frac{(n^2 + 2)^2}{n} f, \tag{1}$$

where n is the refractive index and γ is the half-width of the absorption band (for ZnS, we have $n \approx 2.4$ and $\gamma \sim 0.3$ eV).

If it is assumed that the oscillator strength is $f = 1$, the above relationship yields the experimentally observed values $\sigma_e \approx 10^{-15}$ cm^2. If the optical excitation density is $W_e \approx 10^{-4}$

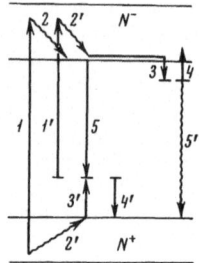

Fig. 1. Simplest energy-
level scheme of a crystal
phosphor. The arrows in-
dicate electrons and hole
transitions.

W/cm^2, which is easily attainable in the case of PRK-4 mercury lamps, the volume density of the excitation of nonequilibrium carriers can reach only $w_e \approx W_e \varkappa / h\nu_e \lesssim 10^{16}$ carriers·cm^{-3}·sec^{-1}. In the case of fundamental absorption by the host lattice (in the case of ZnS this occurs at $\lambda = 313$ nm) the absorption coefficient is 10^4-10^5 cm^{-1} and, consequently, we have $w_e = 10^{18}$-10^{19} pairs·cm^{-3}·sec^{-1}. Thus, the absorption coefficient in the latter case is hundreds or even thousands times higher than in the former case and the excitation density is higher by the same factor. On the other hand, the depth of the active layer of the phosphor, which is governed by the depth of penetration of the exciting light, will be smaller by the same factor than in the first case.

When mixed $SnS_x \cdot CdS_{1-x}$ phosphors are excited with the same wavelength (365 nm), the impurity absorption is observed in the phosphors with low concentrations of CdS and the band— band absorption in the phosphors with high concentrations of CdS. This happens because the forbidden band width decreases from 3.9 eV for pure ZnS to 2.56 eV for pure CdS. This provides a very convenient menas for altering the mechanism of excitation of a crystal phosphor by varying the composition of the host [35, 36]; such variation alters also the color of the emitted luminescence.

The quantum efficiency of the ionization process, i.e., the ratio of the number of nonequilibrium carriers generated by illumination to the number of absorbed quanta, is usually unity. However, in the case of strong excitation this efficiency may be less than unity because of the absorption of the exciting light by localized nonequlibrium carriers. When a phosphor is excited with quanta whose energy exceeds the forbidden band width E_G, the energy efficiency is always less than $E_G/h\nu_e$, because after the ionization of the host lattice the carriers with excess energy (electrons and holes in the allowed bands) are rapidly thermalized as a result of interaction with optical and acoustic phonons (transitions 2 and 2' in Fig. 1) [38].* The duration of the thermalization process, during which the kinetic energy of free carriers decreases from few electron volts to its equilibrium value ($\sim kT$), has been estimated theoretically and experimentally and found not to exceed 10^{-10}-10^{-11} sec. A rough estimate of the duration of the thermalization process can be obtained in the following very simple way. The duration of an elementary event of interaction between a carrier and the phonon field of the lattice is $> 1/\nu_{ph}$, where ν_{ph} is the phonon frequency (10^{13} sec^{-1}), and the number of such interactions during thermalization is a few hundreds. Hence, it follows that the thermalization time is 10^{-11} sec. This time is very short compared with the durations of the subsequent processes. Therefore, we may assume that practically all the free carriers active in the photoluminescence have energies not exceeding a few kT, irrespective of the energy of the incident quanta $h\nu_e$.

Carriers may be localized in vacant trapping levels (processes 3 and 3') or they may participate in radiative or nonradiative recombination with nonequilibrium carriers which are already localized (processes 5 and 5').

We shall now estimate approximately the probabilities of such processes because the nature and the competition between them determine the luminescence kinetics. These processes can be described in terms of the capture and recombination cross sections, the carrier velocity, the free-carrier density, and the density of discrete levels which can be filled or empty. The required probabilities can be written in the form

$$w_{\bar{c}} = N^- \sigma_{\bar{c}} u^- (C_d - n^-), \quad w_{\overline{rr}} = N^- \sigma_{\overline{rr}} \, u^- n^+,$$

$$w_c^+ = N^+ \sigma_c^+ u^+ (C_a - n^+), \quad w_{nr}^+ = N^+ \sigma_{nr}^+ \, u^+ n^-, \tag{2}$$

---

*If the energy of the exciting quanta (or the energy of electrons in cathodoluminescence) exceeds (2-3)E_G, an additional electron—hole pair may be generated and the quantum efficiency of the ionization may increase (this is known as the photon multiplication effect) [38, 39].

where w_c^- and w_c^+ are, respectively, the electron and hole capture probabilities; w_{rr}^- and w_{nr}^+ are the probabilities of radiative and nonradiative recombination of electrons and holes; N^- and and N^+ are the densities of free electrons and holes; C_d and C_a are the concentrations of donor and acceptor impurities; n^- is the density of trapped electrons; n^+ is the concentration of ionized luminescence centers; σ_c^-, σ_c^+, σ_{rr}^- and σ_{nr}^+ are the capture and recombination cross sections of electrons and holes. Since thermalization is rapid, the carrier velocities u^- and u^+ are the thermal velocities: $u_T \approx \sqrt{3kT/m^*}$, where m^* is the effective mass of a carrier. The effective mass of electrons in ZnS phosphors is $m_e \approx 0.28\ m_0$ and the effective masses of holes are $m_{h_1} \approx 0.5\ m_0$ and $m_{h_2} \approx 1.5\ m_0$ (m_0 is the mass of a free electron) so that at T = 300°K the thermal velocities are of the order of $u_T \approx 10^7$ cm/sec and they differ by a factor not exceeding 2.

Densities of the discrete levels C_a and C_d have a lower bound set by the concentration of accidental (residual) impurities in ZnS phosphors, which is $\sim 10^{15}$ cm^{-3}, and an upper bound set by the concentration quenching of the luminescence (which usually occurs at C $\geqslant 10^{18}$ cm^{-3}). In real crystals these level densities vary within two or three orders of magnitude.

The cross sections of the capture (trapping) and recombination processes vary within much wider range. These cross sections are governed by the structure of lattice defects and, particularly, by the charge of these defects. In the case of charged centers the theoretical estimates and the experimental data on semiconductors [39, 40] yield cross sections of 10^{-14}-10^{-12} cm^2; similar estimates give 10^{-16} cm^2 for neutral centers and very low values of 10^{-18}-10^{-20} cm^2 for repulsive centers.

These cross sections are in satisfactory agreement with the experimental values of the frequency factor s (5×10^5 sec$^{-1} \lesssim$ s $\lesssim 1 \times 10^{12}$ sec^{-1}) in the probability of thermal liberation of electrons from various traps in ZnS phosphors, which was obtained by Gobrecht and Hofmann [41] in a study of thermoluminescence.

In view of detailed equilibrium between the processes resulting in the filling (3 and 3') and emptying (4 and 4') of the localization levels, the following equation applies under weak excitation (or high temperature) conditions [32]:

$$\frac{s}{\sigma_c} = N_{eff}\mu_T = \left(\frac{2\pi m^* kT}{h^2}\right)^{3/2} \sqrt{\frac{3kT}{m^*}}, \tag{3}$$

where N_{eff} is the effective density of states in the allowed bands; s is the frequency factor lying within the limits just stated. Substituting the various quantities into Eq. (3) and assuming that $m^* = m$ and T = 293°K, we find that the quoted values of s lead to cross sections in the range from 10^{-14} to 10^{-20} cm^2.

We shall use these values to determine the wide range of the carrier capture times $\tau_c^\pm$ (this range amounts to nine orders of magnitude in the case of ZnS phosphors), which is mainly due to different charge states of the impurities. In fact, it follows from

$$\tau_c^\pm \approx \frac{1}{P_c^\pm} \approx \frac{1}{\sigma_c^\mp u_T C_{a,d}} \tag{4}$$

that for the purest crystal phosphors with the smallest capture centers we have $\tau \leq 10^{-2}$ sec whereas for the phosphors with high impurity concentrations and charged capture centers the value of τ decreases to 10^{-11} sec. We note that the second value is comparable with the thermalization time, i.e., in extreme cases electrons may be captured before thermalization. In view of this wide range of τ an excitation of the same intensity will produce quite different free-carrier densities in different phosphors. On the other hand, it follows from the condition of detailed balance [3] that the free-carrier density $N^\pm$ corresponding to the quoted values of W_e is considerably lower than the density of localized nonequilibrium carriers. The neutrality condition for a phosphor $n^+ + N^+ = n^- + N^-$ leads to equality of the concentrations of the ionized luminescence centers and the density of the localized carriers $n^+ = n^-$, which simplifies the calculations quite considerably.

It should be stressed that the differences due to different values of the parameters of the luminescence centers are greater than the differences due to inequality of their concentrations. In practice, approximate calculations can usually be performed employing the average value of the frequency factor $s \sim 10^8$ sec^{-1} and the corresponding capture cross sections $\sigma_c^{\pm} \sim 10^{-16}$ cm^2 [32].

The recombination cross sections are usually of the same order of magnitude or even smaller: for example, in the case of radiative recombination at neutral centers we have $\sigma_{rr} \sim 10^{-20}$ cm^2 [39]. Thus, the probability of direct recombination may be comparable with the probability of localization only if the concentration of the ionized centers is sufficiently high, i.e., when the excitation intensity is high. Complete filling of the local levels with nonequilibrium carriers is hindered not only by the thermal liberation from these levels but, at high excitation power densities W_e, by the optical liberation of electrons whose probability P_e is proportional to I_e which is the excitation intensity expressed in quanta, i.e.,

$$P_e^{\pm} = \sigma_{el} I_e = \frac{6.3 \cdot 10^{18} \cdot W_e \sigma_{el}}{h\nu_e},$$

where σ_{el} is the absorption cross section of the exciting light by localized carriers.

The liberation of electrons by the exciting light is quite important in the case of deep levels. We find that when $f \approx 1$ and the excitation power density is $\sim 10^{-4}$ W/cm^2, the liberation probability becomes

$$P_e^{\pm} = \sigma_{el} I_e \approx 0.1 \text{ sec}^{-1}.$$

If the frequency factor is $s \sim 10^8$ sec^{-1} and the temperature is T = 300°K, the probability of thermal liberation from levels of depth $E \geq 0.6$ eV is considerably less than the probability of optical liberation by the exciting light and, therefore, such light may determine the population kinetics of the local levels.

The depths of the levels may be even less if the excitation takes place at lower temperatures or when the value of s is less (we have mentioned earlier that this factor can have values down to $\sim 10^6$ sec^{-1}).

One of the most important characteristics among those that govern the intensity and duration of afterglow and the sensitivity to the optical effect of infrared radiation is the optimal light sum stored as a result of prolonged and fairly intense excitation.

When luminescence centers are excited directly ($\varkappa = 100$ cm^{-1}), the conditions for filling the centers with carriers are ideal ($n \sim C_a$), and nonradiative losses are absent, we find that the light sum of the afterglow is $C_a (1/\varkappa) \approx 10^{15}$ photons/cm^2, whereas in the case of band-band excitation ($\varkappa = 10^4$-10^5 cm^{-1}) the light sum is only 10^{12} photons/cm^2. Therefore, in order to obtain a large light sum it is best to use direct excitation of the centers.

We ought to mention, for the sake of comparison, that the threshold sensitivity of photographic films at their sensitivity maxima amount to $\sim 10^8$ photons/cm^2, which is 10^4-10^7 times smaller than the light sums mentioned above. Thus, light sums accumulated in a phosphor film are quite sufficient not only for every strong photographic action but also for multiple photocopying of the image displayed on a luminescence screen and for transmission of images over long distances. These capabilities always exist in the case of ZnS:Cu:Co phosphors in spite of the fact that some liberation of localized carriers occurs during excitation (at low excitation intensities such liberation is a thermal effect whereas at high intensity it is produced by the exciting light itself).

Before we consider the application of crystal phosphors in detection of infrared radiation, we must discuss the efficiency of the optical stimulation and quenching by infrared radiation and the sensitivity of phosphors to such radiation.

Phosphors are sensitive to the optical action of infrared radiation because such radiation is absorbed by and liberates localized carriers (this reduces the density of these carriers) and increases the density of free nonequilibrium carriers. The probability of liberation of localized carriers by infrared radiation $P_{IR}^{\pm}$ is given, like the probability of liberation by exciting light, in terms of a cross section for the absorption of infrared radiation:

$$P_{IR}^{\pm} = \sigma_{IR}^{+} I_{IR} , \tag{5}$$

where the signs $\pm$ refer, respectively, to the liberation of localized holes or electrons, and the maximum value of σ_{IR} is of the order of 10^{-15} cm^2.

In comparisons of the sensitivity to infrared radiation of various classes of phosphors or of the same phosphors but under different excitation conditions (for example at different temperatures) it is usual to employ the concept of the relative stimulation under identical rigorously defined conditions. This quantity, denoted by R, is defined as the ratio of the change in the intensity of afterglow due to infrared radiation $\mathscr{I}_{IR}$ to the intensity of the same afterglow in the absence of such radiation $\mathscr{I}_0$, [32], i.e.,

$$R_{h\nu} = \frac{\mathscr{I}_{IR} - \mathscr{I}_0}{\mathscr{I}_0} , \tag{6}$$

where $\mathscr{I}_{IR}$ is the intensity of the luminescence emitted by a phosphor in the presence of infrared radiation. If the action of infrared radiation is practically instantaneous i.e., it does not have any quenching effect (no liberation of holes takes place), and moreover, only a small fraction of the accumulated light sum is used in the stimulation effect, the change in the intensity of afterglow has a fully defined meaning and it is solely due to a change in the density of free carriers N$^-$ which are liberated by infrared radiation from the traps. In this very simple case the relative change in the free-carrier density is equal to the ratio of the probability of optical liberation of carriers by infrared radiation to the probability of thermal liberation (in the absence of such radiation), i.e.,

$$R_{h\nu_{st}} = \frac{\mathscr{I}_{IR} - \mathscr{I}_0}{\mathscr{I}_0} = \frac{N_{IR}^- - N^-}{N^-} \approx \frac{P_{IR}}{P_T} \approx \frac{\sigma_{IR} I_{IR}}{s e^{-E/kT}} . \tag{7}$$

In the opposite extreme case when infrared radiation exerts only the quenching effect on luminescence centers of depth E, i.e., when such radiation does not affect the localization levels of electrons and, consequently, cannot alter the free-electron density, we can introduce a corresponding concept of the relative sensitivity to the quenching action of infrared radiation:

$$R_{h\nu_q} = \frac{\mathscr{I}_0 - \mathscr{I}_{IR}}{\mathscr{I}_{IR}} = \frac{n^+ - n_{IR}^+}{n_{IR}^+} = \frac{\sigma_{IR} I_{IR}}{s e^{-E/kT}} . \tag{8}$$

These extreme cases are not normally encountered in phosphors, i.e., infrared radiation usually acts simultaneously on localized electrons, giving rise to a stimulation flash, and on localized holes, which gives rise to quenching. In some phosphors (alkaline-earth compounds activated with two rare-earth elements) the stimulation effect predominates, whereas in others (ZnS:Cu:Co) the quenching effect is more important.

Apart from the coexistence of the stimulation and quenching processes, the determination of the relative and absolute sensitivities of phosphors to electromagnetic radiations is additionally complicated by the noninstantaneous nature of the stimulation and quenching processes resulting from infrared radiation. Consequently, the stimulation and quenching effects are functions of time. If low-intensity infrared radiation is used, the weak effect of such radiation may be characterized by a time constant comparable with the duration of afterglow, lasting minutes or even hours. On the other hand, at high infrared radiation intensities, the stimulation and quenching processes are much more rapid but a considerable fraction of the light sum is dissipated and, consequently, the sensitivity of the phosphor to infrared radiation is less. This difference, which is associated with the different methods of comparison of the effects of infrared radiation, may have a considerable influence in comparisons of the optical sensitivity of different classes of phosphors capable of stimulation and quenching (§4, Chap. II). Therefore, the comparison criterion should be selected only when the practical conditions of the application of a given phenomenon are known.

The quantity that is important in practical applications is the quantum effectiveness η'_{IR}, which is governed not by the absorbed but by the incident infrared radiation and, because such radiation is weakly absorbed, it is much smaller than the quantum efficiency η_{IR}. In the case of alkaline-earth phosphors exhibiting the stimulation effect we find that under optimal conditions the quantum efficiency and the quantum effectiveness are $\eta_{IR} \approx 10\%$ and $\eta'_{IR} \lesssim 1\%$ [9], i.e., they differ by one order of magnitude.

The fairly high quantum efficiency of the stimulation effect in these phosphors indicates that in some cases the probability of recombination can be very high (it may be comparable with the capture probability). This is in agreement with the observation that the light sums stored in these phosphors are large because the ratio of the probabilities of these processes is governed by the ratio of free and filled localization levels.

§3. Comparison of the Sensitivities of Crystal Phosphors to the Optical and Thermal Effects of Infrared Radiation

Since the optical and thermal methods of detection of electromagnetic radiation have their advantages and disadvantages, it is essential to compare the sensitivities of these methods so as to make the correct choice under any given conditions. The optical sensitivity of crystal phosphors to infrared radiation has been investigated fairly thoroughly [8-18, 42-46]; however, there are no data on the thermal sensitivity and even theoretical estimates are not available. Therefore, we shall concentrate our attention on the thermal method.

Experimental investigations indicate that the optimal threshold sensitivity to the optical effect of infrared radiation ($10^{-7} W/cm^2$) is considerably higher than the threshold sensitivity to the thermal effect, which lies within the range 10^{-3}-$10^{-2} W/cm^2$ [23, 24, 27-29]. However, the optical sensitivity decreases rapidly with increasing wavelength of infrared radiation and at room temperature it extends no further than $\lambda \sim 2 \mu$. The thermal sensitivity can, in principle, be used to detect fairly strong electromagnetic radiation of any wavelength and it can even be employed in detection of other forms of radiation such as ultrasound [47].

The difference between the optical and thermal effects of infrared radiation on crystal phosphors is due to the fact that in the former case we are dealing with direct selective de-excitation of the localization levels (luminescence centers or carrier traps), whereas in the latter case the infrared radiation energy is first transformed into heat so that the selectivity of the action of this radiation is destroyed. Infrared radiation can transfer localized electrons to the conduction band and localized holes to the valence band provided

$$h\nu_{IR} > E_1. \tag{9}$$

where $h\nu_{IR}$ is the energy of an infrared quantum (photon) and E_1 is the depth of a localization level, measured with respect to the allowed bands. According to this condition, long-wavelength infrared radiation can only liberate carriers from very shallow levels, i.e., the same levels which are very easily emptied by thermal means. Therefore, in order to utilize the optical method of detection of long-wavelength infrared radiation we must have a phosphor which contains very shallow localization levels and the detection process must be carried out at very low temperatures so that

$$\frac{E_1}{kT} \gg 1. \tag{10}$$

This complicates greatly the utilization of phosphors. There are now ZnS-base phosphors which can be operated at liquid nitrogen temperature and which are sensitive to 3-4 μ. Further extension of the sensitivity in the direction of longer wavelengths would be extremely difficult to achieve in the case of ZnS phosphors and it is hardly possible even on theoretical grounds [48].

The use of the second (thermal) method in its pure form, i.e., quenching of the luminescence of a phosphor by infrared radiation absorbed directly by the phosphor, is hindered seriously by the weakness of such absorption. However, this effect can be strongly enhanced by the use of suitable absorbing substrates or coatings, ensuring complete absorption of infrared radiation. For this purpose one can blacken a substrate with gold or soot. A phosphor film on a blackened substrate can be regarded as an enormous system of microscopic nonselective infrared radiation absorbers, which detect the incident radiation and measure its intensity by the degree of quenching of the luminescence. The use of other special substrates makes it possible to extend this effect to radio waves in the millimeter range [28], which is a very desirable feature.

The estimates given below are intended to make more specific the various qualitative comments made above, i.e., it is our intention to compare the sensitivity of the two detection methods at different wavelengths of infrared radiation.

The optical and the thermal action of infrared rays on the photoluminescence of activated crystal phosphors results from the change in the probability of liberation of localized nonequilibrium carriers and a corresponding change in the density of free and localized carriers. This change in the probability can occur (and usually does) simultaneously for several levels. However, we shall simplify our calculations by considering the relative change in the probability of liberation of carriers from the same local level in the optical and thermal action of infrared radiation of the same intensity and wavelength.

The relative change in the probability of liberation of a carrier from a local level as a result of the optical action of infrared radiation of intensity I_{IR} (expressed as the number of quanta) is given by the well-known formula

$$\frac{P}{P_T} = \frac{\sigma_{IR} I_{IR}}{se^{-E_1/kT}} = R_{h\nu}, \tag{11}$$

which specifies the optical sensitivity of a phosphor $R_{h\nu}$ (the notation is the same as in §2).

The change in the probability of liberation of a carrier from a local level in the thermal effect of infrared radiation, which represents — in this simplest case — the thermal sensitivity, can be defined similarly:

$$R_T = \frac{\Delta \mathcal{J}_T}{\mathcal{J}_T} \approx \frac{\Delta P_T}{P_T} = \frac{d\,(se^{-E_1/kT})}{dT\,se^{-E_1/kT}} \Delta T_{IR} = -\frac{E_1}{kT^2} \Delta T_{IR} , \tag{12}$$

where ΔT_{IR} is a small change in the temperature of the surface of a phosphor which is governed by the intensity of the infrared radiation incident on this surface as well as by the conditions of absorption and heat removal from the surface.

In the case of low steady-state infrared radiation intensities, giving rise to small changes in the temperature ($\Delta T_{IR}/T \ll 1$), we have

$$\Delta T_{IR} = \frac{\varkappa_{IR}\,W_{IR}}{\frac{dW_T}{dT}} , \tag{13}$$

where $\varkappa_{IR}$ is the total coefficient of absorption of infrared radiation in a luminescent screen (i.e., the ratio of the absorbed to the incident energy) and dW_T/dT is the change (due to heating by 1 deg C) in the flux of heat removed from the screen.

Temperature changes ΔT_{IR} can be calculated much more simply in the case of optimal conditions of the thermal effect, i.e., when the absorption of infrared ray radiation is sufficiently strong and the thermal contact with the ambient medium is quite poor. This is achieved in practice by blackening the substrate and placing a screen in a thermostat which can be pumped out to 10^{-3} mm Hg (studies of the action of bolometers [49] indicate that under these conditions the convection is eliminated and the sensitivity of the detector increases by about one order of magnitude). Under these optimal conditions the absorption coefficient $\varkappa_{IR}$ can be approximately unity and the heat is lost from the screen solely as a result of thermal radiation emitted from the screen whose blackened substrate emits approximately in accordance with the Stefan–Boltzmann law $W_T = \sigma_{SB} T^4$.

In this case the change in the temperature is given by the simple formula

$$\Delta T_{max} \approx \frac{W_{IR}}{\frac{dW_T}{dT}} = \frac{W_{IR}}{4\sigma_{SB}T^3} , \tag{14}$$

and the thermal sensitivity defined by Eq. (12) becomes

$$R_T = -\frac{E_T}{kT^2} \frac{W_{IR}}{4\sigma_{SB}T^3} = -\frac{E_T}{kT} \frac{W_{IR}}{4W_T} . \tag{15}$$

It is evident from Eq. (15) that the thermal sensitivity is proportional to the depth of the energy level in question divided by kT and to the incident infrared radiation power divided by the power of the radiation emitted by a black body at the average temperature employed in the experiments considered.

Thus, in order to compare the optical and thermal sensitivities we must use Eqs. (11) and (15), and specify the values which occur in these formulas. The thermal sensitivity of Eq. (15) is governed, in the simplest case considered here, by just one characteristic which is the depth of the localization levels. We can use Eq. (15) to find the value of W_{IR} corresponding to a given value of R_T equal to the detection threshold. The depth of the localization levels E or (in the general case which we shall consider in the next section) the activation energy of the thermal quenching E_T lies between 0.1 and 1 eV. If we use the average value $E_T \sim 0.3$ eV, which is typical of ZnS phosphors, we find that at room temperature (kT $\sim$ 0.025 eV, $W_T = 0.045$

W/cm^2) the infrared radiation power corresponding to a change in the probability of thermal liberation by 10% (i.e., that corresponding approximately to the threshold contrast necessary for direct visual observation) amounts to

$$W_{IR}^{th} = 4W_T \frac{kT}{E_T} \left(\frac{\Delta \mathcal{J}_T}{\mathcal{J}_T}\right)_{th} \approx 2 \cdot 10^{-3} \text{ W/cm}^2.$$

This threshold value of W_{IR} is in satisfactory agreement (within the limits of the experimental error) with the results obtained empirically, which confirms the validity of the initial assumptions and of the above elementary analysis of the thermal action of infrared radiation.

The threshold intensity of infrared radiation corresponding to the threshold value of the probability of optical liberation of Eq. (11) can be found if we know not only the depth of the localization levels but also the absorption cross section of infrared radiation $\sigma_{IR}(\nu)$ and the frequency factor in thermal liberation s.

The hydrogen-like model of local centers, which has been employed in studies of shallow donor and acceptor levels in semiconductors [40], yields the following absorption cross section for infrared quanta of energies $h\nu_{IR} \gtrless E$:

$$\sigma_{IR} = \frac{3.5 \cdot 10^{-17}}{E \sqrt{\chi}} \left(\frac{m}{m^*}\right) \left(\frac{E}{h\nu}\right)^{5/2}, \tag{16}$$

where χ is the permittivity; m and m^* are, respectively, the mass of a free electron and the effective mass of a carrier.*

If we consider ZnS and assume that $m^* = 0.28m$ [50] and $\sqrt{\chi} = 2.4$, we obtain the following value of the absorption cross section for $E \approx h\nu_{IR}$: $\sigma_{IR} = 10^{-15}$ cm^2, which corresponds to transitions that are characterized by an oscillator strength $f \sim 1$ and are observed expressed experimentally in semiconducting compounds.

The maximum value of the frequency factor observed for alkali-halide phosphors is close to the frequency of optical phonons (i.e., $\sim 10^{13}$ sec^{-1}) whereas the factor for ZnS lies within a wide range 10^6 sec$^{-1} < s < 10^{12}$ sec^{-1}, which has a considerable influence on the spectral and temperature dependences of the ratio $R_{h\nu}/R_T$.

Assuming optimal conditions for the optical effect of infrared radiation ($f \sim 1$ and $s \sim 10^6$ sec^{-1}) and going over from the energy to the quantum relationships for traps of the same depth as before ($E = 0.3$ eV), we find that Eq. (11) yields the following infrared power density necessary for a 10% change in the liberation probability at room temperature (in the range $h\nu_{IR} \gtrless E$):

$$W_{IR}^{th} = 0.1 \frac{se^{-E/kT}}{\sigma_{IR} \cdot 6.3 \cdot 10^{18}} \approx 0.1 \frac{10^6 \cdot 10^6}{10^{-15} \cdot 6.3 \cdot 10^{18}} \sim 10^{-5} \text{ W/cm}^2.$$

This optimal value of the optical sensitivity is also in satisfactory agreement with the experimental results. The assumed depth of levels is approximate and it occurs also in the argument of the exponential function in Eq. (11). Consequently, the precision of determination of the threshold values of W_{IR} is low. Moreover, not all the infrared energy is utilized in the liberation of carriers (some of it is dissipated by the polarization of the lattice). Consequently, carriers will be liberated not when $h\nu_{IR} > E$ but when $h\nu_{IR} \geq E_T + E_p$, where E_T is the usual thermal depth of a localization level and E_p is the energy lost due to the polarization of the lattice in each liberation event. The polarization losses $E_p = E_{h\nu} - E_T$ are in a sense analogous

---

*The factor $h\nu$ appears because I_{IR} in Eq. (11) is expressed in numbers of quanta whereas W_{IR} in Eq. (15) is the power density in W/cm^2.

to the Stokes losses and they depend strongly on the degree of ionicity of the lattice. In the case of alkali halide compounds the value of E_p may reach 1 eV [51], whereas in the case of semi-conducting covalent crystals it is negligibly small (≤ 0.01 eV) [40]; for II–VI compounds which are characterized by mixed ionic-covalent bonds the value of E_p is intermediate [31, 48]. An exact calculation of the degree of ionicity and an estimate of the value of E_p for ZnS phosphors is quite a difficult problem, which has not yet been solved. Various estimates of the degree of ionicity agree that the ratio of the ionic and covalent components of the bonding in ZnS and CdS is of the order of unity; this is why more precise calculations are difficult. Some workers [40] assume that the bonding is 60% covalent, whereas others [48] estimate the ionic component as 60%. In particular, it is suggested in [48] that polaron states of electrons and holes with energies of 0.27 eV can appear in ZnS and these states govern the difference between the energies of levels active in the thermal and optical liberation methods.*

The difference between the thermal and optical ionization energies is known quite accurately for ZnS phosphors. However, this difference is not just due to the existence of the polaron states but also due to the structure of the energy bands and to the forbiddenness of the direct transitions.

The principal difficulty preventing refinement of the numerical estimates of the optical sensitivity of phosphors to infrared radiation is the lack of an agreed point of view on the reasons responsible for the difference between E_{op} and E_T and on the absolute value of this difference (according to [48], this difference is 0.6 eV, whereas experimental data obtained for shallow traps from the edge luminescence indicate that this difference does not exceed 0.1 eV [52]).

Other important parameters are the relative changes in the luminescence brightness resulting from the optical and thermal effects of infrared radiation, given by the quantities $\Delta P_T / P_T \approx E/kT^2$ and $\sigma_{IR} I_{IR}/P_T$, as well as the absolute quantum sensitivities of the optical and thermal methods of detection of infrared radiation.

The quantum efficiency (sensitivity) of the optical effect is defined as the ratio of the IR-induced change in the number of visible light quanta induced to the number of infrared quanta absorbed by the phosphor in question. The highest published value is 10%: this has been reported for alkaline-earth phosphors and it indicates, in particular, that these phosphors are characterized by high values of the ratio of the probability of direct recombination to the probability of recapture and by fairly high values of the efficiency η_{IR}:

$$\eta_{IR} \leqslant \frac{1}{\dfrac{w_c}{w_{dr}} + \dfrac{1}{\eta}} . \tag{17}$$

In spite of the fact that the mechanism of the thermal effect of infrared radiation is basically different from the corresponding mechanism of the optical effect, we can still introduce the quantum efficiency (sensitivity) of the thermal effect of infrared radiation on the photoluminescence of a phosphor, in full analogy with the optical effect. We shall define the quantum efficiency of the thermal effect η_T as the ratio of the reduction in the number of quanta of the visible luminescence of a phosphor ($\Delta \mathscr{J}_1$) to the number of infrared quanta I_{IR} whose absorption in the phosphor (luminescent screen) is responsible for the observed quenching of the steady-state luminescence. We thus obtain

$$\eta_T = \frac{\Delta \mathscr{J}_1}{I_{IR}} = I_e \frac{d\eta_{lq}}{dT} \frac{\Delta T}{I_{IR}} = \frac{W_e}{h\nu_e} \eta_{lq} \left(\frac{d\eta_{lq}}{dT} \frac{1}{\eta_{lq}} \right) \frac{\Delta T}{I_{IR}} , \tag{18}$$

---

*According to Pekar's theory [53] the difference between the optical and thermal depths amounts to two-thirds of the polaron energy because the energy required for optical annihilation of the polaron state is 3 times as high as the energy for the thermal annihilation.

where I_e is the intensity of the ultraviolet excitation (expressed as the number of quanta) and η_{1q} is the quantum efficiency of the luminescence of the phosphor.

If we use the relationship (14) derived earlier for the optimal heating of a luminescent screen $\Delta T = W_{IR}/4\sigma_{SB}T^3 = I_{IR}h\nu_{IR}/4\sigma_{SB} T^3$ and if we bear in mind that in the region of strong thermal quenching $\eta_{1q} \approx \eta_0 e^{-E_T/kT}$, we find that Eq. (18) leads to

$$\eta_{Tq} \approx \eta_{1q} \frac{W_e}{4W_T} \frac{E_T}{kT} \frac{h\nu_{IR}}{h\nu_{UF}}. \tag{19}$$

For example, if we assume that $\eta_{1q} \sim 0.1$ (strong thermal quenching), $W_e \sim 10^{-4}$ W/cm^2 (which is the usual level of ultraviolet excitation power), $h\nu_{UF} \approx 3$ eV, $h\nu_{IR} = 0.3$ eV, and $E_T \approx 0.3$ eV, we find at room temperature (kT ≈ 0.025 eV, $W_T \approx 0.05$ W/cm^2) the quantum efficiency of the thermal effect of infrared radiation is $\eta_{T_q} \sim 10^{-4}$ quantum/quantum.

Thus, the quantum efficiency of the thermal liberation of carriers at T = 300°K is about 10^{-4} quantum/quantum for $E_T \approx 0.3$ eV, i.e., this efficiency is only two orders of magnitude lower than the highest efficiency of the optical effect.

The difference between the thermal and optical effects of infrared radiation is relatively small due to the more complete absorption of infrared radiation in a blackened substrate (90%) than directly in the phosphor itself (not exceeding 10%) because the quantum efficiency and the quantum effectiveness of the thermal action of infrared radiation are quite similar.

We shall now compare the sensitivities (efficiencies) of luminescent and thermoelectric detectors of infrared radiation. In both cases the incident infrared radiation alters the density of free and localized carriers and the mechanisms of the processes involved are similar. In the case of thermoelectric detectors (for example, bolometers) we measure the change in the current, i.e., the electron flux, resulting from the exponential temperature dependence of the resistivity $\rho_T = \rho_0 e^{E_G/2kT}$, where ρ_0 is a constant and E_G is the forbidden band width. In the case of luminescent-thermal detectors we measure the change in the luminescence brightness, i.e., in the photon flux, whose temperature dependence is again (at least in the simplest case) given by an exponential formula $\mathcal{I}_T = \mathcal{I}_0 e^{E_T/kT}$, where I_0 is a constant and E_T is the activation energy of the thermal quenching. Thus, we can compare directly the relative change in the measured parameter (the current or the luminescence brightness) resulting from the same small change in the temperature caused by direct heating or by the absorption of infrared rays (these rays must be of the same energy in both cases).

The relative value of the change in the luminescence brightness or in the electrical resistivity, resulting from the exponential temperature dependence of the parameter in question, is given by

$$\begin{aligned}
\frac{\Delta \mathcal{I}}{\mathcal{I}} &= \frac{d\mathcal{I}}{\mathcal{I}dT} \Delta T = -\frac{E_T \Delta T}{kT^2}, \\
\frac{\Delta \rho}{\rho} &= \frac{d\rho}{\rho dT} \Delta T = -\frac{E_G}{2kT^2} \Delta T.
\end{aligned} \tag{20}$$

Since the forbidden band width of semiconductors used in bolometers does not exceed 1-2 eV and the activation energy of the thermal quenching of luminescence may reach 1 eV (§4, Chap. I; §2, Chap. III), the changes in the two parameters (expressed in percent) should be similar in the two cases under discussion.

The much higher sensitivity threshold of the luminescent-thermal method is not related to the intrinsic magnitude of the thermal effect but it is simply due to the fact that methods for measuring small differences between optical fluxes have been less developed than the corresponding methods for electron currents. In fact, compensation methods can be used to measure very small differences between currents, amounting to less than 0.01%, whereas direct visual comparison of the brightness can be made to within 2-3% (this applies to visual photometers in which absolute changes of the brightness can be observed to within 10%). Compensation methods and photoelectric cells can be used to detect differences between light fluxes down to 0.3%, i.e., at the present state of development the optical methods are less sensitive than the methods involving measurements of the electric currents.

Further improvements in the methods of detection of small changes in optical fluxes would make it possible to reduce considerably the gap between the values quoted above. However, even at the present stage of development of optical methods we can still use luminescent screens for recording the distribution of the field of laser radiation.

§ 4. Sensitivity of Crystal Phosphors to Changes in Temperature and to the Thermal Effect of Infrared Radiation Corresponding to Different Thermal Quenching Mechanisms

We shall now consider in detail the sensitivity of the photoluminescence of a crystal phosphor to a change in its temperature and the associated sensitivity to the thermal effect of infrared radiation. We shall consider different thermal quenching mechanisms. The simplest case of thermal quenching is that which takes place inside luminescence centers and involves internal nonradiative transitions. This case can be described by the configurational model of luminescence centers (Fig. 2). This quenching mechanism usually occurs in alkali halide compounds and it is described by the well-known Mott formula for the dependence of the luminescence intensities $\mathcal{I}_T$ on the temperature [4] for a given excitation intensity:

$$\mathcal{I}_T = \frac{P_r I_e}{P_r + P_n} = \frac{\mathcal{I}_0}{1 + \dfrac{P_n}{P_r}} = \frac{\mathcal{I}_0}{1 + A e^{-E/kT}}, \tag{21}$$

where I_e is the excitation intensity; $\mathcal{I}_0$ is the luminescence intensity expressed as the number of quanta (in the absence of quenching, we have $I_e = \mathcal{I}_0$); P_r and P_n are, respectively, the probabilities of radiative and nonradiative transitions within a luminescence center (Fig. 2); A and E are the constants of the center, representing the ratio of these transition probabilities.

The quantity E, which is the activation energy of the quenching within a center, is governed by the difference between the energy levels of the excited and unexcited states of the center expressed in configurational coordinates. The value of E for alkali halide compounds is 1-1.5 eV [52].

The probability P_r of the allowed dipole transitions is $\sim 10^9$ sec^{-1}, whereas the experimental value of the probability of magnetic dipole and other forbidden complex transitions is 10^2-10^3 sec^{-1}. The frequency factor s_n, which governs $P_n = s_n \exp(-E_T/kT)$ at high tempera-

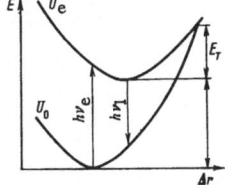

Fig. 2. Energy scheme for internal quenching. U_0 is the energy state of an unexcited center; U_e is the energy state of an excited center.

tures is $\sim 10^{12}\text{-}10^{13}$ sec^{-1} for ionic lattices. Thus, the value of A for different centers may vary from 10^4 to 10^{10} and this, in combination with the different values of E, is responsible for the large differences between the internal thermal quenching in various phosphors.

The knowledge of the range of values of A and E can be used to calculate the values of the thermal sensitivity and its range for different phosphors. Such an analysis of the thermal quenching is useful also in some cases of external quenching, which is of different origin but it is still described by formulas similar to the Mott equation (of course the physical meaning of the quantities in such formulas is different and the numerical values of the parameters A and E are also different).

We shall consider the ratio $\Delta\mathcal{J}/\mathcal{J}$ for the processes described by Eq. (21), which gives the relative change in the luminescence brightness resulting from a small change in the temperature. In practice, the values of $\Delta\mathcal{J}$ should be sufficient to produce a contrast between two parts of the temperature field sufficient for the detection of the temperature difference. The temperature difference producing a given brightness contrast can be regarded as a measure of the thermal sensitivity of a crystal phosphor,

$$\frac{\Delta\mathcal{J}}{\mathcal{J}} = \frac{d\mathcal{J}\Delta T}{\mathcal{J}_T dT} = - \frac{E_T \Delta T}{kT^2}\left(1 + \frac{1}{A}e^{+E_T/kT}\right)^{-1}. \tag{22}$$

Under strong quenching conditions the above formula reduces to the simpler expression

$$\frac{\Delta\mathcal{J}}{\mathcal{J}} = -\frac{E_T \Delta T}{kT^2}, \tag{23}$$

which shows that the temperature drop necessary for the attainment of a specified contrast (for example, the threshold contrast $\Delta\mathcal{J}/\mathcal{J} = 10\%$), is proportional to the square of the temperature, inversely proportional to the activation energy of the thermal quenching E_T (this follows from the elementary discussion in § 3), and is independent of the value of A. During the initial stage of the thermal quenching the brightness contrast increases with increasing T and is proportional to A, i.e.,

$$\frac{\Delta\mathcal{J}}{\mathcal{J}} = -\frac{E_T \Delta T}{kT^2} A e^{-E_T/kT}. \tag{24}$$

The region where a phosphor is most sensitive to any change in temperature is located approximately where $\ln A \sim E_T/kT$. The position of the sensitivity maximum, i.e., the point of steepest slope of the curve $d\mathcal{J}/\mathcal{J}_T dT$, and the amplitude of this maximum can be deduced from the condition $\frac{d}{dT}\left(\frac{d\mathcal{J}}{\mathcal{J}_T dT}\right) = 0$. This condition leads to the following equation which gives the dependence of $(E/kT)_{max}$ on A:

$$\left(\frac{E_T}{kT}\right)_{max} + \ln\left(\frac{1}{2}\frac{E_T}{kT_{max}} - 1\right) = \ln A, \tag{25}$$

which can be employed to find T_{max} by the method of successive approximations. The values of A and of the ratio of the trap depth to kT, corresponding to the temperature of the optimal quenching rate, are listed below:

A	10^{10}	10^8	10^6	10^4	10^2
$\left(\frac{E_T}{kT}\right)_{max}$	21	16.3	12	8	4.3

It is clear from these values that $kT_{max} \sim 1/10\,E_T$ and that it decreases with increasing A, i.e., if the activation energy is fixed, the quenching effect is observed earlier for higher values of A.

In practical situations it is more convenient to calculate the sensitivity by means of the general formula (22), having specified the minimum temperature difference T_{min} corresponding to the threshold value of the brightness contrast, set by the sensitivity of the instrument employed, i.e.,

$$\Delta T = \left(\frac{\Delta \mathcal{J}}{\mathcal{J}}\right)_{th} : \frac{d\mathcal{J}}{\mathcal{J}_T dT} = \left(\frac{\Delta \mathcal{J}}{\mathcal{J}}\right)_{th} \frac{kT^2}{E_T}\left(1 + \frac{1}{A}e^{E_T/kT}\right). \tag{26}$$

Figures 3 and 4 show the results of calculations carried out for the selected value of the contrast $(\Delta \mathcal{J}/\mathcal{J})_{th} = 0.1$, corresponding to reliable visual observations. Figure 3 gives the temperature dependence of ΔT_{min}, corresponding to several values of the parameters E_T and A, whereas Fig. 4 gives the dependence of the same quantity on the activation energy of the thermal quenching for several values of A at room temperature.

It is evident from these figures that in the case of phosphors whose thermal quenching is described by the Mott formula [21, 22] the minimum temperature difference which can be detected visually $(\Delta \mathcal{J}/\mathcal{J} \approx 10\%)$, lies between ~1 and ~10 deg C for a wide range of the constants A and E_T. At low temperatures $(T \approx 100°K)$ the highest sensitivity is obtained for low values of the activation energy $E_T \approx 0.15$ eV. At high temperatures $(T \approx 600°K)$ the maximum sensitivity corresponds to high values of the same energy $E_T \sim 1$ eV.

The highest sensitivity at a given temperature is attained when the energy E_T is high and when the value of A is large so that the quenching in phosphors with high values of E_T is observed earlier. For example, at room temperature (Fig. 4) the optimal value of the activation energy corresponding to the highest value of A $(\approx 10^{10})$ is ≤ 0.6 eV, which ensures detection of a temperature change by ~1.5 deg C, i.e., $d\mathcal{J}/\mathcal{J}dT \approx 8\%$ per degree Celsius.

If we use the relationship between the rise of temperature of a luminescent screen, resulting from the absorption of infrared radiation of power W_{IR} (§ 3) and apply the notation in-

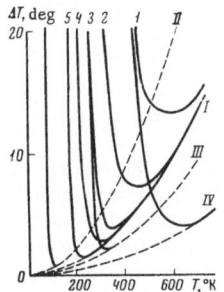

Fig. 3. Temperature dependences of the interval ΔT corresponding to 10% change in the luminescence intensity (threshold contrast), plotted for different values of the parameters E_T and A. I) $E_T = 0.3$ eV: 1) A = 10^3, 2) A = 10^4, 3) A = 10^5, 4) A = 10^6, 5) A = 10^8; II) $E_T = 0.15$ eV, A = 10^8; III) $E_T = 0.45$ eV, A = 10^8; IV) $E_T = 0.9$ eV, A = 10^8.

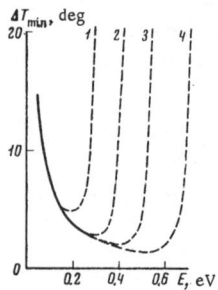

Fig. 4. Dependences of
the temperature interval
ΔT corresponding to 10%
change in the luminescence
intensity (threshold con-
trast) on the activation energy E_T, plotted for T =
300°K and different values
of the parameter A: 1)
10^4; 2) 10^6; 3) 10^8; 4) 10^{10}.

troduced earlier, we find that Eqs. (22) and (23) lead to the general formula

$$\frac{\Delta \mathcal{J}}{\mathcal{J}} = - \frac{E_T}{kT} \frac{W_{IR}}{4\sigma_{SB}T^4} \left(1 + \frac{1}{A e^{\pm E_T/kT}}\right)^{-1}, \tag{27}$$

which reduces to the following expression for the strong quenching region, corresponding to Eq. (23):

$$\frac{\Delta \mathcal{J}}{\mathcal{J}} \approx - \frac{E_T}{kT} \frac{W_{IR}}{4\sigma_{SB}T^4}. \tag{28}$$

These relationships give directly the minimum infrared radiation power which can be recorded for a specified value of the detectable contrast (usually 10%):

$$(W_{IR})_{min} = 4W_T \frac{kT}{E_T} \left(\frac{\Delta \mathcal{J}}{\mathcal{J}}\right)_{min} \left(1 + \frac{1}{A} e^{E_T/kT}\right). \tag{29}$$

Under strong quenching conditions, corresponding to $A > e^{E_T/kT}$, we have

$$(W_{IR})_{min} = 4W_T \frac{kT}{E_T} \left(\frac{\Delta \mathcal{J}}{\mathcal{J}}\right)_{min}. \tag{30}$$

Thus, the threshold infrared radiation power is proportional to the thermal radiation power and to the ratio kT/E_T. Since the thermal radiation power increases rapidly with rising
temperature, the optimal sensitivity to infrared radiation, i.e., the minimum $(W_{IR})_{min}$, is located at somewhat lower temperatures than the optimal sensitivity to the change in temperature
$(\Delta T)_{min}$:

$$\left(\frac{E_T}{kT}\right)_1 + \ln\left[\frac{1}{5}\left(\frac{E_T}{kT}\right)_1 - 1\right] = \ln A. \tag{31}$$

If we use Eqs. (26) and (31) and bear in mind that the shift ΔT between the optimal temperatures is much smaller than temperatures themselves, we find that

$$\Delta T \approx \frac{kT_{max}}{E} \ln \frac{\left(\frac{1}{2}\frac{E}{kT} - 1\right)}{\left(\frac{1}{5}\frac{E}{kT} - 1\right)}. \tag{32}$$

Substituting into this formula the values of the parameters corresponding to T = 300°K, which
are kT = 0.025 eV, E_1 = 0.3 eV, and E_2 = 1 eV, we find that this shift at room temperature can
range from 30 deg (E_1 = 0.3 eV) to 7 deg C (E_2 = 1.0 eV).

In the case of external quenchi~g, which is independent of the excitation density, the temperature dependence of the luminescence brightness is given by the formula

$$\mathcal{J}_T = \frac{I_0}{1 + \dfrac{\beta_1 C_d}{\beta C_a} e^{-\frac{E^- - E^+}{kT}}} \tag{33}$$

where β and β_1 are, respectively, the probabilities of radiative and nonradiative recombination; E^+ and E^- are, respectively, the depths of the luminescence and quenching centers; C_a and C_d are, respectively, the concentrations of these centers.

In this case the formula for the contrast is given by a function which is very similar to the formula in the internal quenching case but the constants now have completely different physical meaning:

$$\frac{\Delta\mathcal{J}}{\mathcal{J}} = \frac{\Delta\mathcal{J}\Delta T}{\mathcal{J}_T dT} = \pm \frac{E^- - E^+}{kT^2} \Delta T \left(1 + \frac{\beta C_a}{\beta_1 C_d} e^{\frac{E^-_1 - E^+}{kT}}\right)^{-1}. \tag{34}$$

The formulas for the optimal sensitivity of a phosphor and its temperature dependence are also similar if we make the substitutions $E \to E^- - E^+$ and $A \to \beta_1 C_d / \beta C_a$. In the absence of concentration quenching the value of A ranges from 10 to 10^4. The concentrations of residual impurities in a phosphor should be quite low, i.e., for $10^{16} < C_a < 10^{18}$ cm^{-3} and $10^{16} < C_d < 10^{18}$ cm^{-3} the ratio of concentrations of these impurities to the total number of atoms should be $10^{-4} - 10^{-6}$.

The difference between the depths of typical acceptor and donor levels is quite small (for example, in the case of ZnS:Cu:Co this difference is of the order of 0.1 eV). Therefore, a redistribution of carriers between these levels as a result of quenching does not lead to a high sensitivity in the region where the luminescence efficiency is independent of the excitation intensity [see Eq. (33)].

Thus, the linear external quenching which is independent of the excitation intensity and which corresponds to a quasiequilibrium of carriers between the localization levels does not provide necessary conditions for a high sensitivity to a change in the temperature because the difference $E^- - E^+$ and the pre-exponential factor $\beta_1 C_d / \beta C_a$ are both small. The nonlinear external quenching (which can be of the first or second order), which has been discussed in [32, 33], can ensure a somewhat higher sensitivity. The formulas for the temperature dependence of the luminescence intensity for the first- and second-order nonlinear external quenching [32] can be written in the form

$$\left.\begin{aligned}
\mathcal{J}_1(T) &= \frac{C_1}{P_T^+}\left(1 + \frac{P_T^-}{P_T^+}\right), \\
\mathcal{J}_2(T) &= \frac{C_2}{(P_T^+)^{1/2}},
\end{aligned}\right\} \tag{35}$$

where C_1 and C_2 are functions which depend weakly on the temperature ($C_1 \propto I_e^2$ and $C_2 \propto I_e^{3/2}$); P_T^- and P_T^+ are, respectively, the probabilities of thermal liberation of localized electrons and holes, given by $s^- e^{-E^-/kT}$ and $s^+ e^{-E^+/kT}$.

Applying the previous method for the determination of the thermal sensitivity and denoting the depths of the radiative and nonradiative recombination centers by E^+ and E^-, respectively, we obtain

$$\frac{\Delta\mathcal{J}_1}{\mathcal{J}} = \frac{d\mathcal{J}_1\Delta T}{\mathcal{J}_1 dT} = -\frac{\Delta T}{kT^2}\left\{E^+ + \frac{E^+ - E^-}{1 + \dfrac{s^+}{s^-}e^{\frac{E^+ - E^-}{kT}}}\right\}$$

or, if $s^-/s^+ \ll 1$ but $E^+ \gg E^-$,

$$\left.\begin{aligned}
\left(\frac{\Delta \mathcal{J}_1}{\mathcal{J}_1}\right)_{\max} &= -\frac{2E^+}{kT^2}, \\
\frac{\Delta \mathcal{J}_2}{\mathcal{J}_2} &= \frac{d\mathcal{J}_2 \Delta_T}{\mathcal{J}_2 dT} = -\frac{E^+}{2kT^2}.
\end{aligned}\right\} \tag{36}$$

Thus, the first- and second-order nonlinear external quenching can lead to somewhat higher values of the thermal activation energy and the corresponding higher rates of thermal quenching than that obtained in the case of linear external quenching independent of the excitation intensity.

Depending on the nature of the quenching processes, the thermal activation energy in the case of nonlinear external quenching lies between half the depth of the luminescence centers to double the depth of the activator levels, which gives about 0.5–0.6 eV. This value of the activation energy can lead to room-temperature quenching rates of $d\mathcal{J}/\mathcal{J}_T\, dT = 8\%$ per degree Celsius, which in the case of threshold contrast $(\Delta \mathcal{J}/\mathcal{J})_{th} = 10\%$ can be used to detect temperature differences down to ~ 1 deg C and infrared radiation powers of $\lesssim 1 \times 10^{-3}$ W/cm^2.

We shall now consider a more complex two-level model of a crystal phosphor in which we shall make an allowance for the saturation, i.e., for the complete filling of the nonradiative recombination centers with nonequilibrium carriers (electrons). This model is shown schematically in Fig. 5. We shall assume that in this phosphor the concentration of the luminescence (radiative recombination) centers is C_a, which is considerably higher than the concentration of the quenching (nonradiative recombination) centers C_d, but most of the recombination flux passes through the latter centers. The intensity of the nonradiative transitions $\mathcal{J}_{nr}^+$, corresponding to these conditions is approximately equal to the excitation intensity I_e and it can be expressed in the following way in terms of the temperature and phosphor parameters:

$$I_e \approx \mathcal{J}_{nr} \approx \sigma_{nr}^+ v N^+ C_d \approx \sigma_{nr}^+ v \frac{C_d^2 N_{eff}^+}{C_a} e^{-\frac{E_a}{kT}}, \tag{37}$$

where σ_{nr}^+, v, and N^+ are, respectively, the nonradiative recombination cross section, the velocity of free holes, and the density of free holes (N_{eff}^+ is the effective density of states in the valence band).

We shall also assume that in spite of saturation (n carriers localized in the C_d centers) of the nonradiative recombination centers, they determine the free-electron density N^-:

$$N^- = \frac{I_e + W_1 n}{\sigma_c C_d v} \; \frac{1}{1 - \dfrac{n}{C_d}\left(1 - \dfrac{\sigma_{rr}^-}{\sigma_c}\right)}, \tag{38}$$

where σ_c is the electron-capture cross section of the nonradiative recombination centers; σ_{rr}^- is the radiative recombination cross section.

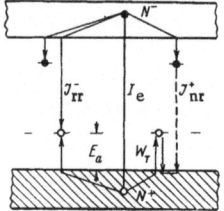

Fig. 5. Energy band scheme of a crystal phosphor exhibiting the highest thermal sensitivity.

It then follows from Eq. (38) that when the cross sections σ_{rr}^- and σ_c differ strongly, the largest relative change in the concentration of ionized centers gives rise to a much larger change in the capture probability $\dfrac{dn}{n}\dfrac{1}{1-\dfrac{n}{C_a}}$ than in the case of a smaller difference between these cross sections and, consequently, the changes in the free-electron density dN^-/N^- and in the luminescence intensity are correspondingly greater. In the first approximation we can ignore the cause of the change in $d\mathcal{J}_{rr}^-/\mathcal{J}_{rr}$, which can be due to a change in the temperature T, in the excitation intensity I_e, or in the composition of a phosphor (for fixed values of T and I_e).

This strong dependence on the excitation conditions may be observed in the following range of the ratio of the free-carrier densities

$$\frac{\sigma_{nr}^+}{\sigma_c^-} \ll \frac{N^-}{N^+} \ll \frac{\sigma_{nr}^+}{\sigma_{rr}^-} \tag{39}$$

and it occurs if the nonradiative and radiative recombination cross sections differ strongly. Under these conditions the thermal sensitivity $R_T = d\mathcal{J}_{rr}^-/\mathcal{J}_{rr}^- dT$ is very high and this is true also of the nonlinearity coefficient $K_{nl} = d \ln \mathcal{J}_{rr}^-/d \ln I_e$, which are related by approximate expression

$$R_T \lesssim \frac{E_a}{kT^2} K_{nl} \cdot 100\% \text{ per degree Celsius.} \tag{40}$$

The quenching conditions described above are difficult to achieve because of the presence of other electron capture and recombination centers which are important in the range $C_d \approx n$. Even in the absence of such centers we must make an allowance for the statistical scatter of the parameters of the various centers and particularly the depth of the luminescence centers (this scatter is not less than 0.03 eV at T = 300°K), which acquire holes from the valence band.

These circumsgances limit the maximum values of R_T and K_{nl} to not more than one order of magnitude in excess of the case of simple linear quenching for which R_T is just equal to $dw/w_T dT < 10\%$.

CHAPTER II

INVESTIGATIONS OF CRYSTAL PHOSPHORS SENSITIVE TO THE OPTICAL EFFECT OF INFRARED RADIATION

§ 1. Selection of Samples and Investigation Method

In the present chapter we shall describe the synthesis and excitation conditions which ensure optimal sensitivity of ZnS:Cu:Co phosphors to the quenching action of infrared radiation. We shall also report measurements of the principal parameters of the kinetics of luminescence of these phosphors and compare the sensitivity of various types of phosphor to the optical action of infrared radiation.

Synthesis. Our phosphors were made from zinc sulfide of the "phosphor" grade to which 5% of sodium chloride flux was added. Copper and cobalt activators were introduced in the form of aqueous solutions of $Cu(NO_3)_2$ and $Co(NO_3)_2$. The charge prepared in this way was dried for 4 h at 120°C and then placed into a crucible which was closed by a tightly-fitting cover and dried again at 400°C applied for 3 h. The second drying was necessary in order to remove the water of crystallization which might otherwise cause ejection of the charge from the crucible when the latter was placed in a furnace heated to the synthesis temperature. The final firing was carried out at temperatures from 900 to 1200°C.

In order to study physicochemical properties of these phosphors, we prepared several batches of samples which differed in the activator concentration and in the firing temperature. The activator concentrations were 1×10^{-7}-1×10^{-4} g of Cu per gram of ZnS and 1×10^{-6}-1×10^{-4} g of Co per gram of ZnS. In these investigations the optimal firing temperature was 1150°C.

We also investigated ZnS:Fe [36] and ZnS:In [54] phosphors, which were prepared by the method described above and which were promising from the point of view of extension of the sensitivity to infrared radiation in the direction of longer wavelengths (up to 2-3 μ).

For the sake of comparison we investigated also the well-known alkaline earth phosphors SrS·CaS:Sm and SrS·CaS:Ce:Sm [9, 10].

The samples investigated were either thick tightly packed layers (d ~ 1 mm) or screens of controlled thickness of density in the range 3-20 mg/cm^2. The grain size in the phosphor powders was < 4-5 μ. The screen thicknesses (0.8-50 μ) were determined by calculation from the weight of the phosphor and were also measured optically with an MIIS-11 microscope. The two methods gave similar results. The thickness of the thinnest screens was considerably less than $1/\varkappa_{365}$ ($\varkappa_{365}$ is the absorption coefficient at 365 nm). Therefore, we could assume that these screens were excited uniformly across the thickness so that we could calculate the volume density of the excitation and the concentration of the ionized centers.

The excitation was provided by the mercury line λ_e = 365 nm emitted by a PRK-4 lamp, which was fitted with filters for removing the other ultraviolet and visible lines and, particularly, the infrared part of the spectrum of this lamp. The excitation power density was varied within three orders of magnitude from 10^{-6} to 10^{-3} W/cm^2. Absolute measurements with a thermocouple were carried out at the ultraviolet power density of $(6.5 \pm 1) \times 10^{-5}$ W/cm^2. The results of these absolute measurements were in satisfactory agreement with the calculations based on the published data on the emission of mercury lamps [55]. In these calculations an allowance was made for the transmission of the filters and for the geometrical position of the lamp relative to the luminescent screen (r ≈ 30 cm, φ = 45°). The excitation power density W_e was varied by means of neutral grids and an additional lens. This power was deduced from the intensity of the red luminescence of a phosphor whose output was constant in a wide range of W_e.

The same red-emitting phosphor was used to determine the efficiency of the phosphors under investigation (an allowance was made for the spectral distributions of the luminescence of these phosphors and for the spectral sensitivity of instruments UM-2 and FÉU-14B). This method was adopted because high-precision data were available on the quantum efficiency of a red-emitting phosphor sample [56] belonging to the same batch as the phosphor used in the present study.

The absolute sensitivity of a photomultiplier was calibrated for a fixed position of this instrument relative to a luminescent screen. Such calibration was necessary in the determination of the absolute light sums stored in a given phosphor and of the absolute sensitivity to infrared radiation. This calibration was carried out by two methods and it was expressed in the form of readings of an indicator M-95, which recorded the photomultiplier current, divided by the number of photons emitted from 1 cm^2 of a phosphor in 1 sec. In the first method we determined the readings of the photomultiplier indicator obtained for the green luminescence of ZnS:Cu phosphors emitted under steady-state excitation conditions. An allowance was made for the absolute intensity of the ultraviolet excitation (measured with a thermocouple) and for the quantum efficiency of the phosphor under investigation (found by comparison with the red-emitting phosphor). The photomultiplier sensitivity found by the first method was given by the formula

$$K_{PM_1} = \frac{\mathscr{J}_{PM}\, h\nu_e}{W_{UF}\, \eta_q \cdot 6.3 \cdot 10^{18}} \left[\frac{\mu A}{quanta/cm^2}\right],$$ (41)

where $\mathscr{J}_{PM}$ represents the readings of the M-95 indicator; W_{UF} is the ultraviolet excitation power density in W/cm^2; 6.3×10^{18} is the number of the luminescence quanta corresponding to an incident power of 1 W and $h\nu = 1$ eV; n_q is the quantum efficiency of the green band of ZnS:Cu.

In the second method we illuminated the photomultiplier with light from a calibrated lamp (colour temperature $T_c = 2850°K$, luminous intensity 7 international candles), whose light was reflected from MgO. The geometry was the same as in the first method and the light incident on the photomultiplier was passed through a Z S8 filter, which was closest to the luminescence emitted by the investigated phosphors. In this case the sensitivity of the photomultiplier was given by

$$K_{PM_2} = \frac{\mathscr{J}_{PM}\, h\nu_1 r^2 \cos\varphi}{\dfrac{\mathscr{J}}{683\, \frac{lm}{W}} \cdot 6.3 \cdot 10^{18}\, \dfrac{quanta}{W.\,eV}} \left[\frac{\mu A}{quanta/cm^2}\right],$$ (42)

where r is the distance from the lamp to the MgO screen; φ is the angle of incident of light from the lamp on the screen (683 is the conversion factor for the change from candles to watts); the rest of the notation is the same as in Eq. (41). The second method yielded a sensitivity which was about 1.5 times higher than the average sensitivity found by the first method.

Infrared radiation was generated by a globar (a Silit rod kept at a color temperature $T_c \approx 1700°K$) or a low-power incandescent lamp ($T_c = 2400°K$, $V_L = 3.6$ V, $\mathscr{J}_{LS} = 0.28$ A). In the first case a spectral interval of $\lambda < 0.1$ μ width was selected by an IKS–12 monochromator in the range between 0.85 and 3 μ; in the second case, we used unresolved infrared radiation from which visible light up to $\lambda = 0.8$ μ was removed.

The spectral distribution of the radiation emitted by the globar was determined by two methods: a photoelectric method in which a PbS photoresistor was used, and a method in which a thermistor was employed as an infrared radiation detector.

The results of our measurements of the spectral distribution of the radiation emitted from the globar are plotted in Fig. 6. The same figure gives also the distribution of the radiation emitted by an absolute black body at the same temperature. The energy of the infrared radia-

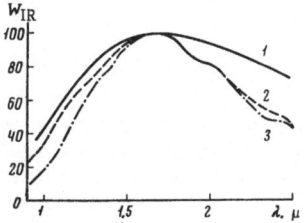

Fig. 6. Normalized spectrum of the radiation emitted by a globar and recorded with an IKS-12 monochromator: 1) black body at a color temperature $T_c = 1700°K$; 2) globar spectrum recorded with a thermistor; 3) globar spectrum measured by a photoelectric method.

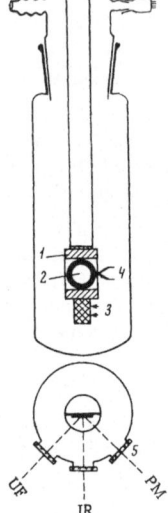

Fig. 7. Dewar flask
used in experiments
involving variation
of temperature: 1)
copper holder; 2)
sample (luminescent
screen); 3) heater; 4)
thermocouple; 5) quartz
windows.

tion emitted from the globar was varied by means of grids and measured with a thermocouple. At $\lambda = 1.3\ \mu$, corresponding to the maximum of the sensitivity of ZnS:Cu:Co to infrared radiation the density of the power reaching this phosphor was usually $(1.8 \pm 0.2) \times 10^{-4}$ W/cm^2.

The arrangement used in the measurements is shown schematically in Fig. 7. A PRK-4 mercury lamp was placed inside a light-tight water-cooled jacket. The ultraviolet radiation emitted by this lamp passed through a window and light filters before it reached cuvettes containing the phosphor under investigation. These cuvettes were replaced on a rotating carriage which was located inside a light-tight chamber with blackened walls. The temperature dependences were determined using a glass Dewar flask with windows transparent to infrared radiation. This flask was located inside a blackened jacket. The rate of heating in investigations of the thermoluminescence was about 10 deg/min.

The radiation from the globar was passed through the IKS-12 monochromator and focused on the whole of the surface of a luminescent screen. The luminescence emitted by screens and powder layers was measured with an FÉU-14B photomultiplier, whose readings were linear throughout the investigated range. An ZhS4 filter and a cuvette filled with a solution of NaNO$_2$ were used to absorb reflected ultraviolet radiation. The flux reaching the photomultiplier was attenuated with a diaphragm placed quite far (of the order of 10 cm) from the multiplier window.

This apparatus enabled us to detect photon fluxes of densities beginning from 10^6 quanta $\cdot$ cm$^{-2}\cdot$ sec^{-1}. Thus, we were able to investigate the later stages of afterglow if the background was eliminated.

The luminescence spectra were determined using different apparatus but keeping the excitation intensity constant.

The absolute values of the ultraviolet excitation intensity, of the infrared radiation, and of the luminescence emitted by the phosphors under investigation were determined with errors not exceeding ± 15-20%. The error in the relative measurements of the luminescence intensity of any one given phosphor was 2-3% throughout the whole intensity range whereas in comparisons of different phosphors the corresponding error was about 5%.

§2. Influence of the Composition of ZnS:Cu:Co Phosphors
and of the Excitation Intensity on Steady-State
Luminescence and Afterglow

The luminescence efficiency and the spectra of the ZnS:Cu:Co phosphors have been investigated before [57–61] and it has been found that when the luminescence centers are ionized directly by exciting light of λ_e = 365 nm wavelength the quantum efficiency is quite high (0.4 ≤ η ≤ 0.9). The luminescence spectra of these phosphors are fairly complex: they consist of two overlapping blue and green bands.

At low and moderate copper concentrations C_{Cu} < 10^{-4} g/g these two bands are due to the radiative recombination at intrinsic lattice defects, usually assumed to be N_{Zn} Cl [34], and at copper ions occupying the cation lattice site (Cu_{Zn}). Our own and the published data indicate that the ratio of the intensity of the green band of copper to the intensity of the blue band increases smoothly with increasing concentration of copper: for C_{Cu} ~ 10^{-5} g/g this ratio is of the order of unity but when the copper concentration is increased to C_{Cu} ~ 10^{-4} g/g only the green band is observed. When the concentration of copper is increased still further, the blue band reappears. Many investigations have established that this blue band is due to the formation of copper centers of different structure. According to Gurvich [34], these centers are ion pairs Cu'_{Zn}-Cu'_{Zn}, located near dislocations.

Introduction of cobalt reduces the steady-state intensities of the green and blue bands. The dependence of the intensity of the green band $\mathscr{I}_{Cu}$ on the concentration of cobalt at three fixed concentrations of copper was investigated in [32, 61]. It was found that the process of quenching of the green band of copper by the presence of cobalt is described by the simple formula

$$\mathscr{I}_{Cu} = \frac{I_e}{1 + s_1 \frac{\nu_1}{\nu} + s_2 \frac{\nu_2}{\nu}}, \tag{43}$$

where I_e is the excitation intensity; ν, ν_1, and ν_2 are, respectively, the concentrations of the copper luminescence centers (E_{Cu}^+ ≈ 0.35 eV), of shallow traps formed by copper (E_{Cu}^- ≈ 0.33 eV), and of deeper cobalt traps (E_{Co}^- ≈ 0.44 eV) which act as nonradiative recombination centers; s_1 and s_2 are determined, in the first approximation, by the exponential factors $\exp\left(\frac{E_{Cu}^- - E_{Cu}^+}{kT}\right)$ and $\exp\left(\frac{E_{Co}^- - E_{Cu}^+}{kT}\right)$.

Equation (43) corresponds to a quasiequilibrium occupancy of the local levels by nonequilibrium carriers and it applies to the region in which the luminescence efficiency is independent of the excitation power density. This region lies between a superlinear region associated with the presence of deeper nonradiative recombination levels (the number of these levels is less than the number of cobalt levels), and a sublinear region, which is associated with the de-exciting action of the ultraviolet light, which results in nonactive absorption and nonequilibrium filling of the local levels [33].

The very slow decay of the luminescence of the ZnS:Cu and ZnS:Cu:Co phosphors (which can last several hours) was investigated by many workers [32, 62–67, 71], who found that during a certain stage of the decay (3 min < t < 30 min) the intensity of the afterglow $\mathscr{I}_t$ obeys the simple law

$$\mathscr{I}_t = \frac{C}{t^\alpha}, \tag{44}$$

where 1 ≤ α ≤ 2, and C is a constant which depends on the composition of the phosphor.

It follows directly from Eq. (44) that the phosphors emitting weak luminescence, which decays slowly (small value of α), emit stronger luminescence at later decay stages than the phosphors characterized by strong emission during excitation and fast decay (large values of α). This has been confirmed in many experiments. In particular, it is reported in [18] that the introduction of cobalt which quenches the luminescence during excitation can increase the luminescence brightness during the later stages of decay, which corresponds to an increase in the constant α in the empirical formula (44).

These characteristics of the steady-state luminescence and afterglow of a phosphor are obviously needed in any determination of the sensitivity to infrared radiation and in refinement of the parameters of the radiative and nonradiative recombination centers participating in these processes. The published data cannot be applied directly to our phosphors, which were made specially sensitive to infrared radiation, because the properties of phosphors depend strongly on the synthesis and excitation conditions.

It is known that the temperature and the duration of firing, the method of introduction of activators and fluxes, and the atmosphere in the growth chamber can alter considerably the concentration of the luminescence, quenching, and capture (trapping) centers, which govern the properties of phosphors.

An important factor which influences the stimulation and quenching effects of infrared radiation is the excitation intensity which is frequently not measured in absolute units (W/cm^2 and photons/cm^2). This makes it difficult to compare the theory and the experimental results and to obtain even quantitative estimates of the cross sections of the various radiative and nonradiative transitions that occur in the ZnS:Cu:Co phosphors.

In view of this we decided to carry out an investigation of the influence of the composition of the ZnS:Cu:Co phosphors on their efficiency and luminescence spectra and on the decay curves recorded at fixed excitation intensities. After some preliminary experiments we selected the excitation intensities which corresponded to maximum values of the stored light sums and to the optimal quantum sensitivity to infrared radiation.

The quantum effectiveness of the steady-state luminescence excited with light of $\lambda_e = 365$ nm at $T = 293°K$ is listed in Table 1 for two batches of the ZnS:Cu and ZnS:Cu:Co phosphors with different concentrations of cobalt but the same concentration of copper ($C_{Cu} \sim 10^{-5}$ g/g). The excitation power density was $(6.5 \pm 1) \times 10^{-5}$ W/cm^2 ($I_e = 1.1 \times 10^{13}$ quanta $\cdot$ cm$^{-2}\cdot$ sec^{-1}), which ensured that the maximum values of the light sums were obtained.

It is evident from Table 1 that an increase in the concentration of copper C_{Cu} from 10^{-7} to 10^{-4} g/g raised the total effectiveness of the steady-state luminescence by a factor of about 5.5. The maximum value of the effectiveness (64%) is in good agreement with the experimental data on the maximum quantum efficiency (40-80%) of ZnS:Cu phosphors [68, 70] obtained later by various investigators in the region where the quantum efficiency is independent of the excitation intensity. It should be noted that the quantum efficiency of these phosphors calculated on the basis of the absorbed energy, is considerably higher than the quantum effectiveness, which is calculated making allowance for the reflected energy which can be quite large. This is particularly true at low copper concentrations. According to the published data [59], the coefficient of reflection of the exciting light ($\lambda_e = 365$ nm) by ZnS:Cu phosphors increases from 20 to 60% when the concentration of copper C_{Cu} is reduced from 10^{-4} to 10^{-7} g/g. In our phosphors the concentration of copper was 10^{-5} g/g and the reflection coefficient was about 40% (we are talking here of the phosphors in which the optimal light sums were stored under the aforementioned excitation conditions). Thus, the quantum efficiency of the luminescence calculated allowing for incomplete absorption reached 60% for the phosphors listed in Table 1 and the probabilities of radiative and nonradiative recombination were comparable.

TABLE 1. Dependence of Quantum
Effectiveness of Steady-State
Luminescence on Composition of
Phosphors

ZnS-Cu		ZnS-Cu, Co	
c_{Cu}, g/g	η, %	c_{Co}, g/g	η, %
10^{-7}	11.5	—	38
10^{-6}	13	10^{-6}	28
$1 \cdot 10^{-5}$	36	$3 \cdot 10^{-6}$	12
$5 \cdot 10^{-5}$	62	$5 \cdot 10^{-6}$	8
$1 \cdot 10^{-4}$	64	$7.5 \cdot 10^{-5}$	5.5

When cobalt was introduced into ZnS:Cu phosphors with copper concentrations $C_{Cu} \sim 10^{-5}$ g/g the effectiveness of the luminescence and, consequently, the quantum efficiency decreased considerably at cobalt concentrations as low as $C_{Co} = 10^{-6}$ g/g. When the concentration of cobalt was increased above 1×10^{-5} g/g, the luminescence of these phosphors decreased strongly in intensity (the quantum efficiency fell to $\eta_q \ll 0.1$). The dependence of the total luminescence efficiency on the concentration of cobalt, plotted on a double logarithmic scale (Fig. 8), was a smooth curve whose slope gradually increased with increasing C_{Co}. The maximum value of the slope, corresponding to high cobalt concentrations, was 2, i.e., the luminescence intensity $\mathcal{I}_1$ was inversely proportional to the square of the concentration of cobalt, rather than to the reciprocal of this concentration $\mathcal{I}_1 < 1/C_{Co}$, predicted by Eq. (43) for strong concentration quenching. Consequently, at high cobalt concentrations there must be a second quenching mechanism, stronger than that discussed in [61]. However, at cobalt concentrations of 10^{-5} g/g, corresponding to the optimal sensitivity to infrared radiation, the model proposed earlier can still be used in an approximate description of the luminescence kinetics.

Investigations of the dependence of the luminescence effectiveness on the composition of the phosphors and on the excitation conditions was complicated by the inconstancy of the ratio of the blue and green band intensities. At copper concentrations $C_{Cu} \sim 10^{-6}$ g/g the luminescence was concentrated mainly in the blue region ($\lambda_{max} = 460$ nm) whereas the phosphors which stored the optimal light sums had steady-state luminescence spectra with comparable contributions of the blue and green ($\lambda_{max} = 525$ nm) bands (Fig. 9).

The recombination interaction between the blue and green luminescence centers was manifested clearly by the dependence of the luminescence spectra on the excitation power density, which became even stronger when cobalt was introduced. The contribution of the blue band increased with increasing excitation power density for all the investigated ZnS:Cu and ZnS:Cu:Co phosphors. The normalized luminescence spectra obtained for the cobalt-free phosphors with $C_{Cu} = 10^{-5}$ g/g and for the phosphors with $C_{Co} = 7 \times 10^{-6}$ g/g are plotted in Figs. 10 and 11 for three excitation power levels. It is evident from these figures that the contribution of the blue

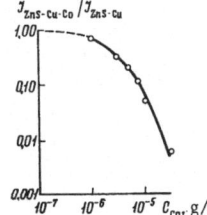

Fig. 8. Quenching of
the luminescence of
ZnS:Cu due to intro-
duction of cobalt.

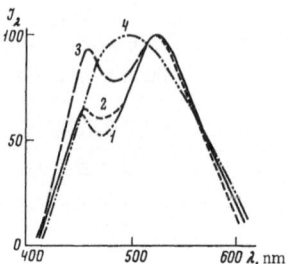

Fig. 9. Influence of the concentration of cobalt on the
spectra of ZnS:Cu:Co phosphors. C_{Co} (g/g): 1) 0; 2)
1×10^{-6}; 3) 5×10^{-6}; 4) 3×10^{-5}.

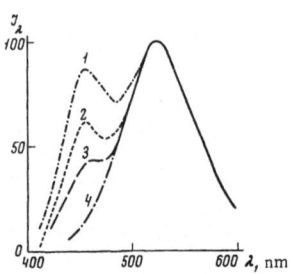

Fig. 10. Influence of the excitation power density W_e on the luminescence spectrum of a ZnS:
Cu phosphor. W_e (W/cm^2): 1)
5×10^{-4}; 2) 5×10^{-5}; 3) 5×10^{-6};
4) green band spectrum.

Fig. 11. Influence of the excitation power density W_e on the luminescence spectrum of a ZnS:
Cu:Co phosphors. W_e (W/cm^2):
1) 5×10^{-4}; 2) 5×10^{-5}; 3) 5×10^{-6}.

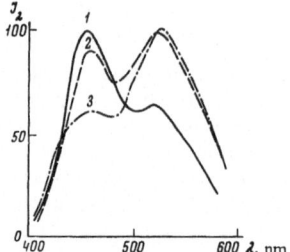

TABLE 2. Influence of Excitation
Power Density on Effectiveness
of Luminescence

W_e, W/cm^2	η, %	
	ZnS-Cu	ZnS-Cu, Co
$5 \cdot 10^{-4}$	29	4,0
$5 \cdot 10^{-5}$	38	5.5
$5 \cdot 10^{-6}$	44	7,1
$5 \cdot 10^{-7}$	46	8

band increased by a factor of about 3.5, ranging from 20% for low values of W_e to 70% for high values of W_e. In the case of the cobalt-activated phosphors the luminescence observed at high excitation intensities was dominated by the blue band. The total effectiveness of both bands in the range $W_e > 10^{-6}$ W/cm² decreased with increasing excitation power density (Table 2): the blue band increased almost linearly in intensity (at low values of W_e the rise was superlinear) whereas the green band increased sublinearly and became saturated more rapidly than the rest of the luminescence.

The saturation of the green band is due to the quenching effect of the exciting light, which liberates holes on the ionized copper centers. In this range of intensities the exciting light does not yet affect the blue band and the absolute losses because of inactive absorption are still small. If we use the excitation intensity $I_e \approx 10^{14}$ quanta·cm⁻²·sec⁻¹ and the depth of the centers formed by copper, we can estimate the frequency factor of the thermal liberation process. The probabilities of the optical P_{opt} and the thermal P_t liberation of carriers from the green luminescence centers are comparable and, therefore,

$$se^{-E/kT} \sim \sigma_{el}\, I_e \approx 10^{-15} \mathrm{cm^2} f \cdot 10^{14}\ \mathrm{quanta \cdot cm^{-2} \cdot sec^{-1}} \ \lesssim 0.1\ \mathrm{sec^{-1}}$$

for $f \sim 1$ (f is the oscillator strength for the absorption of excited light and σ_{el} is the cross section for the absorption of exciting light by local centers). Thus, we find that the upper limit of the frequency factor is 10^4-10^5 sec⁻¹. These values indicate that the probability of thermal liberation of carriers from the copper centers is extremely low, which is in good agreement with the slow decay of the afterglow.

We have mentioned in Chap. I that the low value of the frequency factor s provides favorable conditions for the storage of large light sums and for a high sensitivity of the ZnS:Cu:Co phosphors to the optical effect of infrared radiation at room temperature.

The frequency factor s = 10^5 sec⁻¹ obtained above can be used also to estimate the maximum value of the cross section for the capture of holes by the copper luminescence centers, as given by Eq. (3). If we assume that $N^+_{eff} = 3 \times 10^{19}$ cm⁻³ and $u_T = 10^7$ cm/sec, we obtain

$$\sigma^+_c = \frac{s}{N^+_{eff} u_T} \leqslant \frac{10^5}{3 \cdot 10^{19} \cdot 10^7} < 10^{-21}\ \mathrm{cm^2}.$$

The values of σ^+_c and of s are probably underestimated (not more than one order of magnitude) because multiple scattering of light in powder films has been ignored. However, even when the necessary correction is made, the value of σ^+_c remains small ($\sigma^+_c \lesssim 10^{-20}$ cm²). Therefore, we may assume that the copper centers are repulsive with respect to holes.

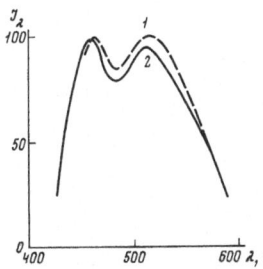

Fig. 12. Influence of infrared illumination on the relative intensities of the luminescence bands emitted by a ZnS:Cu:Co phosphor during ultraviolet excitation: 1) no infrared illumination; 2) infrared illumination.

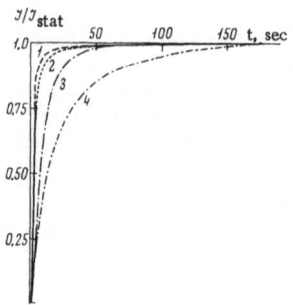

Fig. 13. Influence of the concentration of cobalt on the stimulation of a ZnS:Cu:Co phosphor
($C_{Cu} = 1 \times 10^{-5}$ g/g). C_{Co} (g/g):
1) 0; 2) 1×10^{-6}; 3) 5×10^{-6}; 4)
1×10^{-5}.

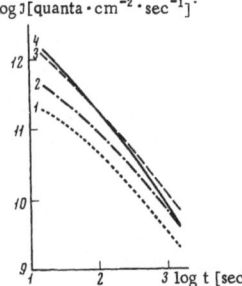

Fig. 14. Decay of the luminescence of a ZnS:Cu:Co
phosphor ($C_{Co} = 7.5 \times 10^{-6}$
g/g) with different concentrations of copper (g/g): 1)
1×10^{-7}; 2) 1×10^{-4}; 3) 1×10^{-5}; 4) 5×10^{-5}.

Infrared radiation of 1.8×10^{-4} W/cm² power density, which reduces the steady-state
luminescence of the ZnS:Cu:Co phosphors by a factor of 2-3, has hardly any effect on the ratio
of the band intensities: its action on the green band is only about 10% stronger than on the blue
band (Fig. 12). This provides further evidence for the absence of an equilibrium population of
the green and blue luminescence centers at the moment of excitation because otherwise we would
observe a much stronger influence of infrared radiation on centers of greater depth, i.e., on the
blue luminescence centers. Thus, all the above experimental data on the luminescence of the
ZnS:Cu and the ZnS:Cu:Co phosphors indicate that large light sums can be stored (i.e., a considerable fraction of the green luminescence centers can be ionized) even when the carrier population of the local levels in the green and blue luminescence centers does not have equilibrium values.

The process of buildup of luminescence of these phosphors if fairly slow and when cobalt
is introduced the time needed to reach the steady-state brightness increases considerably (Fig.

Fig. 15. Decay of the luminescence of a ZnS:Cu:Co
phosphor ($C_{Cu} = 1 \times 10^{-5}$
g/g) with different concentrations of cobalt (g/g): 1)
0; 2) 1×10^{-7}; 3) 1×10^{-6};
4) 1×10^{-5}; 5) 1×10^{-4}.

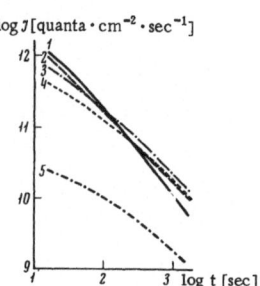

13). When the excitation power density is 5×10^{-5} W/cm², this time exceeds 2 min for the optimal ZnS:Cu:Co phosphors. The number of quanta of the exciting light which reaches a phosphor during this interval is 10^{16} cm⁻², which gives the upper limit of the possible values of the light sum. In spite of the fact that during continuous excitation the blue and the green bands are of comparable intensity, the green band is much stronger than the blue band during the afterglow.

The decay curves obtained for two batches of phosphors over time intervals from 10 sec to 20 min are plotted on a double logarithmic scale in Figs. 14 and 15.

The decay curves satisfy the law (44) beginning from $t \geq 3$ min and the highest afterglow intensity during this stage of decay is observed for phosphors with $C_{Cu} = 10^{-5}$ g/g and $C_{Co} = 5 \times 10^{-6}$ g/g. The minimum values of the slopes α of the lines representing the decay of the luminescence of the ZnS:Cu and the ZnS:Cu:Co phosphors are $\alpha = 1.35$ for $C_{Cu} = 1 \times 10^{-5}$ g/g (Co-free phosphors) and $\alpha = 0.95$ for $C_{Cu} = 1 \times 10^{-5}$ g/g and $C_{Co} = 5 \times 10^{-6}$ g/g, i.e., these values are much smaller than $\alpha = 2$ which is predicted by the simplest model of recombination luminescence.

It is usually assumed that the values of α are low because of an inhomogeneous excitation of the phosphor film across its thickness or because of a complex system of electron traps. The use of very thin films ($\alpha \leq 1/\varkappa_e$), ensuring homogeneous excitation, increased only slightly the value of α: this value rose from 0.97 to 1.03 in the case of phosphors with the optimal concentration of cobalt and from 1.36 to 1.43 in the case of cobalt-free phosphors.

The influence of thickness of a luminescent screen on the value of α for the ZnS:Cu:Co

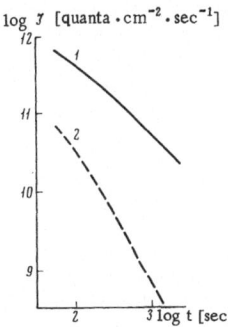

Fig. 16. Decay of the luminescence of a ZnS:Cu:Co phosphor ($C_{Cu} = 1 \times 10^{-5}$ g/g), $C_{Co} = 7.5 \times 10^{-6}$ g/g): 1) natural decay; 2) decay in the presence of 1.3 μ infrared illumination.

Fig. 17. Influence of temperature on the decay of the luminescence of a ZnS:Cu phosphor ($C_{Cu} = 1 \times 10^{-5}$ g/g): 1) −60°C; 2) 0°C; 3) 20°C; 40 50°C; 5) 100°C.

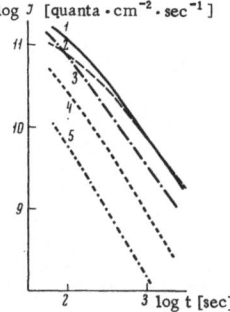

TABLE 3. Influence of Temperature on Decay Factor α (film
thickness d = 20 μ)

T, °C	α	
	ZnS - Cu·10⁻⁵ g/g	ZnS - Cu·10⁻⁵·Co 7,5 × × 10⁻⁶ g/g
100	1.64	1.72
75	1.50	1.51
50	1.59	1.14
35	1.54	1.12
20	1.43	1.04
0	0.85	0.84
−30	0.51	0.63
−65	1.35	1.02

phosphors (C_{Cu} = 1 × 10⁻⁵ g/g, C_{Co} = 7.5 × 10⁻⁶ g/g) at T = 20°C can be judged from the following data:

d, μ . .	8	20	30	1000 (tightly packed powder)
α	1.06	1.02	0.96	0.92

Thus, even in the case of thin screens the value of α is considerably less than 2, i.e., the
presence of layers with different degrees of excitation and the possibility of reabsorption of
luminescence cannot explain the low values of α.

A considerably greater increase in the decay factor α was achieved by illuminating a
luminescent screen with infrared radiation (λ_{IR} ~ 1.3 μ, W ~ 2 × 10⁻⁴ W/cm²) during afterglow.
Such illumination equalized the probability of liberation of carriers from various local levels
(Fig. 16). Under the influence of further illumination the values of α rose to 1.85 for a thick
powder layer and 1.9 for thin screens. These experiments indicated the existence of a complex
system of local levels in our phosphors, which was supported also by the structure of the thermoluminescence curves.

The value of the decay factor α depends strongly on the temperature. The data on the influence of temperature on the decay of the luminescence of the ZnS:Cu and the ZnS:Cu:Co phosphor are given in Table 3 and in Figs. 17 and 18. A gradual acceleration of the decay was observed for the ZnS:Cu phosphors when the temperature was raised from −30 to 100°C; in the
range from +20 to 100°C the value of α varied slowly. When the temperature was lowered, the
decay factor α first decreased strongly (it even became smaller than ~1) but this was followed
by a rapid rise in a narrow temperature range between −35 and −65°C, so that this factor
reached 1.35 at T = −65°C (this value was close to that obtained at T = 20°C).

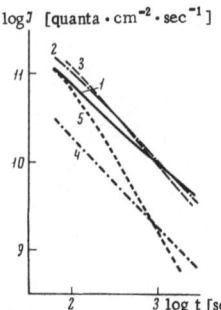

log J [quanta · cm⁻² · sec⁻¹]

Fig. 18. Influence of temperature on the decay of the
luminescence of a ZnS:Cu:Co
phosphor (C_{Cu} = 1 × 10⁻⁵
g/g), C_{Co} = 7.5 × 10⁻⁶ g/g):
1) −60°C; 2) 0°C; 3) 20°C;
4) 50°C; 5) 100°C.

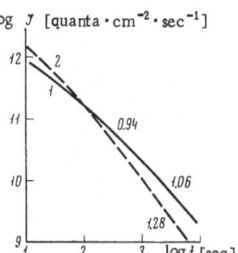

Fig. 19. Later stages of the decay of the luminescence of two phosphors: 1) ZnS:Cu:Co ($C_{Co} = 1 \times 10^{-5}$ g/g, $C_{Cu} = 7.5 \times 10^{-6}$ g/g); 2) ZnS:Cu ($C_{Cu} = 1 \times 10^{-5}$ g/g).

In the case of the ZnS:Cu:Co phosphors a stronger increase of α from 1.04 to 1.72 was observed between 20 and 100°C, the strongest rise occurring in the range 50 < T < 75°C. At 75 and 100°K the decay of the luminescence of the cobalt-activated and the cobalt-free phosphors could be described approximately by Eq. (44) with $\alpha \sim 1.7$. At −65°C the cobalt-activated phosphors had the same value of α as at room temperature. Thus, the greatest difference between the decay curves of the ZnS:Cu and ZnS:Cu:Co phosphors was observed in the temperature range 20-50°C, i.e., the range preceding the thermal de-excitation of the cobalt levels (T ≈ 70°C). The approximation of the decay curve by $\mathcal{J} = Ct^{-\alpha}$ was very convenient in comparisons of the decay of different phosphors, but this law was applicable only to a narrow range of luminescence intensities.

The decay factor α gradually increased at later stages of the afterglow (Fig. 19), which was expected because the retention of $\alpha = 1$ would have implied an infinite light sum stored in the phosphors.

§ 3. Light Sums and Sensitivity of Infrared Radiation of ZnS:Cu:Co Phosphors

The de-exciting action of infrared radiation consists of a change in the number of carriers which have become localized at capture (trapping) levels during excitation. When crystal phosphors are used in investigations of infrared radiation fields it is convenient to employ changes in the light sums of excited phosphors, which are related directly to changes in the number of localized carriers. Therefore, we shall consider the sensitivity of the ZnS:Cu:Co phosphors to infrared radiation as judged by changes in the light sums stored in the phosphors.

The light sums were determined by two methods. In the first method we used thick screens (d = 1 mm) and measured the area under the curve representing the decay of the luminescence during the first 20 min from the end of excitation. In the second method we recorded the thermoluminescence spectra, which were then used to determine the light sums of thin screens.

In both cases the excitation was provided by the 365 nm line emitted by a PRK-4 lamp. The excitation power density was 8×10^{-5} W/cm²; the duration of the excitation was 10 min; the temperature of the phosphor was 20°C.

In the first method we studied the influence of the concentration of copper on the light sums stored in the ZnS:Cu phosphors, the influence of the concentration of cobalt in the ZnS:Cu:Co phosphors with the optimal concentration of copper (10^{-5} g/g), and the changes in the light sums of these phosphors during prolonged decay.

The influence of the concentration of copper on the light sums S stored in the ZnS:Cu phosphors can be judged from the following table:

C_{Cu}, g/g	10^{-7}	10^{-6}	10^{-5}	$5 \cdot 10^{-5}$	10^{-4}
$S \cdot 10^{-13}$, quanta/cm²	1.6	3.4	8.1	8.7	9.0

TABLE 4. Depletion of Light Sums Stored in ZnS:Cu and ZnS:Cu:Co
Phosphors during Various Stages of Natural Decay

Δt, min	ZnS - Cu			ZnS - Cu, Co		
	ΔS, rel. units	ΔS, %	$\Sigma \Delta S$, %	ΔS, rel.units	ΔS, %	$\Sigma \Delta S$, %
0—0.25	800	28.5	28.5	450	12.5	12.5
0.25—1.0	800	29	57.5	600	16.5	2.9
1—10	232	8	65.5	280	7.5	36.5
10—20	266	9	74.5	494	13.5	50
20—60	310	11	86	700	19.5	69.5
60—120	156	6	91.5	408	11.5	81
120—480	208	8	100	648	19	100
0—480	2772			3580		

It is evident from these values that when the concentration of copper is increased, the light sums increase at first but, beginning from $C_{Cu} = 10^{-5}$ g/g$_{ZnS}$, the value of S remains practically constant right up to copper concentrations $C_{Cu} = 10^{-4}$ g/g$_{ZnS}$. Therefore, we prepared ZnS:Cu:Co phosphors with copper concentrations of 10^{-5} g/g$_{ZnS}$.

The dependence of S on the concentration of Co in the ZnS:Cu:Co phosphors was as follows:

C_{Co}, g/g ZnS	0	10^{-6}	$3 \cdot 10^{-6}$	$5 \cdot 10^{-6}$	$7.5 \cdot 10^{-6}$	10^{-5}	$3 \cdot 10^{-5}$
$S \cdot 10^{-13}$, quanta/cm^2	7.5	7.0	7.1	6.6	5.2	4.5	0.35

It is clear from this table that the light sums which can be extracted from the ZnS:Cu:Co phosphors during the first 20 min are somewhat smaller than the light sums which can be extracted from the ZnS:Cu phosphors, which is due to the slower de-excitation of the latter phosphors.

The influence of the rate of decay was investigated by measuring the light sums extracted during different stages of the luminescence decay.

The results of this investigation are given in Table 4.

It follows from Table 4 that the total light sum stored is a ZnS:Cu phosphor in 2772 units whereas the corresponding sum stored in a ZnS:Cu:Co phosphor is 3580 units, i.e., is the latter case the light sum represents ~129% of the light sum in the former case. During the first 20 min of the decay a ZnS:Cu phosphor loses 74.5% of the light sum whereas a ZnS:Cu:Co phosphor loses 50%. Therefore, a comparison of the light sums stored in the two types of phosphor can be made provided we increase by a factor of 1.5 the light sums extracted from the ZnS:Cu:Co phosphors extracted during the first 20 min of the decay.

The relationship between the number of defects generated by the introduction of an activator and the maximum number of localized electrons was determined by measuring the light

TABLE 5. Influence of Concentration of Copper on Light Sums of
Thin ZnS:Cu Screens

C_{Cu}, g/g	$C_{Cu} \cdot 10^{-12}$, ion/cm^2	$S \cdot 10^{-12}$, quanta/cm^2	$\dfrac{S_q}{C_{Cu}}$	η	$\dfrac{S_q}{C_{Cu}} \dfrac{1}{\eta}$
$1 \cdot 10^{-7}$	0.8	1	1.2	0.11	10
$1 \cdot 10^{-6}$	8	2.3	0.3	0.13	2.3
$1 \cdot 10^{-5}$	80	6.8	0.08	0.36	0.2
$5 \cdot 10^{-5}$	400	9.3	0.02	0.62	0.03
$1 \cdot 10^{-4}$	800	8.6	0.01	0.64	0.01

TABLE 6. Influence of Concentration of Cobalt on Light Sums of Thin ZnS:Cu:Co Screens with $c_{Cu} = 10^{-5}$ g/g

c_{Co}, g/g	$c_{Co} \cdot 10^{-14}$, ion/cm²	$S \cdot 10^{-14}$, quanta/cm²	$\Delta S \cdot 10^{-14}$, quanta/cm²	$\dfrac{\Delta S}{c_{Co}}$	η	$\dfrac{\Delta S}{c_{Co}} \dfrac{1}{\eta}$
0	—	0.66			0.38	
$1 \cdot 10^{-6}$	0.8	0.95	0.3	0.37	0.28	1.3
$3 \cdot 10^{-6}$	2.4	1.8	1.1	0.46	0.12	3.8
$5 \cdot 10^{-6}$	4	2.0	1.3	0.3	0.08	3.7
$7.5 \cdot 10^{-6}$	6	1.9	1.2	0.2	0.055	3.6
$1 \cdot 10^{-5}$	8				0.02	
$3 \cdot 10^{-5}$	12	0.17			0.002	

sums of thin films ($d = 20 \pm 3\,\mu$, $M = 6$ mg/cm²) of the ZnS:Cu and the ZnS:Cu:Co phosphors excited homogeneously across their thickness. We determined the absolute values of the number of defects in these phosphors, the number of the activator atoms introduced, and the quantum values of the light sums deduced from the thermoluminescence spectra. The number of stored quanta was then divided by the steady-state effectiveness of the luminescence (Table 2) and this enabled us to determine the total number of excited sites in a phosphor (the number of localized carriers).

In the case of the ZnS:Cu:Co phosphors it seemed desirable to find the ratio of the change in the light sum ΔS (expressed in quanta) resulting from the introduction of cobalt atoms to the number of such atoms c_{Co}, and to compare this ratio with one obtained making allowance for the quantum efficiency of the steady-state luminescence (Table 1).

The results obtained for the thin screens are given in Tables 5 and 6.

A comparison of the data given in § 2 and those given in Tables 5 and 6 shows that the total number of quanta emitted in the thermoluminescence of thin screens is approximately half the number of quanta emitted by an infinitely thick layer during afterglow. Since our experiments on the ZnS:Cu phosphors screens have indicated that the light sums are independent of the de-excitation method, these results show that under our conditions thin films are excited almost homogeneously across their thickness.

The number of quanta emitted from a thin film of ZnS:Cu with a low concentration of copper is close to or even higher than the number of copper atoms (10^{-7} g/g). Bearing in mind the effectiveness (0.11) of the steady-state luminescence, we find that the stored light sums are at least 9 times (i.e., almost an order of magnitude) higher than the concentration of the copper luminescence centers. Hence, it follows that in the case of phosphors with low concentrations of copper a considerable fraction of holes is localized during excitation at other hole traps, which can be the blue luminescence centers associated with intrinsic lattice defects.

When the concentration of copper is increases from 1×10^{-5} to 5×10^{-5} g/g, the stored light sum increases considerably but the ratio of the number of quanta emitted by a phosphor to the concentration of copper decreases strongly and becomes $\sim 8\%$ for $c_{Cu} = 10^{-5}$ g/g or, if the effectiveness of the luminescence is taken into account, about 15-20%.

Introduction of cobalt in concentrations up to 5×10^{-6} g/g increases the light sum of the ZnS:Cu:Co screens and the absolute increase in this sum represents a considerable fraction (25-50%) of the number of cobalt atoms introduced.

However, if an allowance is made for the effectiveness of the luminescence, which decreases when cobalt is introduced, we find that the increase in the stored light sum (expressed in quanta) proceeds at a faster rate than the increase in the number of electron localization levels formed by cobalt atoms.

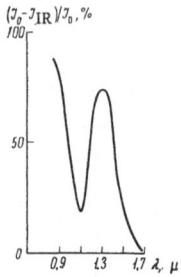

Fig. 20. Spectrum of the sensitivity of ZnS:Cu:Co phosphors to the quenching action of infrared radiation.

Since such a difference between the rates of increase is unlikely, it follows that the quantum efficiency of the thermoluminescence of the ZnS:Cu:Co phosphors is at least 3-5 times larger than the quantum effectiveness of the photoluminescence of the same phosphors, even if we assume minimum values of the absolute light sums which, according to our calculations, are 1.5 times smaller than the average tabulated values.

Comparison of the Sensitivities to Infrared Radiation

An investigation of the influence of infrared radiation on the afterglow of the ZnS:Cu:Co phosphors with different concentrations of the components indicated that the spectrum of the sensitivity to infrared radiations of different wavelengths was not greatly affected by changes in the relative concentrations of the activators. Figure 20 shows the sensitivity spectrum of a ZnS:Cu:Co phosphor, optimal from the point of view of the quenching effect. The curve plotted in Fig. 20 is normalized to unit incident energy of the infrared radiation.

It is evident from Fig. 20 that, in agreement with the published data [42], there are two sensitivity maxima: one at short wavelengths in the region of 0.8-0.9 μ ($h\nu_{IR}$ = 1.4 eV) and another at longer wavelengths in the region of 1.3 μ ($h\nu_{IR}$ = 0.95 eV).

The second of these sensitivity maxima is of greatest practical interest because it lies beyond the long-wavelength edge of the sensitivity of photographic emulsions. In the 1.1 μ region, corresponding to the output radiation of $CaWO_4:Nd^{3+}$ lasers (1.06 μ) there is a 15% dip of the sensitivity below the maximum value, but the sensitivity is still sufficient for recording the radiation field of such lasers.

The long-wavelength edge of the sensitivity spectrum extends to about 1.7-1.8 μ (0.7 eV), which is quite close to the theoretical limit (Chap. I) of the optical sensitivity at room temperature ($\sim 2 \mu$), which is set by the thermal liberation of carriers from traps.

Since the sensitivity spectrum depended weakly on the concentrations of Co and Cu, we compared the sensitivities of different phosphors and estimated the absolute value of the sensitivity of the 1.3 μ maximum, especially as this maximum was of greatest practical importance.

In the first series of experiments the phosphors were first excited for 5 min with ultraviolet light of $W_e = 5 \times 10^{-5}$ W/cm^2 power density and then illuminated with infrared radiation after 10-min decay, i.e., we used that part of the decay curve which obeyed the law $\mathcal{J}_t = \mathcal{J}_0/t$. The power density of the 1.3 μ infrared radiation (1.2 × 10^{-5} W/cm^2) was such that the rate of de-excitation by this radiation was approximately an order of magnitude faster than the rate of natural decay. It was found that illumination with infrared radiation first enhanced the intensity of the afterglow but this was followed by quenching (compared with the parts of the phosphor which were not illuminated in this way). These results were observed for all the ZnS:Cu phosphors (Fig. 21) and also, but to a lesser degree, for the cobalt-activated phosphors (Figs. 22 and 23). Thus, all the investigated phosphors exhibited the stimulating and the quenching effects

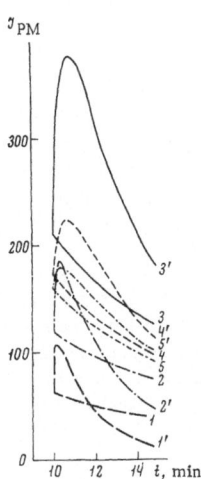

Fig. 21. Influence of infrared radiation ($\lambda = 1.3$ μ) on the afterglow of ZnS:Cu phosphors with different concentrations of copper (g/g): 1)-1') 1×10^{-7}; 2)-2') 1×10^{-6}; 3)-3') 1×10^{-5}; 4)-4') 5×10^{-5}; 5)-5') 1×10^{-4}. Curves 1-5 represent natural decay and curves 1'-5' represent decay in the presence of infrared illumination.

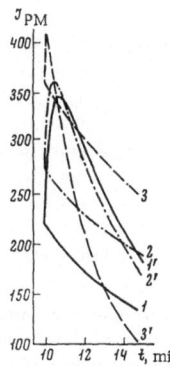

Fig. 22. Influence of infrared radiation on the afterglow of ZnS:Cu:Co phosphors with different concentrations of cobalt (g/g): 1)-1') 0; 2)-2') 1×10^{-6}; 3)-3') 3×10^{-6}. Curves 1-3 represent natural decay and curves 1'-3' represent decay in the presence of infrared illumination.

of infrared radiation and the ratio of these effects was a function of the concentrations of copper and cobalt. Table 7 lists the maximum values of the luminescence intensity $\mathcal{I}_{max}$, the time taken to reach this maximum t_{max}, the difference between the maximum intensity in the presence and absence of infrared illumination $\Delta \mathcal{I}_{max}$, the same difference $\Delta \mathcal{I}/\mathcal{I}_0$ expressed in percent, and the time t_a in which the afterglow intensity returned to that which would be observed in the absence of infrared illumination.

TABLE 7. Comparison of Natural and Infrared-Influence Decay
of Luminescence of ZnS:Cu Phosphors

$c_{Cu} \cdot$ g/g	t_{max}, sec	$\mathcal{I}_{max}$	$\Delta \mathcal{I}_{max} = \mathcal{I}_{max} - \mathcal{I}_0$	$\frac{\Delta \mathcal{I}}{\mathcal{I}_0}$, %	t_a, min (for $\mathcal{I}_{IR} = \mathcal{I}_0$)
10^{-7}	8	110	50	83	1.75
10^{-6}	15	185	60	50	2.5
10^{-5}	45	380	190	100	15
$5 \cdot 10^{-5}$	40	225	65	40	10
10^{-4}	20	180	30	20	10

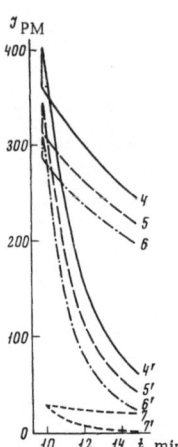

Fig. 23. Influence of in-
frared radiation on the
afterglow of ZnS:Cu:Co
phosphors with different
concentrations of cobalt
(g/g): 4)-4') 5×10^{-6};
5)-5') 7.5×10^{-6}; 6)-6')
1×10^{-5}; 7)-7') 3×10^{-5}.
Curves 4-7 represent nat-
ural decay and curves 4'-7'
represent decay in the presence of infrared illumination.

It is evident from Table 7 that the optimal values of the relative and absolute stimulating effect of infrared radiation on the ZnS:Cu phosphors corresponds to copper concentrations of 10^{-5} g/g (4×10^{17} cm^{-3}). The same concentration of copper corresponds to the longest time needed to reach the maximum intensity (t_{max} = 45 sec) and to the longest time during which the intensity in the presence of infrared illumination exceeds the afterglow intensity under natural decay conditions (15 min). However, the absolute quantum efficiency of the stimulation flash of the ZnS:Cu phosphors is low: in the optimal case it reaches 0.03-0.05%. The light sum extracted by infrared illumination is much smaller than the total sum stored in the phosphors and liberated during natural decay or by thermal means. This means that even in the case of the ZnS:Cu phosphors the stimulating effect of infrared radiation is simultaneous with the quenching action of the same radiation, the latter resulting in the liberation of holes from copper, which then undergo nonradiative recombination with nonequilibrium electrons localized at the quenching centers.

Introduction of cobalt (Figs. 22 and 23, and Table 8) does not reduce greatly the maximum value of the luminescence intensity but the actual amplitude of the flash $\Delta \mathcal{J}$ becomes much smaller. However, the time t_a necessary to reach the same intensity as during natural decay of the afterglow becomes shorter, i.e., quenching effect of infrared radiation in the presence of cobalt is stronger. We can see from Table 8 that t_a decreases with increasing C_{Co}.

If we define the sensitivity to infrared radiation as $\left(\dfrac{\mathcal{J}_0 - \mathcal{J}_{IR}}{\mathcal{J}_0} \right)$ % measured 1 min after the beginning of infrared illumination, we find that this sensitivity increases almost linearly

TABLE 8. Quenching Effect of Infrared Radiation on Luminescence
of ZnS:Cu:Co Phosphors

c_{Co}	t_{max}, sec	$\mathcal{J}_{max}$	$\Delta\mathcal{J}_{max}$	$\Delta\mathcal{J}/\mathcal{J}_0$, %	t_a, min
	45	345	145	72	15
10^{-6}	30	360	100	39	4
$3 \cdot 10^{-6}$	8.5	410	60	17	0.75
$5 \cdot 10^{-6}$	2	400	40	11	0.35
$7.5 \cdot 10^{-6}$	1	345	30	9	0.25
$1 \cdot 10^{-5}$	1	315	20	7	0.15
$3 \cdot 10^{-5}$	1	30	2	6	0.05

with increasing cobalt concentration. Since the intensity of the afterglow increases with increasing coblat concentration, the quantum sensitivity to the quenching effect of infrared radiation has a maximum in the region of cobalt concentrations between 5×10^{-5} and 7.5×10^{-5} g/g.

Estimates of the Absolute Sensitivities of ZnS:Cu:Co Phosphors

to Infrared Radiation

A study of the influence of an increase in the concentration of copper indicated that such an increase does not raise the quantum sensitivity to infrared radiation and reduces considerably the relative sensitivity. An analysis of the influence of the thermal conditions during firing made it possible to raise the intensity of the afterglow of phosphors by about 15% and the absolute sensitivity to infrared radiation by about 20%. The absolute value of the quantum sensitivity, defined as the difference between the areas under decay curves in the presence and absence of infrared radiation, is small. This absolute value becomes much larger when an allowance is made for the reduction in the light sum which is extracted after the end of infrared illumination. Moreover, one must add the aftereffect of infrared radiation, i.e., the strong reduction in the intensity of the afterglow immediately after the end of infrared illumination. When these allowances are made the difference between the light sums, expressed in quanta, for the phosphors not illuminated and illuminated with infrared radiation, divided by the number of incident infrared quanta, is 0.3–0.5%. These values increase by a factor of 20, reaching 7–10%, if the absorbed rather than the incident energy is used in the calculations because the reflection coefficient for infrared radiation incident on the surface of a phosphor is 92 ± 3% and, moreover, about half of the infrared radiation entering a thin phosphor film emerges on its other side.

Estimates of the Parameters of the Luminescence Kinetics

of ZnS:Cu:Co Phosphors

The experimental data on the intensity of the ultraviolet excitation (I_e, quantum/cm^2), the effectiveness of the steady-state luminescence η_{stat}, the absolute values of the light sums stores in screens S, and the sensitivity to infrared radiation η_{IR} can be used to estimate various parameters of the synthesized phosphors.

The intensity of the luminescence of thin screens ($\mathcal{J}_s$, quanta $\cdot$ cm$^{-2} \cdot$ sec^{-1}) can be used to determine the number of radiative (w_{rr}^-, cm$^{-3} \cdot$ sec^{-1}) and nonradiative (w_{nr}^+, cm$^{-3} \cdot$ sec^{-1}) transitions occurring per unit volume in unit time during excitation:

$$w_{rr}^- = \beta n^+ N^- = \frac{\eta I_e}{d} \left(\frac{\mathcal{J}_s}{\mathcal{J}_p} \right),$$

$$w_{nr}^+ = \beta_1 n^- N^+ = \frac{(1-\eta) I_e}{d} \left(\frac{\mathcal{J}_s}{\mathcal{J}_p} \right), \tag{45}$$

where $\beta = u_T \sigma_{rr}^-$, $\beta_1 = u_T \sigma_{nr}^+$ (u_T is the thermal velocity of carriers; σ_{rr}^- and σ_{nr}^+ are the radiative and nonradiative recombination cross sections); N^- and N^+ are the free-carrier densities; $d = 20 \mu$ is the thickness of a screen; $\mathcal{J}_s/\mathcal{J}_p$ is the ratio of the intensities of the luminescence emitted by a screen and a powder, n^+ is the concentration of ionized centers; n^- is the density of localized electrons ($n^+ \approx n^-$) deduced from the absolute values of the light sums S.

If we assume that $W_e = 6 \times 10^{-5}$ W/cm^2 (1.1×10^{14} quanta/cm^2), which is the excitation intensity corresponding to the optimal values of the light sums, we obtain the following transition rates or luminescence intensities for the ZnS:Cu screens ($C_{Cu} = 10^{-5}$ g/g) and the ZnS:Cu:Co screens ($C_{Cu} = 10^{-5}$ g/g and $C_{Co} = 5 \times 10^{-6}$ g/g: $\mathcal{J}_{rr,Cu} = 1.5 \times 10^{16}$, $\mathcal{J}_{nr,Cu} = 1.0 \times 10^{16}$,

$\mathcal{I}_{rr.Cu.Co} = 2.5 \times 10^{15}$, and $\mathcal{I}_{nr.Cu.Co} = 2.2 \times 10^{16}$ (all these values give the number transitions per cubic centimeter per second).

If we use the absolute values of the light sums extracted from such screens (the relevant formulas are $\beta N^- = w_{rr}^-/n$, and $\beta N^+ = w_{nr}^+/n$), we can find the rates of recombination of localized electrons and holes, expressed as the number of transitions per center per second: $\beta N_{Cu}^- \sim 1$ and $\beta N_{Cu}^+ \sim 1$ for ZnS:Cu; $\beta N_{Cu}^- \sim 0.1$ and $\beta N_{Co}^+ \sim 1$ for ZnS:Cu:Co.

These data can be used to estimate the upper limit of the recombination cross section bearing in mind that because of the stimulating and quenching effects of the exciting ultraviolet light the density of free carriers is higher than the quasiequilibrium value (N_{eff}^+ is the effective density of states in the allowed bands, which is $\sim 10^{18}$ cm^{-3}), i.e., that

$$N^{\pm} \geqslant N_{eq}^+ = \frac{n^{\pm} N_{eff}^+}{C_{a,d}} e^{-E\pm/kT} \tag{46}$$

The nonradiative and radiative recombination cross sections can then be found from

$$\sigma_{nr,\,rr} \leqslant \frac{\beta N^{\pm}}{u_T N_{eq}^+}, \tag{47}$$

where u_T is the thermal velocity of carriers (1×10^7 cm/sec).

Calculations made using the depths of the Cu (0.33 and 0.35 eV) and Co (0.44 eV) levels give very low values of these cross sections. If we substitute the experimental data, we obtain the following equilibrium carrier densities:

in ZnS:Cu phosphors

$$N_{eq}^+ \approx \frac{3 \cdot 10^{19} \cdot 2 \cdot 10^{16}}{3 \cdot 10^{17}} \cdot 10^{-6} = 2 \cdot 10^{12} \text{ holes/cm}^3,$$
$$N_{eq}^- \approx \frac{3 \cdot 10^{19} \cdot 2 \cdot 10^{16}}{3 \cdot 10^{17}} \cdot 10^{-6} = 2 \cdot 10^{12} \text{ electrons/cm}^3;$$

in ZnS:Cu:Co phosphors

$$N_{eq}^- \approx \frac{3 \cdot 10^{19} \cdot 6 \cdot 10^{16}}{1.5 \cdot 10^{17}} \cdot 10^{-8} = 1.2 \cdot 10^{11} \text{ electrons/cm}^3,$$
$$N_{eq}^+ \approx \frac{3 \cdot 10^{19} \cdot 6 \cdot 10^{16}}{3 \cdot 10^{17}} \cdot 10^{-6} = 6 \cdot 10^{12} \text{ holes/cm}^3.$$

It is evident from Eq. (47) that the nonradiative and radiative recombination cross sections of the various levels are within the limits

$$\left(\frac{1 - 0.1}{10^{11} - 10^{12}}\right) \cdot 10^{-7} = 10^{-18} - 10^{-22} \text{ cm}^2.$$

These cross sections are very small, in qualitative agreement with the estimates of the same cross sections obtained by comparing the probabilities of the thermal liberation of carriers and of the stimulation by the exciting light (§ 2, Chap. II).

Thus, the small values of the nonradiative and radiative recombination cross sections, like the small values of the frequency factors of thermal liberation of carriers, are in a basic agreement with the slowness of the afterglow of the ZnS:Cu:Co phosphors. This is why large light sums can be stored in these phosphors and why they can have fairly high quantum sensitivity to infrared radiation.

§ 4. Comparison of the Optical Sensitivities of Various

Types of Phosphor to Infrared Radiation

The experimental data on the absolute quantum sensitivity of our ZnS:Cu:Co phosphors to infrared radiation must be compared with the optimal sensitivity of other crystal phosphors. It should be particularly interesting to make a comparison with alkaline-earth crystal phosphors such as CaS·SrS:Eu:Sm [9-10] and CaS·SrS:Ce:Sm (these phosphors exhibit stimulation flash), because the reported values of the absolute sensitivity of these phosphors are two or three orders of magnitude higher than the sensitivities of all other phosphors. We carried out this comparison under the same conditions which were employed in the main experiments, described in the preceding chapter. The exciting light was provided by a PRK-4 lamp (λ_e = 365 nm, $W_e = 5 \times 10^{-5}$ W/cm^2, $I_e \sim 1 \times 10^{14}$ quanta·cm^{-2}·sec^{-1}). Illumination with infrared radiation (8×10^{-5} W/cm^2) of wavelengths 1.3 and 1.0 μ, coinciding with the sensitivity maxima of the ZnS:Cu:Co phosphors and the alkaline earth phosphors, respectively, was started after 2 min after the end of the ultraviolet excitation.

During these first 2 min the afterglow intensity decreased to $\sim 2 \times 10^{12}$ quanta·cm^{-2}·sec^{-1} for the ZnS:Cu:Co phosphors and to a value approximately an order of magnitude lower (10^{11} quanta·cm^{-2}·sec^{-1}) for the alkaline earth phosphors. In both cases only a small fraction of the total light sum (not exceeding 20%) was lost during the first 2 min.

Since illumination with infrared radiation quenched the afterglow of the ZnS:Cu:Co phosphors and stimulated the afterglow of the CaS·SrS:Eu:Sm and CaS·SrS:Ce:Sm phosphors (Fig. 24), a direct comparison of the sensitivities could not be made. Such a comparison would have been ambiguous because of the difference between the quenching and stimulation mechanisms

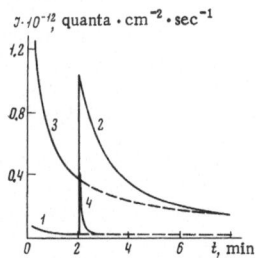

Fig. 24. Comparison of the optical sensitivities of ZnS: Cu:Co and SrS · CaS:Ce:Sm phosphors subjected to infrared illumination (λ_{IR} = 1.3 μ) 2 min after beginning of decay: 1) SrS · CaS:Ce:Sm in the absence of infrared illumination; 2) same phosphor in the presence of infrared illumination; 3) ZnS: Cu:Co in the absence of infrared illumination; 4) same phosphor in the presence of infrared illumination.

and the difference between the optimal conditions of practical applications of these phosphors.
For example, in the case of stimulation the enhancement of the afterglow intensity, which produced a positive image, was observed only during infrared illumination; in the case of quenching, the reduction in the afterglow intensity and the corresponding negative image were retained
after the end of infrared illumination and could be observed as long as the brightness of the un-
quenched regions was sufficient for the detection of luminescence.

It follows from the above discussion that for practical purposes it is often necessary to
compare not a change in the afterglow intensity but the light sum which is extracted or reduced
depending on whether stimulation or quenching occurs in a given phosphor. This definition of
the sensitivity increases considerably the values obtained for the ZnS:Cu:Co phosphors, compared with the $CaS \cdot SrS$:Eu:Sm and $CaS \cdot SrS$:Ce:Sm phosphors and at the same time it provides
a more natural measure of the processes occurring in these phosphors. However, the comparison should be carried out for a relatively short exposure to infrared radiation because the
sensitivity and the contrast, especially for the phosphors exhibiting the stimulation effect, decrease as the light sum is depleted. On the other hand, the exposure must not be too short because all the phosphors exhibiting an overall quenching effect, produce a flash during the initial
stage of infrared illumination.

The total light sum extracted from $CaS \cdot SrS$:Ce:Sm by a 15-min infrared illumination
($\lambda_{IR} = 1.0\ \mu$, $I_{IR} = 4 \times 10^{14}$ quanta/cm^2) was 1.5×10^{14} quanta/cm^2, i.e., it was of the same order of magnitude as for ZnS:Cu:Co. The optimal quantum effectiveness was 0.40-0.45% for
$I_{IR} = 4 \times 10^{14}$ quanta/cm^2. This value is in order-of-magnitude agreement with the published
data [9-15] but is approximately half the published values, which may be due to differences in
the excitation conditions.

The sensitivity to $\lambda = 1.3$ radiation was approximately 1.5 times lower than the optimal
sensitivity to the $\lambda = 1.0\ \mu$ rays, in agreement with the sensitivity spectrum given in [9].

The absolute sensitivity of ZnS:Cu:Co determined under the same conditions during infrared illumination was an order of magnitude lower, i.e., about 0.1%. The sensitivity deduced
from the difference between the light sums extracted during 15 min in the presence and absence
of infrared illumination was 0.5-1%. Thus, in contrast to the values reported earlier, the absolute sensitivity of the ZnS:Cu:Co phosphors was not inferior to the sensitivity of the alkaline-
earth phosphors. Therefore, the selection of either type of phosphor for detection of infrared
radiation should be governed by the actual problem which was to be tackled. In particular, when
the combined effect of several repeated signals has to be determined, it will be preferable to
use ZnS:Cu:Co phosphors.

It must be stressed particularly strongly that the absolute sensitivity of the ZnS:Fe phosphors, determined at 77°K under the excitation conditions given above, was 0.4% for $\lambda_{IR} = 1.3\ \mu$,

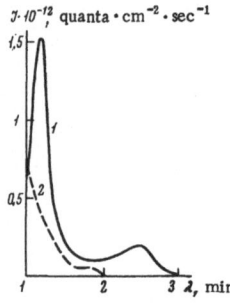

Fig. 25. Sensitivity spectra
of two phosphors at $T = 77°K$:
1) ZnS:Fe; 2) ZnS:In.

in close agreement with the published data [25]. This agreement confirmed that our measurements were quite accurate (these measurements were carried out on the same ZnS:Fe phosphors as were used in [25]) and it showed that at room temperature it would be preferable to use the ZnS:Cu:Co phosphors in investigations of the near infrared region. However, at low temperatures the ZnS:Fe phosphors would be preferable at longer wavelengths because of the presence of a second maximum in the infrared sensitivity, located at 2.5 μ (the sensitivity at this maximum is about 0.1%, as shown in Fig. 25).

The quantum sensitivity of the ZnS:In phosphors exhibiting the stimulation effect (these phosphors were recommended in [56] because of its high sensitivity in the region of 2 μ) was found to be much lower than that of ZnS:Fe.

Thus, the absolute quantum sensitivity of the best currently available phosphors, normalized to the incident flux, is 0.1-1% which is quite sufficient for recording laser radiation fields of 10^{-3}-10^{-4} J/cm^2 energy density. The greatest losses in the incident infrared radiation (> 90%) result from incomplete absorption of this radiation in an excited phosphor and the quantum efficiency of the stimulation and quenching effects can reach 10%. This is evidence of a large value of the ratio of the probabilities of recombination and recapture in phosphors which can store large light sums. Further development of these phosphors should be aimed not so much in the direction of increasing quantum efficiency of the stimulation and quenching effects as in the direction of enhancement of the absorption of infrared radiation by the phosphor in question.

This leads us to a discussion of the considerable advantages of the utolization of the thermal effect of infrared radiation because in this case the radiation is absorbed not in the phosphor itself but in its blackened substrate.

CHAPTER III

INVESTIGATION OF CRYSTAL PHOSPHORS EXHIBITING HIGH THERMAL SENSITIVITY

§ 1. Synthesis of Crystal Phosphors
and Investigation Method

In the present chapter we shall consider the luminescence of phosphors sensitive to small changes in the temperature and we shall establish experimentally the chemical compositions and the excitation conditions ensuring maximum thermal sensitivity.

We shall consider the following two promising systems:

1) ZnS · CdS:Ag:Ni phosphors, which exhibit strong quenching of the room-temperature luminescence when the temperature is raised [17]; these phosphors have used in black-and-white thermography [23, 24, 27, 28] and should be capable of producing a negative contrast in investigations of temperature and electromagnetic fields;

2) ZnS:Ag:Sm phosphors, which exhibit a strong temperature dependence of the ratio of intensities of two activator bands of different colour: the change in color of the luminescence near room temperature can be used in color thermography [26].

These phosphors were prepared from zinc sulfide and cadmium sulfide of the "phosphor" grade.

The ZnS·CdS:Ag:Ni phosphors were prepared as follows. The Ag and Ni activators were introduced into the charge in the form of solutions of their nitrates, whereas the flux (NaCl) was in the salt form. The firing was carried out in air in quartz ampoules. The charge was first dried at 400°C for 2 h in order to remove the water of crystallization. Next, a crucible was rapidly transferred to a furnace heated to 900°C and kept at this temperature for 30 min.

The following batches were prepared for the purpose of investigating the influence of physicochemical properties on the luminescence: a batch with different concentrations of Ni (from 1×10^{-7} to 1×10^{-4} g/g) and a constant concentration of Ag (1×10^{-4} g/g) and of CdS (30%); a batch of phosphors with different concentrations of Ag (from 5×10^{-5} to 5×10^{-4} g/g), a constant concentration of Ni (6.0×10^{-6} g/g) and a variable concentration of CdS (0-70%).

The ZnS:Ag:Sm phosphors were prepared from a charge which included $MgCl_2$ as a flux. The Ag and Sm activators were introduced as solutions of nitrates. The charge was dried at 120°C and poured into a quartz boat. This boat was then placed in a closed quartz tube through which hydrogen sulfide was passed at a rate of 60 liters/min.

The charge in the tube was first heated for 2 h at 400°C in order to remove zinc oxide and water of crystallization. Then, the tube containing the boat and the charge was placed in a furnace heated to 1200°C. This temperature was applied for 30 min in a stream of hydrogen sulfide. The phosphor formed in this way was rapidly cooled to room temperature in a stream of hydrogen sulfide.

The phosphors were investigated in the form of thin screens (≤ 6 mg/cm^2) deposited on Dural substrates. The excitation and the measurement conditions were the same as the preceding chapters. The stray light and the reflected exciting radiation were eliminated by blackening the jacket of a Dewar flask and by using combined filters. This made it possible to investigate the thermal quenching over a range of intensities differing by a factor of 1000. The total temperature interval was about 300 deg C (From −100 to 200°C). The most detailed investigations, in which the luminescence intensity was measured at intervals of 1 deg, were carried out in region of the highest quenching rate. The curves were recorded several times and a check was made that the dependences obtained during heating and cooling agreed.

The relative values of the luminescence intensity were measured to within ± 1.5%. Small temperature intervals of 5 deg C were determined to ± 10% and the relative thermal sensitivity was found to within ± 12% of the measured value.

§ 2. Thermal Quenching of $ZnS_x \cdot CdS_{1-x}$:Ag:Ni Phosphors

The quantum efficiency of the luminescence of $ZnS_{70} \cdot CdS_{30}$:Ag(10^{-4} g/g) at room temperature was quite high: a comparison of this luminescence with that emitted by a standard red-emitting phosphor gave a value of about 75 ± 5%, which was in agreement with the optimal values of the efficiency of the ZnS·CdS:Ag phosphors reported in the literature (~80-90%) [35]. A relatively wide luminescence band of silver was observed in the spectra of the $ZnS_x \cdot CdS_{1-x}$:Ag phosphors. When the concentration of CdS was increased, this band gradually shifted toward longer wavelengths because of reduction in the forbidden band width. The maximum of the luminescence band emitted by the ZnS:Ag phosphors was located at $\lambda = 435$ nm, whereas in the case of $ZnS_{70} \cdot CdS_{30}$:Ag phosphors the maximum was at $\lambda_{max} \approx 510$ nm ($h\nu_1 \approx 2.4$ eV). These phosphors exhibited a relatively weak thermal quenching: the luminescence decreased by a factor of 10 only at T ~ 120°C, i.e., when the temperature rose by 100 deg C.

The thermal quenching of the luminescence of the ZnS·CdS:Ag:Ni phosphors depended very strongly on the composition of these phosphors and the excitation conditions, particularly on the excitation intensity.

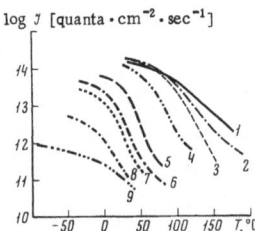

Fig. 26. Temperature dependence of the absolute intensity
of the luminescence of ZnS_{70} ·
CdS_{30}:Ag:Ni phosphors with
different concentrations of
nickel C_{Ni} (g/g): 2) 0; 3) 1 ×
10^{-7}; 4) 1 × 10^{-6}; 5) 5 × 10^{-6};
6) 8.5 × 10^{-6}; 7) 1 × 10^{-5}; 8)
2 × 10^{-5}; 9) 5 × 10^{-5}. Curve
1 represents a ZnS:Ag phos-
phor.

The temperature dependences of the luminescence intensity obtained for a batch of ZnS_{70} ·
CdS_{30}:Ag:Ni phosphors are given in Fig. 26 for samples containing 1 × 10^{-4} g/g (3 × 10^{-18} cm^{-3})
of silver and various amounts of nickel ranging from 10^{-7} to 5 × 10^{-5} g/g.

In many practical applications the thermal stability of the luminescence is a favorable
property. In our case it was necessary to produce phosphors which would be sensitive to
changes in the temperature and thus more suitable for thermal detection of long-wavelength infrared radiation. This could be done very effectively by introducing nickel quenching impurities, which not only reduced the quenching threshold to lower temperatures but ensured strong
quenching. Even small amounts of Ni (less than 10^{-7} g/g) altered the nature of the quenching
effect, which indicated that the amount of nickel present as an accidental impurity in the original zinc sulfide was even lower than assumed (according to factory data the concentration of nickel was 10^{-8}-10^{-7} g/g). Introduction of nickel in amounts of 10^{-6} g/g resulted in a fall of the
luminescence intensity by a factor of 2 at T = 20°C and by more than two orders of magnitude
at 120°C. The optimal quenching rate was observed in samples with Ni concentrations from
3.5 × 10^{-6} to 7 × 10^{-6} g/g: the quenching in such samples amounted to more than one order of
magnitude over a temperature range of only 20 deg C. For these phosphors the luminescence
intensity below 0°C was only 10-20% of the intensity of the luminescence emitted from the
nickel-free phosphors.

The region of the optimal sensitivity, i.e., of the steepest temperature dependences,
shifted toward lower temperatures when the concentration of nickel was increased. For the excitation intensities employed in the present study the optimal sensitivity near room temperature
was obtained for C_{Ni} = 7 × 10^{-6} g/g.

Further increase in the concentration of Ni was undesirable: it reduced strongly the intensity of the luminescence observed at low temperatures without increasing the steepness of
the temperature dependences (in fact, in the range $C_{Ni} \geq 3.5 \times 10^{-5}$ g/g the quenching rate decreased with increasing nickel concentration). Moreover, the range of temperatures corresponding to the maximum thermal sensitivity did not shift any further in the direction of lower
temperatures.

Fig. 27. Temperature de-
pendences of the absolute
intensity of the lumines-
cence of $ZnS_{70} \cdot CdS_{30}$:Ag:Ni
phosphors with different concentrations of silver C_{Ag} (g/g):
1) 5×10^{-5}; 2) 1×10^{-4}; 3)
5×10^{-4}. In all cases $C_{Ni} =$
6.0×10^{-6} g/g.

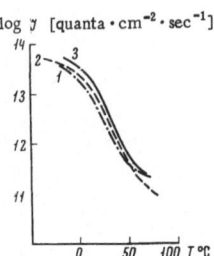

$\log \mathcal{J}$ [quanta $\cdot$ cm$^{-2} \cdot$ sec^{-1}]

Thus, the range of nickel concentrations in which the thermal sensitivity of the $ZnS \cdot CdS$: Ag:Ni phosphors could be increased effectively was 10^{-6}–10^{-5} g/g and the presence of nickel in such concentrations shifted the region of maximum sensitivity from 120°C to 20°C.

The lower limit of this range of concentrations was set by the presence of other quenching centers at concentrations of ~10^{-7} g/g and the upper limit by the formation of larger quenching centers such as nickel and silver complexes.

When the concentration of silver was reduced to 5×10^{-5} g/g or increased to 5×10^{-4} g/g, the rate of quenching of the $ZnS \cdot CdS$:Ag:Ni phosphors did not change significantly compared with that obtained for 1×10^{-4} g/g of silver (Fig. 27). When the concentration of silver, i.e., the concentration of the luminescence centers, was increased the luminescence intensity at low temperatures rose and the thermal quenching was observed 10-15 deg C higher. However in the strong-quenching region where T > 50°C and $\mathcal{J}/\mathcal{J}_0 > 10$, the luminescence emitted by the phosphors with higher and lower silver concentrations was slightly higher than for the phosphors with the "normal" concentration of 1×10^{-4} g/g. It should be stressed that the temperature interval in which the luminescence intensity changed by a factor of 100 was 60 deg C. The rate of quenching was lower at temperatures below this range (due to the weak thermal sensitivity) than at higher temperatures (strong general quenching).

Extension of the range of high thermal sensitivity is a very important task in the development of phosphors suitable for practical applications such as infrared detection and temperature measurement. Therefore, we carried out additional investigations of the influence of the composition and the synthesis conditions of the phosphors on their thermal sensitivity.

When the composition was varied (Fig. 28) it was found that the thermal quenching of the $ZnS \cdot CdS$:Ag:Ni phosphors depended strongly on the concentration of CdS and the greatest changes occurred up to $C_{CdS} \approx 30\%$.

At these concentrations narrowing of the forbidden band width altered the nature of the excitation process: direct absorption by the silver centers in ZnS:Ag:Ni gave way to band–band

$\log \mathcal{J}$ [quanta $\cdot$ cm$^{-2} \cdot$ sec^{-1}]

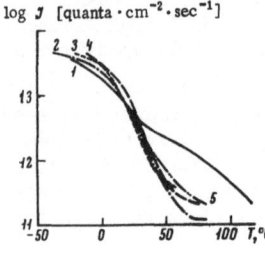

Fig. 28. Temperature dependences of the absolute intensity of the luminescence of
$ZnS_x \cdot CdS_{1-x}$:Ag:Ni phosphors
with different concentrations
of CdS ($C_{Ag} = 1 \times 10^{-4}$ g/g,
$C_{Ni} = 6.0 \times 10^{-6}$ g/g): 1) 0;
2) 20%; 3) 30%; 4) 40%; 5) 50%.

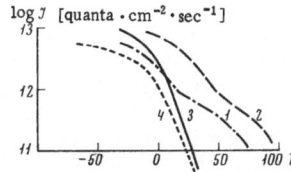

Fig. 29. Temperature dependences of the absolute intensity of the luminescence of ZnS:Ag:Ni $(1-\lambda_e = 365$ nm, $2-\lambda_e = 313$ nm) and $ZnS_{70} \cdot CdS_{30}$:Ag:Ni $(3-\lambda_e = 365$ nm, $4-\lambda_e = 313$ nm) phosphors.

transitions in the host substance at concentrations of CdS in excess of 30% (these results were obtained for $\lambda_e = 365$ nm). At a fixed value of the ultraviolet excitation intensity $(W_e = 1 \times 10^{-5}$ W/cm^2) the nature of the quenching of the ZnS:Ag:Ni phosphors was found to be quite different for $\lambda_e = 365$ and 313 nm, whereas in the case of the $ZnS_{70} \cdot CdS_{30}$:Ag:Ni phosphors with the same concentrations of silver $(10^{-4}$ g/g) and nickel $(6.0 \times 10^{-5}$ g/g) the influence of the excitation wavelength on the quenching was less (Fig. 29). At a fixed value of the excitation intensity the thermal quenching of the ZnS:Ag:Ni phosphors started much earlier if the excitation was provided by the $\lambda_e = 365$ nm line whereas the quenching of the ZnS $\cdot$ CdS:Ag:Ni phosphor started earlier if the excitation was provided by the $\lambda_e = 313$ nm line. In the strong quenching region of the phosphors containing cadmium sulfide results obtained for both excitation wavelengths were similar but the curves were considerably steeper than for the CdS-free phosphors. It is worth mentioning that the direct excitation of the silver centers in ZnS $\cdot$ CdS:Ag:Ni by $\lambda_e = 365$ nm was postulated in [72], but the results reported in that paper actually indicated excitation via the host substance.

At a fixed value of the excitation intensity the volume density of the excitation of the host substance was considerably higher than the volume density of the excitation resulting in the ionization of the luminescence centers. This was a very important point because the ZnS $\cdot$ CdS:Ag: Ni phosphors were nonlinear and the intensity of the luminescence $\mathcal{J}$ emitted by them was related to the excitation intensity I_e by $\mathcal{J} = CI_e^n$, where n > 1, and it depended on the composition of the phosphor as well as on the value of the excitation intensity. According to the results reported in [72], the highest value of n could exceed 3, i.e., these phosphors exhibited a dependence of the luminescence intensity on the excitation intensity which was stronger than cubic whereas external quenching would usually correspond to n $\leq$ 2-2.5.

Our measurements indicated that the nonlinearity of our phosphors was somewhat weaker but we still found that n > 2. For example, a reduction in the excitation intensity by a factor of 4.5 reduced the luminescence intensity of the ZnS $\cdot$ CdS:Ag $(10^{-4}$ g/g):Ni $(6.0 \times 10^{-6}$ g/g) phosphor by a factor of 25, whereas an increase in the excitation intensity by a factor of 5 raised the luminescence intensity by a factor of 16. These results were obtained at room temperature and the excitation intensity was deduced from the intensity in the red luminescence emitted by a standard phosphor.

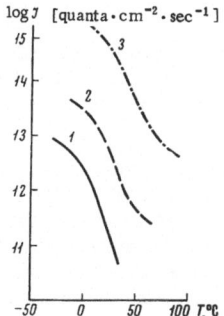

Fig. 30. Temperature dependences of the absolute intensity of the luminescence of a $ZnS_{70} \cdot CdS_{30}$:Ag:Ni phosphor recorded at different excitation power densities W_e (W/cm^2): 1) 2×10^{-5}; 2) 1×10^{-4}; 3) 6×10^{-4}.

The thermal quenching curves recorded at different excitation intensities were significantly different and a reduction in the excitation intensity led to an earlier and a more pronounced thermal quenching. Figure 30 shows the thermal quenching curves of the ZnS · CdS:Ag
(10^{-4} g/g):Ni (6.0×10^{-6} g/g) phosphor obtained at three excitation power densities (2.5×10^{-3},
5×10^{-4}, and 1.1×10^{-4} W/cm^2). It is evident also from this figure that the quenching curves
shifted along the temperature scale when the excitation intensity was raised. When the excitation power density was increased by a factor of 25 the shift of the quenching curves was about
70 deg. At low excitation intensities the highest rate of the thermal quenching was observed
near room temperature.

Thus, variation of the volume density of the excitation made it possible to shift quite considerably the thermal quenching curves and this could be used to achieve the maximum sensitivity at a given temperature, including room temperature. On the other hand, the strong dependence of the thermal quenching curves on the excitation power density was evidence of the
recombination mechanism of the quenching which occurred in the absence of quasiequilibrium
population of the electron (quenching centers) and the hole (luminescence centers) localization
levels. The region of maximum sensitivity to temperature coincided with the region of maximum nonlinearity of the dependence on the excitation density.

§ 3. Thermal Sensitivity of ZnS · CdS:Ag:Ni Phosphors

In the strong-quenching region ($\mathcal{I}/\mathcal{I}_0 \leq 0.1$) the thermal quenching of these phosphors
does not follow a strictly exponential dependence and varies with the excitation intensity I_e.
However, the dependences $\ln \mathcal{I} = f(1/T)$ can be approximated by straight lines to within one
order of magnitude of the luminescence intensity. The activation energy of the thermal quenching, deduced from the slope of such straight lines, is 1.1 ± 0.1 eV. This value of the activation
energy E_T is considerably greater than the depth of the acceptor levels of the silver luminescence centers and the depth of the donor levels of the nickel quenching centers (in the case of
ZnS this depth is $E_{Ni} \sim 0.56$ eV [32], whereas in the case of CdS it is somewhat lower, $E_{Ni} \leq$
0.5 eV). Therefore, a simple model of external quenching involving recombination interaction
between the luminescence and quenching centers cannot yield this high value of the activation
energy of the thermal quenching or the high rate of such quenching at room temperature (§ 4,
Chap. I).

The cited value of E_T is more than 4 times as large as the difference between the depths
of the donors and the acceptors, which should govern the temperature dependence of the luminescence intensity in the case of external linear second-order quenching. This value is almost
twice as large as the difference between the forbidden band width E_G and the sum of the energy
of the silver luminescence quanta $h\nu_1$ and the depth of the silver acceptor levels E_a, i.e., it is
twice as large as the maximum possible value of the activation energy of internal quenching
E_T^*. For example, in the case of ZnS$_{70}$CdS$_{30}$:Ag:Ni, we have $E_G \approx 3.2$ eV, $h\nu_1 \approx 2.4$ eV, $E_a \approx 0.2$ eV,
$E_T^* \approx 0.6$ eV.

All these various quenching mechanisms are based on the assumption that the luminescence intensity is a linear function of the excitation power density, which is in conflict with the
experimental data showing that this dependence is strongly nonlinear in the region where thermal quenching is strong.

Nonlinear second-order quenching, for which the activation energy is half the depth of the
luminescence centers (i.e., 0.1-0.15 eV), also does not agree with the experimental values of
E_T. Moreover, the external nonlinear first-order quenching mechanism, which is characterized
by a stronger temperature dependence than that obtained for other mechanisms and for which
E_T reaches twice the depth of the luminescence centers, is characterized by activation energies
which are about four times smaller than those found experimentally.

Thus, in order to explain the very high value of the activation energy and the corresponding high rate of thermal quenching we must make allowance for the influence of deeper levels (for example, silver and zinc complexes) on the temperature dependence of the ratio of the probability of radiative recombination of electrons at the ionized luminescence centers (Ag) and the probability of nonradiative recombination of holes at the electron-filled quenching centers (Ni). The most probable maximum value of the activation energy, governed by the difference between the quasi-Fermi levels of holes and electrons, amounts to one half of the forbidden band width, i.e., not more than 1.6 eV (§ 4, Chap. I), which is only 30% greater than the experimental value.

It is pointed out in Chap. I that in recording infrared, submillimeter, and other radiations (for example, ultrasound) on the basis of the thermal quenching effect, it is convenient to define the thermal sensitivity of a phosphor as $R_T = d\mathcal{J} \cdot 100/\mathcal{J}(T)\ dT$ i.e., as the ratio of the change in the luminescence intensity as a result of the change in the temperature by 1 deg C to the luminescence intensity at a given temperature (this ratio is expressed in percent). The thermal sensitivities of our phosphors, deduced in this way from the experimental data, are plotted as a function of temperature in Figs. 31-35.

It is evident from Fig. 31 that the maximum sensitivity of the ZnS · CdS:Ag phosphors free of nickel (curve 7) is located at relatively high temperatures (~125°C) and it amounts to 5% per degree Celsius. The values of the sensitivity of these phosphors cover a wide range but at room temperature they do not exceed 1.5%. Introduction of small amounts of Ni (10^{-7} g/g) improves considerably the optimal sensitivity (to 10%) but it does not shift it toward lower temperatures. Consequently, the sensitivity at room temperature is still quite low. When more Ni is added (10^{-6} g/g) the optimal sensitivity increases to 11.5% and the sensitivity curve shifts significantly in the direction of lower temperatures so that $T_{max} = 85°C$. Consequently, the room-temperature sensitivity rises somewhat.

Further increase in the concentration of Ni shifts continuously the quenching curve in the direction of low temperatures and improves the maximum sensitivity which reaches 15% per degree Celsius for $C_{Ni} = 6.0 \times 10^{-6}$ g/g and this maximum is found at about 40°C. The highest sensitivity at room temperature (12% per degree Celsius) is exhibited by the ZnS · CdS:Ag:Ni phosphors with 7×10^{-6} g/g of Ni. It is undesirable to increase the concentration of nickel still further because this reduces the sensitivity quite significantly and results in only a small shift along the temperature scale (up to 10 deg C) and, moreover, the luminescence intensity is

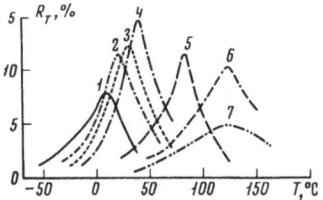

Fig. 31. Temperature dependences of the thermal sensitivity of $ZnS_{70} \cdot CdS_{30}$:Ag:Ni phosphors with different concentrations of nickel C_{Ni} (g/g): 1) 1.5×10^{-5}; 2) 7×10^{-6}; 3) 6×10^{-6}; 4) 3.5×10^{-6}; 5) 7×10^{-7}; 6) 1×10^{-7}; 7) 0. In all cases $C_{Ag} = 1 \times 10^{-4}$ g/g.

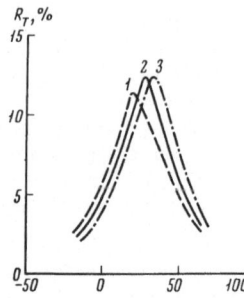

Fig. 32. Temperature dependences of the thermal sensitivity of $ZnS_{70} \cdot CdS_{30}$:Ag:Ni phosphors with difference concentrations of silver C_{Ag} (g/g): 1) 5×10^{-5}; 2) 1×10^{-4}; 3) 5×10^{-4}. In all cases $C_{Ni} = 6.0 \times 10^{-6}$ g/g.

reduced considerably if $C_{Ni} > 10^{-5}$ g/g (this is due to other quenching mechanisms of chemical nature, as manifested by the precipitation of Ni in the form of a black deposit). It follows that it is undesirable to use nickel concentrations higher than 10^{-5} g/g.

According to the experimental data given in the preceding section the concentration of silver can be varied near its optimal value because such variation alters only slightly the amplitude and the position of the thermal sensitivity maximum (Fig. 32). On the other hand, introduction of CdS into ZnS:Ag:Ni is a necessary condition for the attainment of a high thermal sensitivity, which increases sharply when the concentration of CdS is raised to 30% and has its maximum and constant value when the concentration of CdS is 45-55%. The temperature dependence of the thermal sensitivity of the phosphors containing at least 30% of CdS is practically the same irrespective of whether $\lambda_e = 313$ or 365 nm excitation lines are used (Fig. 33). This observation confirms that the optical width of the forbidden band of $ZnS_{70} \cdot CdS_{30}$:Ag:Ni is less than 3.4 eV and that the exciting light of both wavelengths is absorbed by the host substance. The temperature dependences of the thermal sensitivity obtained for these two excitation wavelengths are close only if the excitation intensity is the same. This intensity governs the shape of the thermal quenching curves and, consequently, it alters the thermal sensitivity attained at different temperatures (Fig. 34). For example, when the excitation power density is increased by a factor of 15 (from $W_e = 2 \times 10^{-5}$ to 6×10^{-4} W/cm²), the thermal sensitivity maximum is shifted by 26 deg C in the direction of higher temperatures and the amplitude of the sensitivity maximum is reduced somewhat.

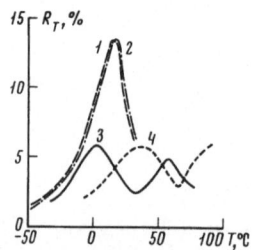

Fig. 33. Temperature dependences of the thermal sensitivity of two phosphors ($C_{Ag} = 1 \times 10^{-4}$ g/g, $C_{Ni} = 6.0 \times 10^{-6}$ g/g) plotted for different excitation wavelengths λ_e. $ZnS_{70} \cdot CdS_{30}$:Ag:Ni: 1) $\lambda_e = 365$ nm; 2) $\lambda_e = 313$ nm. ZnS:Ag:Ni: 3) $\lambda_e = 365$ nm; 4) $\lambda_e = 313$ nm.

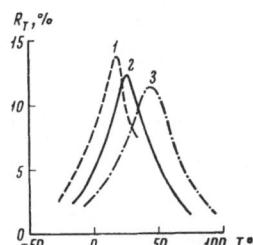

Fig. 34. Temperature dependences of the thermal sensitivity of a $ZnS_{70} \cdot CdS_{30}$:Ag:Ni
phosphor ($C_{Ag} = 1 \times 10^{-4}$ g/g,
$C_{Ni} = 6.0 \times 10^{-6}$ g/g recorded
for different excitation power
densities W_e (W/cm^2): 1) 2 ×
10^{-5}; 2) 1 × 10^{-4}; 3) 6 × 10^{-4}.

Improvements in the conditions of preparation of the $ZnS_{0.5} \cdot CdS_{0.5}$:Ag:Ni phosphors made
it possible to raise the thermal sensitivity to ~25% per degree Celsius (Fig. 35), i.e., the activation energy E_T was increased to 2.1 eV and the maximum nonlinearity K_{nl} to 3.5. These
values of the thermal sensitivity and nonlinearity can be explained only if we assume that the
Ni centers are saturated (§ 4, Chap. I). The shift of the sensitivity maxima along the temperature scale resulting from variation of I_e, C_{Ag}, and C_{Ni} yields the depth of the silver centers
$E_{Ag} \approx 0.6$ eV, which is in agreement with the approximate relationship $\Delta E_T = \Delta E_{Ag} K_{nl}$.

Thus, the maximum thermal sensitivity, expressed as the percentage change in the luminescence intensity resulting from a change in temperature by 1 deg C, is considerably higher
than the thermal sensitivity of bolometers which is 4-6% per degree. The greater capabilities
of bolometers are simply due to the ease with which small differences between currents can be
measured.

We shall now consider the problem of the quantum efficiency of the thermal method for
detection of infrared radiation under optimal conditions using ZnS · CdS:Ag:Ni phosphors developed by us.

The luminescence intensity at which the thermal quenching is strongest (up to ~25% per
degree Celsius) amounts to about 5×10^{13} quanta · cm^{-2} · sec^{-1}, i.e., the change in the luminescence intensity resulting from the temperature change of 1 deg C is 10^{12} quanta · cm^{-2} · sec^{-1} ·
deg^{-1}.

The infrared radiation power which can alter the temperature luminescence screen by
1 deg C is, under optimal absorption and heat removal conditions, $4\sigma_T T^3 \approx 6 \times 10^{-4}$ W · cm^{-2} ·
deg^{-1} at room temperature. When the energy of the infrared quanta is $h\nu \approx 0.3$ eV, which corresponds to the threshold of the optical detection methods, the infrared radiation intensity is
$I_{IR} \approx 1 \times 10^{16}$ quanta · cm^{-2} · sec^{-1}.

Thus, the quantum effectiveness of the thermal action of infrared radiation is 0.01%, i.e.,
it is smaller by up to two orders of magnitude than the quantum effectiveness of the optical ac-

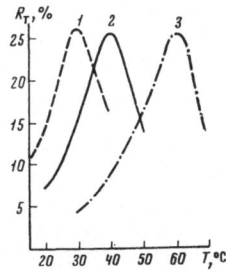

Fig. 35. Temperature dependences of the thermal sensitivity of a ZnS · CdS:Ag:Ni phosphor with an optimal composition, recorded for different excitation power densities We (W/
cm^2): 1) 5 × 10^{-5}; 2) 2 × 10^{-4};
3) 5 × 10^{-4}.

tion. This shows that the temperature-sensitive phosphors can be used in investigations of the spatial distribution of laser radiation of wavelengths lying in the middle and far parts of the infrared range.

§4. Change in Color of Luminescence of Two-Activator ZnS:Ag:Sm Phosphors Resulting from Changes in Temperature and Optical Excitation Intensity

In phosphors with one type of luminescence center a change in the temperature causes energy transfer from the luminescence (Ag) centers to nonradiative recombination centers. In phosphors with two interacting luminescence centers a change in the temperature can alter the ratio of the luminescence band intensities. If the luminescence bands are located in different spectral regions, a change in the temperature may alter the color of the luminescence and this can be used in color thermography of temperature fields and color visualization of electromagnetic radiations.

The most favorable conditions for this kind of effect are found in phosphors which exhibit stimulation of one of the luminescence bands, as a result of energy transfer from the other activator, rather than in phosphors with different thermal quenching of the bands. For such phosphors the color of the luminescence will vary faster with the temperature than the intensities of the band separately and this can be used as a means for increasing the sensitivity of the thermographic method of recording infrared radiation.

Thermal quenching of the luminescence of one activator (I), accompanied by stimulation of the luminescence of another activator (II) whose band lies in a different spectral region, is exhibited by the ZnS:Ag:Sm phosphors. We investigated in detail these interesting phosphors. Their luminescence spectrum was found to consist of several groups of bands between 400 and 700 nm (Fig. 36). The interaction between a heavy metal (Ag) and a rare earth (Sm) in the ZnS lattice was first analyzed for various temperatures in [26]. It is shown in that paper that, in contrast to the normal mutual quenching of two activators, a stimulation of the luminescence of Sm^{3+} is observed in a certain range of activator concentrations and temperatures. Moreover, the influence of temperature on the properties of this two-activator phosphor is quite different from the influence of temperature on the properties of each of these activators separately in one-activator phosphors. Thermal quenching of the blue luminescence of Ag ($\lambda = 430$ nm) in the ZnS:Ag:Sm phosphors was observed when the temperature was increased only slightly up to 50°C.

Thermal quenching of the red luminescence bands of Sm, the main of which was located at $\lambda = 650$ nm, was observed at somewhat higher temperatures whereas in the thermal quenching region of the luminescence of the silver centers, the luminescence of Sm was practically inde-

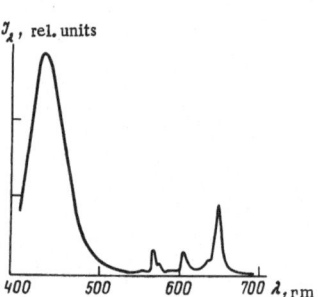

Fig. 36. Luminescence spectrum of a ZnS:Ag:Sm phosphor ($C_{Ag} = 1 \times 10^{-4}$ g/g, $C_{Sm} = 1 \times 10^{-5}$ g/g) excited with $\lambda_e = 365$ nm.

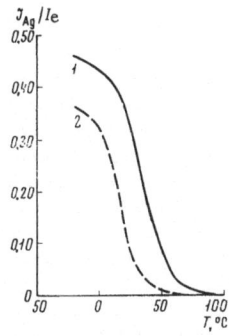

Fig. 37. Thermal quench-
ing of the Ag luminescence
of an ZnS:Ag:Sm phosphor
at two excitation power densities: 1) $W_0 \sim 10^{-3}$ W/cm^2;
2) $W_e = 0.125W_0$.

pendent of the temperature. When Sm was introduced into ZnS:Ag phosphors, thermal quenching
of the silver luminescence was enhanced, as in the case of addition of Ni. However, a new effect was observed: the luminescence of Sm was stimulated in the thermal quenching region of
the silver band and this indicated an interaction between the two activators. In spite of the
weakening of the luminescence of Sm at low temperatures, the presence of Ag enhanced not only the relative [26] but also the absolute intensity of the room-temperature luminescence of Sm,
i.e., Ag acted as a sensitizer on the luminescence of Sm. This effect of silver can be explained
bearing in mind that when excitation occurs in the activator absorption region (λ_e = 364 nm) the
absorption by silver centers (allowed transitions) is much more likely than by rare-earth ions
(forbidden transitions). This establishes conditions favorable for the accumulation of electrons
and holes at the Ag centers rather than at the nonradiative recombination centers. The carriers
localized at the Ag centers can be liberated thermally and transferred to the rare-earth centers.
Such a recombination mechanism involving an interaction between the two activators was confirmed by our investigations of the dependence of the thermal redistribution of the energy between the Ag and Sm bands on the excitation intensity. The temperature dependences of the intensities of the Sm and the Ag luminescence bands, recorded at different excitation intensities
(λ_e = 365 nm), are plotted in Figs. 37 and 38. It is evident from these figures that a reduction
in the excitation power density enhanced thermal quenching of the silver luminescence and
stimulated the samarium luminescence. The shift of the samarium luminescence band amounted
to about 30 deg C when the excitation power density was increased by one order of magnitude.
The position of the samarium luminescence maximum on the temperature scale T_{max} was simply given by the empirical dependence $1/kT_{max} \approx C \log I_e$, where I_e is the excitation intensity.

Consequently, the value of $1/kT_{max}$, plotted as a function of the logarithm of the excitation
intensity, was represented by a straight line (Fig. 38b) whose slope yielded an activation energy

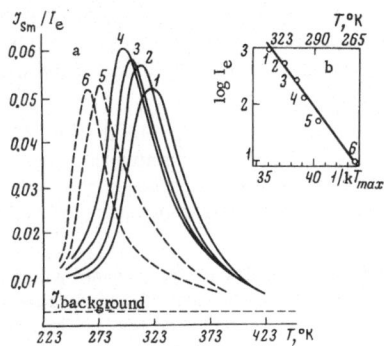

Fig. 38. Temperature dependences of the parameters of the
Sm luminescence of a ZnS:Ag:
Sm phosphor obtained at different excitation intensities: a) luminescence efficiency; b) position of the luminescence maximum (the coordinates in Fig.
38b correspond to the range of
I_e in Fig. 38a).

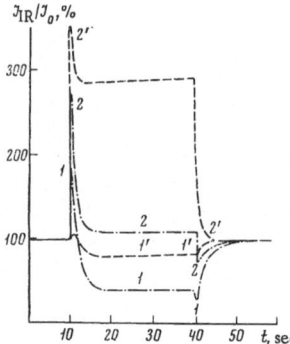

Fig. 39. Time dependences of the intensities of the blue and red luminescence bands of a ZnS:Ag:Sm phosphor excited with λ_e = 365 nm and subjected to infrared illumination: 1)–1') luminescence of silver centers; 2)–2') luminescence of samarium centers; 1), 2) weak ultraviolet excitation; 1'), 2') strong ultraviolet excitation.

of about 0.63 eV. This energy was approximately equal to the depth of the silver centers, deduced for the ZnS·CdS:Ag:Ni phosphors. The strong dependence of the relative intensities of the Ag and Sm bands on the excitation intensity could be used to shift the region of the greatest change in the color of the luminescence to room temperatures, which would be very convenient in color thermography.

In the temperature range in which the samarium luminescence was strongest the intensity of this luminescence increased 5–5.5 times compared with the almost temperature-independent intensity of the luminescence of Sm at low temperatures. In the same range of temperatures the luminescence of silver became weaker by approximately the same factor (5.5–6). This redistribution of the luminescence intensities resulted in visible change in the color of the luminescence from dark blue to orange-red. The maximum rate of change of the intensities of the bands of both activators was observed approximately in the middle of this temperature range and it amounted to 3–4% per degree Celsius, which corresponded to a change of 8% per degree in the ratio of the intensities. This sensitivity was only about half the optimal thermal sensitivity of the ZnS·CdS:Ag:Ni phosphors to infrared radiation but it could be used to detect changes in temperature of the order of 1–2 deg and, as demonstrated in special experiments, it was possible to record infrared radiation of power densities not exceeding 10 mW/cm².

The energy efficiency of the Ag luminescence of the ZnS:Ag:Sm phosphors increased with increasing excitation intensity and reached 30–40% at room temperature, which was approximately equal to the efficiency of the luminescence of the ZnS:Ag phosphors prepared under the same conditions. The energy efficiency of the Sm luminescence was much lower and it amounted to 3–5% in the maximum stimulation region.

Thus, in spite of the strong interaction between the Ag and Sm centers induced by temperature changes, a large fraction of the energy was lost by nonradiative processes. A much stronger energy transfer from the Ag to the Sm centers was achieved by illumination with infrared radiation.

Our experiments showed that the optical effect of infrared radiation on the ZnS:Ag:Sm phosphors gave rise, like the thermal effect, to a redistribution of the luminescence energy in the direction of enhancement of the Sm luminescence. Infrared illumination (with a sensitivity maximum in the region of 1.3 μ) gave rise to simultaneous stimulation and quenching of both luminescence bands (Fig. 39).

The relative importance of the stimulation and the quenching processes was a function of the ultraviolet excitation density. When a phosphor was illuminated with infrared radiation after a weak ultraviolet excitation, large stimulation flashes were recorded for the Sm and the Ag bands. After a few seconds the intensity of the Ag luminescence decayed to a value much lower than the intensity in the absence of infrared illumination. On the other hand, the intensity of the Sm luminescence observed during continuous infrared illumination was slightly higher than the intensity observed in the absence of such illumination.

After the end of infrared illumination the intensities of both types of luminescence decreased at first, which indicated that infrared radiation both stimulated and quenched the two luminescence bands. This effect was much weaker for the Ag centers in ZnS:Ag. When the intensity of the ultraviolet excitation was increased, the relative amplitude of the stimulation flash and of the aftereffect of infrared illumination increased strongly for the silver and samarium luminescence. Such an increase in the excitation intensity enhanced the coefficient of steady-state energy transfer from silver to samarium, i.e., it enhanced the quantity $(\mathcal{J}_{IR} - \mathcal{J})\,Sm/(\mathcal{J} - \mathcal{J}_{IR})Ag$, where $\mathcal{J}$ and $\mathcal{J}_{IR}$ are, respectively, the luminescence intensity in the absence and in presence of infrared illumination.

An important result of this investigation was the observation of complete transfer of the excitation energy from the Ag to the Sm centers, which occurred at high intensities of ultraviolet excitation and infrared illumination. The quantum efficiency of the samarium luminescence in the presence of infrared radiation could reach 20-30%. This was established by a direct comparison of the intensity of the Sm luminescence under ultraviolet excitation and in the presence of infrared illumination and the intensity of the red luminescence of a standard phosphor whose quantum efficiency was 50%.

Such an efficient energy transfer from Ag to Sm, exceeding considerably the efficiency of the excitation of Sm in the ZnS lattice, and the enhancement of the quenching effect of infrared radiation on the silver centers in the presence of Sm are evidence of adjacent positions of these centers.

Apart from practical applications of these effects in optical detection of infrared radiation, the results obtained are also important in the interpretation of the nature of the interaction between rare-earth and heavy metals in the lattice of ZnS.

The results obtained are in good agreement with the hypothesis [26] of the formation of large complexes consisting of trivalent rare earths and monovalent metals (these arise because of the principle of compensation applicable to divalent lattices of II–VI compounds). The formation of these complexes is manifested by the appearance of new deep localization levels and by changes in the fine structure of the Sm luminescence. A redistribution of energy between these complexes can give rise to efficient utilization of the Ag luminescence in the excitation of the rare-earth element. In this respect the two-activator phosphors of the ZnS:Ag:Sm type are not inferior to the two-activator alkaline-earth phosphors.

The quantum effectiveness of the stimulation of Sm could reach 1% and could be even higher than the quantum effectiveness of the alkaline-earth phosphors in our possession.

CONCLUSIONS

The present paper reports theoretical and experimental investigations of the optical and thermal effects of long-wavelength electromagnetic radiations on the photoluminescence of specially prepared zinc sulfide phosphors. The purpose of these investigations was to determine the potentialities of these phosphors in recording the spatial distributions of laser radiation fields. We also found the optimal conditions for such applications. The following results were obtained.

1. A comparison was made of the change in the probability of liberation of carriers from local levels in phosphors as a result of direct optical and thermal action of long-wavelength

(infrared) radiation. It was found that high–contrast images could be obtained of infrared radiation fields and that the optical and thermal methods of investigating laser radiation were com-The direct optical effect was more suitable to near radiation whereas at longer infrared wavelengths the thermal method was more effective.

2. ZnS:Cu:Co phosphors capable of storing light sums close to the theoretical limit were developed. The number of photons emitted by these phosphors during afterglow was comparable with the concentration of the luminescence centers at the depth of penetration of the exciting light. This made it possible to make a practically unlimited number of photocopies of the images of electromagnetic fields obtained on screens covered with the ZnS:Cu:Co phosphors. The quantum efficiency of the reduction in the light sum under the action of infrared radiation was up to 10%. Estimates were obtained of the parameters describing the kinetics of the luminescence of the ZnS:Cu:Co phosphors under conditions of optimal sensitivity of these phosphors to infrared radiation.

3. A theoretical analysis was made of the rate of thermal quenching corresponding to different quenching mechanisms. It was found that the thermal sensitivity of the investigated phosphors was governed directly by the activation energy of the thermal quenching process. An experimental study was made of the dependence of the thermal quenching of $ZnS_x \cdot CdS_{1-x}:Ag:Ni$ phosphors on their composition and on the excitation conditions. The relative rate of change of the luminescence intensity with temperature, which governed the contrast in the temperature fields, could be as high as 25% per degree Celsius for these phosphors. This thermal sensitivity was the highest ever reported and much higher than the sensitivity of other types of phosphor.

4. A calculation was made of the threshold values of the power of electromagnetic radiations needed in optical and thermal recording of these radiations. It was found that a temperature-sensitive phosphor could be used, under optimal absorption and heat removal conditions, to record visually and photographically temperature changes amounting to less than 1 deg C and electromagnetic radiation powers down to 1×10^{-3} W/cm^2. The quantum effectiveness of the thermal action of infrared radiation was two orders of magnitude lower than the corresponding effectiveness of the optical action but the former had the advantage of nonselectivity. When absorbing substrates were used, such thermophotoluminescent detectors could be used to visualize with equal effectiveness fields of electromagnetic sources ranging from 10^{-3} to 10 W/cm^2.

5. The thermal colorimetric method for investigating of electromagnetic fields was developed. This method was based on a thermally induced redistribution of the composition (color) of the luminescence emitted by a two-activator phosphor. The color contrast established between parts of a field which were at different temperatures could be used to study electromagnetic radiation fields. A ZnS:Ag:Sm phosphor was found to be suitable for this purpose. This phosphor exhibited a high efficiency of energy transfer from the blue (Ag) to the orange-red (Sm) luminescence centers. The conditions were determined under which the rate of change of the composition of the luminescence was greatest at room temperature.

The current and future developments in quantum electronics, construction of new lasers emitting infrared radiation of various wavelengths, and other applications make it imperative to develop efficient luminescent detectors of infrared radiation.

The authors are grateful to M. V. Fok for discussing some of the problems, to Yu. V. Voronov for his help in several experiments, and to Yu. P. Pikin for his participation in the measurements.

LITERATURE CITED

1.   A. É. Becquerel, La Lumière, Paris (1867).

2. P. Lenard, Handbuch der Experimental Physik (ed. by W. Wien and F. Harms), Vol. 23, Akademische Verlagsgesellschaft, Leipzig (1928), p. 2.
3. J. W. Draper, Phil. Mag., 11:160 (1881).
4. E. Lommel, Ann. Phys. Chem., 3:251 (1878).
5. A. Dahms, Ann. Phys. (Leipzig), 13:425 (1904).
6. A. Güntz, Ann. Chim. (Paris), 5:363 (1926).
7. L. Vanino and S. Rotschild, Chem. Zentralb., 50:45 (1926).
8. V. L. Levshin, V. V. Antonov-Romanovskii, and L. A. Tumerman, 4:1033 (1934); Phys. Z. Sowjetunion, 5:811 (1934).
9. V. V. Antonov-Romanovskii, V. L. Levshin, Z. L. Morgenshtern, and Z. A. Trapeznikova, Dokl. Akad. Nauk SSSR, 54:19 (1946).
10. V. L. Levshin, V. V. Antonov-Romanovskii, Z. L. Morgenshtern, and Z. A. Trapeznikova, Zh. Eksp. Teor. Fiz., 17:949 (1947).
11. V. L. Levshin, Zh. Eksp. Teor. Fiz., 18:149 (1948).
12. V. V. Antonov-Romanovskii, V. V. Levshin, Z. L. Morgenshtern, and Z. A. Trapeznikova, Izv. Akad. Nauk SSSR, Ser. Fiz., 13:76 (1949).
13. V. L. Levshin, Photoluminescence of Liquids and Solids [in Russian], Gostekhteoretizdat, Moscow (1951).
14. Z. L. Morgenshtern, Dokl. Akad. Nauk SSSR, 54:79 (1946).
15. B. O'Brien, J. Opt. Soc. Amer., 36:351, 369 (1946).
16. B. O'Brien, in: Preparation and Characteristics of Solid Luminescent Materials, (Proc. Symp. at Cornell University, Ithaca, N. Y., 1946), Wiley, New York (1948), p. 190.
17. F. Urbach, N. R. Nail, and D. Pearlman, J. Opt. Soc. Amer., 39:1011 (1949).
18. S. A. Fridman and A. A. Cherepnev, Dokl. Akad. Nauk SSSR, 59:53 (1948).
19. V. V. Antonov-Romanovskii, I. B. Keirim-Markus, M. S. Poroshina, and Z. A. Trapeznikova, Soviet Contributions at Intern. Conf. on Peaceful Uses of Atomic Energy [in Russian], Moscow (1955), p. 342.
20. Z. L. Morgenshtern, Dokl. Akad. Nauk SSSR, 74:493 (1950).
21. L. A. Vinokurov, Zh. Eksp. Teor. Fiz., 21:338 (1951).
22. B. A. Lengyel, Lasers, 1st ed., Wiley, New York (1962).
23. J. D. McGee, and L. J. Heilos, IEEE J. Quantum Electron., QE-3:31 (1967).
24. T. J. Bridges and E. G. Burkhardt, IEEE J. Quantum Electron., QE-3:168 (1967).
25. E. Ya. Arapova, V. L. Levshin, N. V. Mitrofanova, T. S. Reshetina, V. F. Tunitskaya, S. A. Fridman, and V. V. Shchaenko, Izv. Akad. Nauk SSSR, Ser. Fiz., 30:573 (1966).
26. V. L. Levshin, Yu. V. Voronov, V. B. Gutan, S. A. Fridman, and V. V. Shchaenko, Izv. Akad. Nauk SSSR, Ser. Fiz., 25:392 (1961).
27. V. A. Arakelyan, N. V. Karlov, and S. A. Fridman, Prib. Tekh. Eksp., No. 2, p. 186 (1969).
28. A. P. Bazhulin, E. A. Vinogradov, N. A. Irisova, and S. A. Fridman, ZhETF Pis. Red., 8:261 (1968).
29. L. L. Alt, in: Proc. Intern. Conf. on Luminescence, Budapest, 1966 (ed. by G. Szigeti), Vol. 2, Akadémiai Kiádo, Budapest (1968), p. 2085.
30. E. I. Adirovich, Some Topics in The Theory of Luminescence of Crystals [in Russian], Gostekhizdat, Moscow (1951).
31. D. Curie, Luminescence in Crystals, Wiley, New York (1963).
32. M. V. Fok, Introduction to the Kinetics of Luminescence of Crystal Phosphors [in Russian], Nauka, Moscow (1964).
33. V. V. Antonov-Romanovskii, Kinetics of Photoluminescence of Crystal Phosphors [in Russian], Nauka, Moscow (1966).
34. A. M. Gurvich, Usp. Khim., 35:1495 (1966).
35. L. Ya. Markovskii, F. M. Pekerman, and L. N. Petoshina, Phosphors [in Russian], Khimiya, Moscow-Leningrad (1966).

36. V. L. Levshin, N. V. Mitrofanova, and L. A. Pakhomycheva, Yu. P. Timofeev, S. A. Frid-
     man, and V. V. Shchaenko, Izv. Akad. Nauk SSSR, Ser. Fiz., 33:852 (1969).
37. V. L. Levshin, E. Ya. Arapova, A. I. Blazhevich, Yu. V. Voronov, I. G. Voronova, V. B.
     Gutan, A. V. Lavrov, Yu. M. Popov, S. A. Fridman, V. A. Chikhacheva, and V. V. Shchaen-
     ko, Tr. Fiz. Inst. Akad. Nauk SSSR, 23:64 (1963).
38. É. R. Il'mas, G. G. Liid'ya, and Ch. B. Lushchik, Tr. Inst. Fiz. Astron. Akad. Nauk Est.
     SSR, 26:213 (1964).
39. A. Rose, Concepts in Photoconductivity and Allied Problems, Interscience, New York
     (1963).
40. R. H. Bube, Photoconductivity of Solids, Wiley, New York (1960).
41. H. Gobrecht and D. Hofmann, J. Chem. Phys. Solids, 27:509 (1966).
42. N. T. Melamed, J. Electrochem. Soc., 97:33 (1950).
43. E. F. Daly, Proc. Roy. Soc. London, 196:554 (1949).
44. F. Stöckmann, Naturwiss., 39:226, 246 (1952).
45. K. Przibram, Naturwiss., 39:425 (1952).
46. F. Urbach, J. Opt. Soc. Amer., 36:372 (1946).
47. L. C. Bradley, C. C. van Voorhis, and D. Bershader, Bull. Amer. Phys. Soc., 28:345
     (1953).
48. M. V. Fok, Fiz. Tverd. Tela, 5:1489 (1963).
49. G. Rosenthal, Z. Instrumentenk., 59:432, 437 (1939).
50. J. C. Miklosz and R. G. Wheeler, Phys. Rev., 153:913 (1967).
51. Ch. B. Lushchik, N. E. Lushchik, and K. K. Shvarts, Tr. Inst. Fiz. Astron. Akad. Nauk
     Est. SSR, 8:3 (1958).
52. Yu. V. Voronov, Author's Abstract of Thesis for Candidate's Degree [In Russian], P. N.
     Lebedev Physics Institute, Academy of Sciences of the USSR, Moscow (1969).
53. S. I. Pekar, Untersuchungen über die Elektronentheorie der Kristalle, Akademie Verlag,
     Berlin (1954).
54. H. Koelmans, J. Phys. Chem. Solids, 17:69 (1960).
55. E. A. Bozhevol'nov, Luminescence Analysis of Inorganic Substances [In Russian], Khimi-
     ya, Moscow (1966).
56. Z. L. Morgenshtern, V. B. Neustruev, and M. I. Épshtein, Zh. Prikl. Spektrosk., 3:49
     (1965).
57. S. A. Fridman and A. A. Cherepnev, Compositions Emitting Continuous and Transient
     Luminescence [in Russian], Izd. AN SSSR, Moscow (1945).
58. S. A. Fridman and A. A. Cherepnev, Dokl. Akad. Nauk SSSR, 57:563 (1947).
59. M. N. Alentsev and A. A. Cherepnev, Zh. Eksp. Teor. Fiz., 26:473 (1954).
60. M. N. Alentsev, Opt. Spektrosk., 1:260 (1956).
61. L. A. Vinokurov and M. V. Fok, Opt. Spektrosk., Suppl. No. 1, (Luminescence), p. 263
     (1963).
62. V. V. Antonov-Romanovskii, Zh. Fiz. Khim., 6, 1022 (1935).
63. V. V. Antonov-Romanovskii, Tr. Fiz. Inst. Akad. Nauk SSSR, 1:35 (1937).
64. V. V. Antonov-Romanovskii, Dokl. Akad. Nauk SSSR, 17:95 (1937).
65. L. A. Vinokurov, Dokl. Akad. Nauk SSSR, 85:529 (1952).
66. M. V. Fok, Opt. Spektrosk., 2:475 (1957).
67. E. S. Krylova, Zh. Eksp. Teor. Fiz., 20:905 (1950).
68. M. N. Alentsev, V. V. Antonov-Romanovskii, and L. A. Vinokurov, Dokl. Akad. Nauk SSSR,
     96:1133 (1954).
69. M. N. Alentsev, Opt. Spektrosk., 1:240 (1956).
70. F. M. Pekerman and L. N. Petoshina, Collection of Abstracts of Papers on Chemistry
     and Technology of Phosphors [in Russian], No. 28, Izd. GIPKh, Moscow (1958).
71. S. A. Fridman and A. A. Chereinev, Dokl. Akad. Nauk SSSR, 59:53 (1948).
72. N. R. Nail, F. Urbach, and D. Pearlman, J. Opt. Soc. Amer., 39:690 (1949).

INVESTIGATION OF GALLIUM ARSENIDE LASERS PUMPED BY ELECTRON-BEAM BOMBARDMENT

B. M. Lavrushin*

An investigation was made of the principal characteristics (power, efficiency, laser threshold, and emission spectrum) of semiconductor lasers made of gallium arsenide crystals with different types and concentrations of dopants. The investigation was made in the temperature range 80-300°K. A relationship was established between the degree of doping, the laser characteristics, and the influence of temperature on the emission spectrum and the laser threshold. An investigation was made of the influence of the excitation inhomogeneity, typical of the electron-beam pumping, on the laser threshold and the angular distribution of the output radiation. Pulsations of the emission intensity with a period of 0.3 nsec were observed. The investigations reported here made it possible to increase the output power and to reach the theoretical efficiency (~30%) of the laser.

§1. Formulation of the Problem

The use of an electron beam in the pumping of semiconductor lasers extends considerably the range of applications of these devices.

The widely used method of pumping semiconductor lasers by injection of nonequilibrium carriers across a p—n junction has a number of disadvantages which limit the practical applications of these lasers. The main disadvantage is the small volume of the active region.† In the final analysis this limits the output power that can be obtained with the aid of diode lasers. Moreover, this pumping method imposes severe restrictions on possible improvements in the angular distribution and the coherence of the output radiation.

The active region can be increased in volume by optical pumping of semiconductor lasers but this is not the ideal solution because light fluxes of 1 MW/cm^2 density are needed to reach the laser threshold. In practice, light fluxes of this density can be generated only by means of other lasers (for example, the ruby laser can be used). Consequently, an optically pumped semiconductor laser acts as a frequency converter because it effectively alters the wavelength of coherent optical radiation. One must bear in mind that under normal conditions the angular distribution and coherence of the output radiation are considerably poorer for a semiconductor laser than for a ruby laser. Moreover, optical pumping gives rise to additional difficulties as-

---

*Thesis for the Degree of Candidate of Physicomathematical Sciences. Defended in 1969 at the P. N. Lebedev Physics Institute, Academy of Sciences of the USSR, Moscow.

†The thickness of the active region is governed by the thickness of the p—n junction and amounts to 1-2 μ. For a number of reasons it would be undesirable to increase the junction thickness.

sociated with the need to select the optimal wavelength of the pumping source for a given semiconductor crystal. From the point of view of efficiency, the optimal conditions are achieved
when the energy of the pumping radiation photons is approximately equal but slightly higher than
the energy represented by the forbidden band of the semiconductor.

The use of an electron beam in pumping a semiconductor laser has several advantages
over the injection and optical pumping methods. The most important advantage is the considerably greater size of the active region. For example, when the electron energy is 200 keV,
the thickness of the active region can be 50-60 μ. Therefore, the output power of an electron-
beam-pumped semiconductor laser can be several orders of magnitude higher than the output
power under injection pumping conditions. The efficiency of conversion of the electron-beam
energy into the energy of coherent radiation is quite high and can reach about 30%. Moreover,
the electron energy is considerably higher than the energy of the output quanta and, therefore,
this conversion efficiency is independent of the forbidden band width of the semiconductor used
as the active substance. This makes it possible to construct high-efficiency coherent-radiation
sources operating in a wide range of wavelengths from infrared to ultraviolet: the wavelength
can be varied simply by using semiconductors or dielectrics with different widths of the forbidden band.

Finally, one must mention that the methods of generation of high power electron beams
and of the effective control of these beams are highly developed. All the conditions are, therefore, favorable for the practical realization of the advantages of electron-beam-pumped semiconductor lasers.

The possibility of using an electron beam in pumping a semiconductor laser was first
suggested by Basov in 1961 [1]. The first experiments aimed at constructing an electron-beam-
pumped laser were carried out at the Lebedev Physics Institute in 1961-1963 [2]. In these experiments the crystals used were gallium arsenide and germanium but the laser action was not
achieved. The first successes were reported in 1964: Basov et al. [3] were able to reduce the
width of the output radiation spectrum of cadmium sulfide and observed other signs that the
laser threshold was reached at ~20°K [3].

The laser action in gallium arsenide was first reported in [4-8]. Initially, Hurwitz and
Keyes [4] achieved laser action in p-type GaAs at liquid helium temperature. Subsequently,
Cusano and Kingsley [5, 6] reached the laser threshold in n-type GaAs at liquid nitrogen temperature. Kurbatov et al. [7] were able to generate coherent radiation at room temperature in
undoped samples and similar results were reported by Basov et al. [8]. Basov et al. were the
first to measure the output power and efficiency of a gallium arsenide laser. However, the values of these parameters were much lower than those expected theoretically or those achieved
in injection lasers. Similar low values of the output power and efficiency were subsequently
also reported for other semiconductor lasers.[*]

In view of this situation, it seemed very desirable to study the principal properties of
electron-beam-pumped semiconductor lasers and to make an attempt to improve the characteristics of these lasers. This was the purpose of the investigation reported in the present paper.
The active material was gallium arsenide since its basic electrical and optical properties have
been thoroughly investigated and the crystal growth technology of GaAs has been mastered on
an industrial scale.

---

[*]Apart from cadmium sulfide and gallium arsenide, the laser action has been achieved in the
following semiconducting materials: InSb [9], InAs [10], GaSb [11], CdTe [12, 13], CdSe [14],
GaSe [15], ZnO [16], ZnS [17], ZnTe [18[, ZnSe [19], Te [20], PbS, PbTe, PbSe [21], and
CdSnP$_2$ [22]. Moreover, coherent radiation has been generated in CdS$_x$Se$_{1-x}$ and GaP$_x$As$_{1-x}$
solid solutions. In the first case, i.e., in CdS$_x$Se$_{1-x}$, the laser action was achieved by Hurwitz
[23] and, in the second case, this action was obtained by the present author [24]. However,
one should point out that only the laser threshold and few data are known for most of these crystals.

§ 2. Energy Spectrum and Some Optical Properties

of Gallium Arsenide

Gallium arsenide is a "direct" III-V semiconducting compound. Its energy band structure [27, 28] is such that the lowest minimum in the conduction band and the highest maximum in the valence band are located at the same point in the reciprocal lattice, i.e., at k = 0. Near this point the valence band consists of three subbands: a heavy-hole band, a light-hole band, and a split-off band. The first two subbands are degenerate at the point k = 0 and the third is shifted downward by the spin-orbit splitting, which is about 0.35 eV at k = 0.

At excitation levels typical of semiconductor lasers the nonequilibrium carriers have a narrow energy distribution of the order of kT. Under these conditions the existence of the split-off valence subband can be ignored.

The forbidden band width Δ of gallium arsenide, deduced from the absorption spectra of the purest available samples, has the following values [29]:

T, °K . . .	21	55	90	185	294
Δ, eV . . .	1.521	1.518	1.511	1.479	1.435

However, as in the case of other semiconductor crystals, the fundamental absorption edge is somewhat broad. Therefore, the forbidden band width is not defined very accurately.*

The broadening of the fundamental absorption edge of semiconductor crystals is known to be due to the interaction of carriers and with impurity ions.

The simplest manifestation of the interaction between the carriers is the existence of exciton states. The presence of such states is manifested by one or more peaks in the absorption spectra and these peaks correspond in energy to the ground or excited states of excitons. In the case of gallium arsenide, the ground-state exciton absorption peak is observed only for the purest crystals in which the free-carrier density is of the order of 10^{10} cm^{-3} [29]. The ionization energy of the ground state of excitons in gallium arsenide is about 3 MeV [29]. However, at the excitation levels encountered in semiconductor lasers the carrier density exceeds 10^{17} cm^{-3} and the exciton states are unlikely to form at these densities. This is supported by estimates of the type given in [30] and by the fact that the exciton absorption peak is not observed in samples with carrier densities of $\gtrsim 10^{16}$ cm^{-3}.

The energy spectrum of gallium arsenide near the fundamental absorption edge is affected most strongly by group II, IV, and VI impurities (Te, Se, Ge, Si, Cd, Zn, etc.), which can act as donors or acceptors in III-V semiconducting compounds. It is known that when an atom in the host lattice is replaced by a donor or acceptor impurity, the valence bonding gives rise to an electron or hole and an impurity ion of charge +e or −e. The interaction of a carrier with an impurity ion may give rise to a bound state whose energy corresponds to a discrete level in the forbidden band of the host crystal.

When the concentration of impurities is increased the discrete levels corresponding to isolated impurity ions are shifted because of the interaction between electrons or holes localized at neighboring impurity atoms. Since the distribution of impurities is random, this shift differs from one impurity to another. This gives rise to an impurity energy band whose width increases with increasing impurity concentration.†

---

*The method used in [29] to determine the forbidden band width from the absorption spectra involved measurements of the absorption coefficient of photons of energy $\hbar\omega = \Delta$, which amounted to about 9000 cm^{-1}.

†This picture of the formation of an impurity band is not complete. In particular, it is known that a random distribution of impurities is not a necessary condition for the formation of an impurity band: it is also observed in the case of a fully ordered impurity distribution [31, 32].

We can distinguish light, intermediate, and heavy doping conditions, depending on the impurity concentration.

A semiconductor in which the impurity concentration is so low that the width of the impurity band is much less than the ionization energy of isolated impurity atoms is regarded as lightly doped. When these quentities are comparable, so that the impurity band may overlap partly the nearest allowed band (which may be the conduction or the valence band), a semiconductor is regarded as having intermediate doping. Finally, in a heavily doped semiconductor the impurity band overlaps completely the nearest allowed band and it is no longer possible to distinguish the two bands.

The formation and evolution of a spectrum of impurity states with increasing concentration of impurities in a semiconductor can be studied by recording the luminescence spectra. According to [5, 33], the light doping conditions in gallium arsenide are attained for the following values of the donor and acceptor concentrations:

$$N_d < 10^{17}\,\text{cm}^{-3}, \qquad N_a < 10^{18}\,\text{cm}^{-3}\,,$$

whereas heavy doping corresponds to

$$N_d > 10^{18}\,\text{cm}^{-3}, \qquad N_a > 10^{19}\,\text{cm}^{-3}.$$

The ionization energies of donors and acceptors can be deduced from luminescence spectra:

$$E_d = 6\ \text{MeV}, \quad E_a = 30\ \text{MeV}.$$

Measurements of the electrical conductivity and the Hall effect at low temperatures [34] yield lower values of $E_{d,\,a}$ and $N_{d,\,a}$ corresponding to light and heavy doping.

There are reasons for assuming that, under intermediate doping conditions, the center of the impurity band shifts with the concentration of impurities [33, 35].

The influence of doping on the energy spectrum of a semiconductor is not limited to the formation of impurity levels and impurity bands: doping also alters the spectrum of the energy states of the host crystal. A calculation of the density of states in a doped semiconductor is a problem in the many-electron theory. This problem is considered in [36–39] in the limiting case of heavy doping. The main conclusions reached in these papers are as follows.

1. The doping of a semiconductor alters the density of states of free electrons and holes in such a way that a density-of-states tail forms in the forbidden band and the densities of states within the allowed bands are also affected. The changes in the allowed bands are small and simply reduce to a small correction to the effective mass. The length of a density-of-states tail in the case of strong carrier degeneracy ($F_{e,h} \gg kT$, where $F_{e,h}$ is the Fermi level of electrons or holes measured from the relevant band edge) is given by the following order-of-magnitude relationship [36]:

$$E_t \sim E_{d,\,a}\,(n_{e,h}\,a_{d,\,a}^3)^{1/2}, \tag{1}$$

where $n_{e,\,h}$ is the density of electrons or holes and $a_{d,\,a}$ is the Bohr radius of a bound state of an electron or hole, given by

$$a_{d,\,a} = \frac{\varepsilon_0 m_0}{m_{e,h}}\,a_H. \tag{2}$$

Here, ε_0 is the static permittivity; $m_{e,\,h}$ is the effective mass of an electron or hole in a crystal; m_0 is the mass of an electron in free space; a_H is the Bohr radius of the ground state of the hydrogen atom. In the case of GaAs, we have $\varepsilon_0 = 12.5$ [40], $m_e = 0.07m_0$, $m_h = 0.5m_0$ [41, 42], so that $a_d = 95$ Å and $a_a = 13$ Å.

2. The total number of states in a tail is small compared with the impurity concentration, i.e., compared with the total number of carriers. This means that the total carrier density and the positions of the quasi-Fermi levels can be estimated ignoring the existence of tails.

3. Since carriers interact with one another (in the case we are considering the main effect is the exchange interaction), the gap between the allowed bands decreases by an amount

$$E_{exc} = 2 \frac{e^2}{\varepsilon_0} \left(\frac{3}{\pi} n_{e,h} \right)^{1/3}, \tag{3}$$

but one must bear in mind that this decrease is of a very formal nature because of the existence of tails.

4. One-electron states with a definite electron or hole quasimomentum are not stationary because of scattering processes. Therefore, it is not meaningful to speak of the conservation of quasimomentum in a time interval of the order of the nonequilibrium carrier lifetime ($\sim 10^{-9}$–10^{-10} sec). In this case, the only important characteristic of the system under investigation is the density of states, which still retains its physical meaning.

In the case of gallium arsenide, the orders of magnitude of the tail length E_t and of the exchange interaction energy E_{exc}, defined by Eqs. (1) and (3), are as follows

$$\left. \begin{array}{l} E_t \sim 6 \text{ MeV} \\ E_{exc} \sim 23 \text{ MeV} \end{array} \right\} \quad \text{for} \quad n_e = N_d \approx 10^{18} \text{ cm}^{-3},$$

$$\left. \begin{array}{l} E_t \sim 23 \text{ MeV} \\ E_{exc} \sim 100 \text{ MeV} \end{array} \right\} \quad \text{for} \quad n_h = N_a \approx 10^{20} \text{ cm}^{-3}.$$

Quantitative expressions are available for the densities of states in heavily doped semiconductors. However, these expressions are usually unsuitable for the quantitative interpretation of the experimental results because they are valid in a very narrow range of conditions (this point is discussed in [36]). Stern [43] made an attempt to extend the theoretical treatment to lower dopant concentrations and to avoid the strong degeneracy condition, which might not be satisfied in semiconductor lasers. However, the results of Stern's numerical calculations on a computer are applicable only to the case of strong impurity compensations and are of limited validity.

The direct manifestation of the existence of density-of-states tails is the experimentally observed broadening of the fundamental absorption edge. The experimental data on the dependence of the absorption coefficient α of gallium arsenide on the photon energy $\hbar\omega$, obtained in the range $\alpha(\hbar\omega) \propto \exp(\hbar\omega/E_0)$. The constant E_0, which represents the broadening of the edge, increases with increasing impurity concentration. This is in basic agreement with the theoretical predictions and the order of magnitude agrees with Eq. (1). However, these experimental data give only qualitative information on the energy spectrum of gallium arsenide near its fundamental absorption edge. This is due to the fact that the absorption coefficient depends not only on the density of states but also on the positions of the quasi-Fermi levels, which usually cannot be determined independently. Moreover, there is no universal relationship between the absorption coefficient and these quantities: the relationship varies with the actual model used to represent radiative transitions.

If it is assumed that the stationary states of an electron or hole have a definite quasimomentum and the interaction with an optical electromagnetic field is regarded simply as a

perturbation (all other interactions, such as scattering by impurity ions, interaction with phonons, etc., are ignored), one should be able to derive a quasimomentum selection rule: the
matrix element of a transition does not vanish only for those transitions in which this momentum is conserved [48]. This model has been considered in earlier investigations such as that
reported in [9]. It is now known that the model is inapplicable to a detailed analysis of the optical properties of semiconductors near the fundamental absorption edge and, in particular, it
is unsuitable for the analysis of the spectral characteristics of semiconductor lasers.

We have already mentioned that the state of an electron or a hole with a definite quasimomentum is not stationary because of the scattering by impurities. Even if a semiconductor
is free of impurities, this state is not stationary because of the scattering by optical and acoustic phonons. If the impurity or phonon scattering time is much shorter than the nonequilibrium
carrier lifetime, the direction of the quasimomentum is completely indeterminate (it is averaged out during the lifetime and no quasimomentum selection rule can be derived). In this case,
the absorption coefficient or the gain α for a system with a density of states $\rho_{e,h}$ ($E_{e,h}$) and
with quasi-Fermi levels $F_{e,h}$ * is described by the relationship [50]

$$\alpha(\hbar\omega) = B \int_{-\infty}^{+\infty} \rho_e(E_e)\rho_h(\hbar\omega - \Delta - E_e) [f_e(E_e) + f_h(\hbar\omega - \Delta - E_e) - 1] \, dE_e. \tag{4}$$

Here, B is a constant which includes the matrix element of the transition (assumed to be independent of the energies of the initial and final states) and $f_{e,h}(E_{e,h})$ is the distribution function of electrons or holes, given by

$$f_{e,h} = \left\{ \exp\left(\frac{E_{e,h} - F_{e,h}}{kT}\right) + 1 \right\}^{-1}. \tag{5}$$

The model of optical transitions in which quasimomentum is not conserved and the corresponding expression (4) which allows for the existence of density-of-states tails are used widely in the analysis of the spectral characteristics of semiconductor lasers and they allow us to
interpret (at least qualitatively) a considerable proportion of the experimental observations [51].

In particular, it follows from Eqs. (4) and (5) that optical radiation of frequency ω is amplified ($\alpha > 0$) if

$$F_e + F_h > \hbar\omega - \Delta. \tag{6}$$

The relationship between the gain and the nonequlibrium carrier density in Eq. (4) is implicit
in the quasi-Fermi levels. In the simplest case of parabolic energy bands the relationship between the quasi-Fermi levels, the carrier densities $n_{e,h}$, and the temperature is given by the
formula [52]

$$n_{e,h} = N_{c,v} \mathscr{F}_{1/2}\left(\frac{F_{e,h}}{kT}\right), \tag{7}$$

where

$$N_{c,v} = 2\left\{\frac{m_{e,h}kT}{2\pi\hbar^2 m_0}\right\}^{3/2},$$

---

*The quasi-Fermi levels $F_{e,h}$ and the energy of an electron or a hole $E_{e,h}$ are measured from
the edges of the conduction and the valence bands, respectively. The direction used in these
measurements is selected so that these energies have negative values in the forbidden band.

and

$$\mathscr{F}_{1/2}(x) = \frac{2}{\sqrt{\pi}} \int\limits_{0}^{\infty} \frac{\sqrt{x'}dx'}{1 + e^{x'-x}}$$

is the Fermi−Dirac integral.

The corrections to the total carrier density resulting from the presence of tails are small. Therefore, Eq. (7) can be used even if such tails do exist. In particular, the application of Eq. (7) allows us to calculate the nonequilibrium carrier density n_0 (T) corresponding to the condition $F_e + F_h = 0$. In the case of undoped or compensated gallium arsenide ($n_e = n_h$ and $m_e/m_h = 7.05$), the results of calculations can be written in the form

$$n_0(T) = 2 \left(\frac{m_e kT}{2\pi \hbar^2 m_0} \right)^{3/2} \mathscr{F}_{1/2}(1.9). \tag{8}$$

Hence, we find that, for example, $n_0 \approx 3.5 \times 10^{17}$ cm^{-3} at T = 80°K.

An analysis of the experimental data relating to the dependence of the laser threshold under injection conditions on the losses in the optical resonator (see, for example [51]) shows that the gain is proportional to the volume density of the pumping power (in the case of injection lasers, this quantity is proportional to the current density across the p−n junction). This conclusion is supported by direct measurements of the gain reported, for example, in [53].

§ 3. Electron Beam Pumping of Semiconductor Lasers

It is known [54] that electrons of energies of the order of 10^3-10^5 eV penetrating into a crystal lose practically all their energy in the ionization of the lattice atoms. This process can be described as follows. In the first interaction with a lattice atom, a primary electron loses some of its energy to a secondary electron and a hole which are formed as a result of the ionization of the atom. The primary and secondary electrons and holes have sufficiently high energies for the further ionization of the lattice atoms. In this way, tertiary, quaternary, etc. electrons and holes are formed. This avalanche-like multiplication of electrons and holes stops at the stage when the energy of all the electrons and holes becomes less than the energy needed to form electron−hole pairs. It has been shown theoretically and experimentally [55-57] that the energy of formation of an electron−hole pair is about three times as large as the forbidden band width Δ of a semiconductor.

In this case, the rate of generation of electron−hole pairs in unit volume g (x) is given by

$$g(x) = \frac{1}{3\Delta} \frac{j}{e} \left(-\frac{dE(x)}{dx} \right), \tag{9}$$

where E (x) is the average energy of a primary electron at a distance x from the surface of a crystal; j is the electron-beam current density; e is the charge of an electron.

The distribution (−dE/dx) for electron energies $E_0 = 50, 250,$ and 500 keV was found by Yurkov [58] for germanium and silicon by numerical integration of the appropriate transport equation. This was done using the Spencer method [59], which made it possible to obtain an approximate solution of this equation with any desired degree of precision. The results of these calculations were in good agreement with the experimental data [56, 60, 61].

The distribution (−dE/dx) is usually called the "ionization curve." This curve depends on the atomic number and the density of the substance being considered. Since the atomic number and the density are practically the same for germanium and gallium arsenide, we can use the ionization curve of germanium when dealing with GaAs.

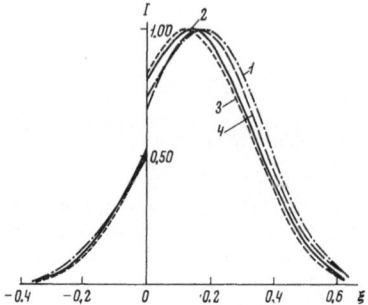

Fig. 1. Ionization curve of germanium [58], plotted for E = 50, 250, and 500 keV (curves 1, 2, and 3, respectively). Curve 4 is a theoretical approximation based on Eq. (11) with $a = 0.16s_0$ and $b = 0.26s_0$.

The ionization curve can be represented in the simplest form by introducing the following dimensionless quantities

$$I\,(\xi) = \frac{-\,dE/dx}{\left(-\dfrac{dE}{dx}\right)_{max}}\,, \qquad \xi = \frac{x}{s_0}\,,$$

where s_0 is the total range of a primary electron traveling on a curvilinear trajectory.* The dependence $I\,(\xi)$ for germanium is plotted in Fig. 1 for the three electron energies mentioned above. It is evident from Fig. 1 that the distribution $(-dE/dx)$ can be approximated by the following expression, which is independent of the initial electron energy:

$$I\,(\xi) = \exp\left\{-\frac{(\xi - \xi_1)^2}{\xi_2^2}\right\}. \tag{10}$$

If $\xi_1 = 0.16$ and $\xi_2 = 0.26$, the precision of the above approximation is about 15% in the energy range 50–500 keV.

If we transform Eq. (10) by replacing ξ, and $I\,(\xi)$ with x and $(-dE/dx)$, we finally obtain the following approximate expression for the ionization curve:

$$-\frac{dE}{dx} = \frac{E_0'}{x_0}\exp\left\{-\left(\frac{x - a}{b}\right)^2\right\} \quad \text{for} \quad x \geqslant 0, \tag{11}$$

where $a = 0.16s_0$, $b = 0.26s_0$, and the effective depth of penetration of electrons x_0 is found by normalization to the total energy E_0' lost in the ionization of the lattice atoms:†

$$x_0 = \int\limits_0^\infty \exp\left\{-\left(\frac{x - a}{b}\right)^2\right\} dx \approx 0.37s_0. \tag{12}$$

The values of the parameters of the ionization curve are listed in Table 1 for different electron energies. These parameters are calculated using the results given in [58] for $E_0' = 0.85E_0$.

In some cases, the ionization curve can be usefully approximated by the following expression

$$-\frac{dE}{dx} = \begin{cases} \dfrac{E_0'}{x_0} & \text{for} \quad 0 \leqslant x \leqslant x_0, \\[2mm] 0 & \text{for} \quad x < 0,\ x > x_0. \end{cases} \tag{13}$$

---

*The dependence of s_0 on E_0 is tabulated in [58].

†Some of the primary electrons are "reflected" from the surface of a crystal, i.e., they emerge from a crystal through its x = 0 surface because of multiple scattering [57, 58, 60]. Therefore, $E_0' < E_0$.

TABLE 1. Parameters of the Ionization Curve

E_0, keV	a, μ	b, μ	x_0, μ	$\frac{E_0'}{x_0}$, keV/μ	E_0, keV	a, μ	b, μ	x_0, μ	$\frac{E_0'}{x_0}$, keV/μ
10	0.1463	0.2192	0.3384	25.1	150	13.44	21.84	31.08	4.10
20	0.4541	0.7380	1.050	16.2	200	21.21	34.47	49.06	3.47
30	0.9146	1.486	2.115	12.1	250	29.92	48.62	69.19	3.07
40	1.480	2.405	3.4225	9.94	300	39.34	63.92	90.97	2.80
50	2.161	3.512	4.998	8.50	400	59.70	97.01	138.1	2.46
60	2.941	4.780	6.802	7.50	500	81.44	132.3	188.3	2.26
70	3.817	6.202	8.826	6.74	600	104.0	169.0	240.4	2.12
100	6.926	11.26	16.02	5.31					

The average kinetic energy of the electrons and holes generated by the ionization of the lattice atoms is of the order of the forbidden band width [62]. This means that the effective temperature of the electron—hole gas is much higher than the lattice temperature. Therefore, the recombination of electrons and holes should occur simultaneously with the establishment of a thermodynamic equilibrium between the electron—hole gas and the lattice: the electrons and holes should slow down because of the interaction with the lattice. The theoretical calculations reported in [63] show that the second process should be much faster than the first: the time needed for slowing down to the "thermal velocities" because of the interaction of electrons and holes with phonons should be much shorter than the carrier lifetime. This conclusion was proved experimentally by Basov and Bogdankevich [2], who showed that the width of the spontaneous radiation line emitted from Ge and GaAs excited by electron bombardment (the excitation level was close to that encountered in semiconductor lasers) was approximately 2kT (T is the lattice temperature. Later investigations of the spontaneous emission spectra of gallium arsenide excited by electron bombardment (see, for example, [5, 33]) also provided no evidence of any difference between the temperature of the electron—hole gas and the lattice temperature in the range $T \leq 4.2°K$.

We may thus regard it as established that the energy distribution of the nonequilibrium electrons and holes is described by the Fermi—Dirac function of Eq. (5), in which T is the lattice temperature.

The spatial distribution of the nonequilibrium carriers is an important point in the analysis of the operation of an electron-beam-pumped semiconductor laser. In principle, this distribution may differ from the ionization curve because of the drift and diffusion of electrons and holes during their slowing down. An analysis of this question [64] shows that when the energy of the incident electrons is sufficiently high (≥ 50 keV), this difference can be neglected. Therefore, the steady-state spatial distribution of the density n (x) of the electron—hole pairs is given by the relationship

$$n(x) = g(x)\tau, \tag{14}$$

where τ is the carrier lifetime.*

It follows, in particular, that in order to establish a population inversion in gallium arsenide by bombardment with 50 keV electrons at T = 80°K, we need an electron current whose density is 0.3-3 A/cm².

---

*When the nonequilibrium carrier density is n $\approx 10^{17}$-10^{18} cm⁻³, their lifetime is of the order of 10^{-8}-10^{-9} sec [65, 66].

CHAPTER I

APPARATUS AND EXPERIMENTAL METHOD

§ 1. Apparatus Used in Electron-Beam Bombardment

of Semiconductors

Figure 2 shows schematically the apparatus which was used in most of our measurements.[*]

The electron gun employed in our studies was a three-electrode electrostatic system with a spherical cathode of 3.6 mm in diameter. The cathode was made of lanthanum hexaboride and was heated indirectly. The cathode produced pulses of about 0.5 A at the exit from the anode aperture when the accelerating voltage pulses were of 50 kV amplitude.

The basic circuit of the high-voltage pulse generator is shown in Fig. 3. A generator of this type was described in [67]. The parameters of this generator were as follows: pulse amplitude 10 to 50 kV; bell-shaped pulses of 0.2 μsec duration, measured along the base; repetition frequency up to 1000 pulses/sec.

The high-voltage pulse generator 12 (Fig. 2) was triggered by an external pulse generator 13, used also in the synchronization of all the other instruments employed in our measurements. The repetition frequency of the high-voltage pulses was governed by the frequency of the master (trigger) generator 13 and it usually amounted to 50 Hz. The pulse voltage was taken from the output of the generator 12 and applied to the cathode 10 and (via a capacitor C_1) to the control electrode 9 of the electron gun (the anode 7 was usually at zero potential). The electron beam was formed in the accelerating field applied between the cathode and anode. A magnetic lens ML-1 (6 in Fig. 2) was placed directly behind the anode aperture in a vacuum chamber 5. This lens reduced the scattering of electrons emerging from the anode aperture. A second magnetic lens ML-2 (4) focused the electron beam in the plane of the sample. The beam could be deflected slightly relative to the sample by the application of a homogeneous magnetic field generated in the deflection coils 3. The sample 2 was attached to a heat sink of a cryostat 1. Figure 2 shows also the power supply of the heater (11) and of the control electrode (9).

The total electron-beam current in the plane of the sample was governed primarily by the first magnetic lens. This current was 250-300 mA under optimal conditions. If this lens was switched off, the total current decreased to about 100 mA. The current density in the plane of the sample was governed primarily by the second magnetic lens (ML-2) and under optimal conditions its value was 10-12 A/cm^2. Small deflections of the beam within ± 5 mm in the plane of the sample did not alter significantly the density of the current or its distribution over the beam cross section.

The electron-beam current density was regulated by varying the current in the winding of the magnetic lens ML-2. Therefore, the beam diameter was minimum at the maximum current density and it increased when the current density was reduced. The minimum diameter of the beam in the plane of the sample was 0.6-0.8 mm (at the 0.5 level of the current density at the center of the beam). The diameter of the electron beam could be reduced with the aid of the control electrode 9: when the negative potential of this electrode was varied relative to the cathode, the beam diameter could be reduced without a significant change in the current density. The control voltage was generated by charging the capacitor C_1 via a choke (L = 2.4 mH) and the winding of the high-voltage pulse generator supplied by a rectifier 14.

---

[*]We shall mention specially those cases when the experimental results were obtained using apparatus differing significantly from that shown in Fig. 2.

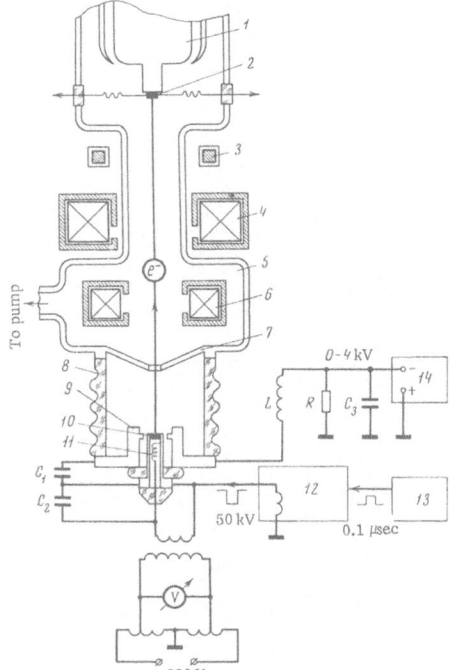

Fig. 2. Schematic diagram of the apparatus used in electron-beam pumping of semiconductors: 1) cryostat; 2) semiconductor; 3) deflection coils; 4), 6) magnetic lenses; 5) vacuum chamber; 7)-11) parts of the electron gun (anode, insulator, control electrode, cathode, and heater, respectively); 12) high-voltage pulse generator; 13) trigger-pulse generator; 14) control-voltage source.

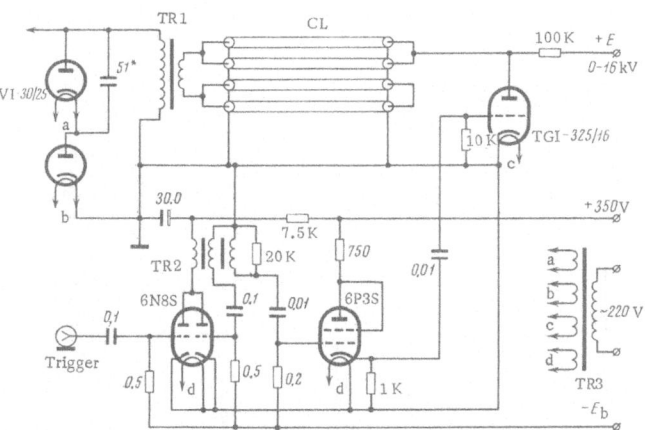

Fig. 3. Basic circuit of the high-voltage pulse generator: CL is the charging line (four sections of an RK-101 coaxial cable, each 10 m long); TR1 is a cable transformer [67].

TABLE 2. Some Parameters of Gallium Arsenide Crystals Used
in Present Investigation

Crystal No.	Ingot designation	Growth method	$n_{e,h}$, cm^{-3}	μ, cm$^2\cdot$V$^{-1}\cdot$sec^{-1}
		Zinc-doped crystals		
14	№ 29	B	$(2.0-2.3)\cdot10^{18}$	124—100
			$(2\cdot10^{18})$	(120)
22	№ 268	B	$(2.0-3.6)\cdot10^{18}$	113
23	№ 250	B	$(1.3-2.3)\cdot10^{19}$	75—56
26	№ 645	C	$(1.1-2.2)\cdot10^{18}$	161—150
65	№ 150	B	$(1.7-2.4)\cdot10^{19}$	69—63
66	№ 137	C	$(1.2-2.6)\cdot10^{18}$	134—85
		Tellurium-doped crystals		
19	№ 138	B	$1.0\cdot10^{18}$	3000—2720
27	AGB 87T (1)	F	$3.6\cdot10^{17}$	4200
28	AGB 87T (2)	F	$3.2\cdot10^{17}$	4800
30	№ 239 s	B	$1.5\cdot10^{17}$	3700
31	№ 57	B	$2\cdot10^{18}$	2450
32	—	B	$7\cdot10^{17}$	3000
34	№ 289	C	$(0.9-2)\cdot10^{18}$	3100
39	№ 128	C	$(1-4)\cdot10^{18}$	2800—2200
42	1811 + 1871	B	$4.7\cdot10^{18}$	2200
43	№ 36 — C	C	$6.8\cdot10^{18}$	1200
49	AGB — 108T — 1	F	$8\cdot10^{17}$	3200
52	73 s	—	$(2\cdot10^{17})$	(4000)
54	AGB — 109T — 1	F	$1.0\cdot10^{18}$	2900
55	AGB — 110T	F	$1.5\cdot10^{18}$	2500
56	AGB — 111T	F	$3.8\cdot10^{18}$	2400
57	AGB — 112T	F	$3.0\cdot10^{18}$	2400
61	C — 161	C	$2\cdot10^{18}$	3000
			$(2\cdot10^{18})$	(2950)
62	AGB — 113T	F	$3.3\cdot10^{18}$	2100
		Undoped crystals		
1	Light (England)	—	$2\cdot10^{15}$	6000
6	Semielements (England)	—	$10^{15}-10^{16}$	6000
			$(3\cdot10^{15})$	(6100)
10	№ 303	B	$1\cdot10^{16}$	5200
			$(1.0-1.5)\cdot10^{16}$	(5000)
13	№ 753	B	$5\cdot10^{16}$	3800
25	16 f	C	$(7.8-2.4)\cdot10^{15}$	6000—6050
			$(1\cdot10^{16})$	(4800)
40	—	C	$5\cdot10^{15}$	5500
41	AGB —60—3	F	$6.4\cdot10^{7}$	980
44	—	E	10^{15}	6000
51	—	—	$3\cdot10^{16}$	5500—5800
58	VPK 163/2	E	$7\cdot10^{15}$	7200
59	VPK 166	E	$9\cdot10^{14}$	7050
60	VPK 172/2	E	$1.8\cdot10^{15}$	6350
63	321 1	B	$6.8\cdot10^{15}$	5700
			$(4\cdot10^{15})$	(4900)
64	318 1	B	$5.7\cdot10^{15}$	6600
			$(8\cdot10^{15})$	(6500)

Notation. Growth method: B denotes the Bridgman method, C the
Czochralski method, F the floating-zone method, and E the vapor-phase
epitaxy. The equilibrium densities of electrons and holes are denoted
by $n_{e,h}$ and the mobility at T $\approx$ 300°K by μ. The values in parentheses
represent control measurements carried out by A. V. Dudenkova.

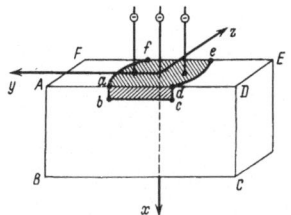

Fig. 4. Shape of the sample and
experimental geometry.

The electron current pulses were shorter than the high-voltage pulses and their duration was in the range 0.05-0.1 μsec.[*]

§ 2. Semiconductor Materials and Fabrication of Lasers

In the experimenta described below we used gallium arsenide single crystals whose main properties are listed in Table 2.

The samples were rectangular parallelepipeds of the shape shown in Fig. 4. The ADEF face, which was bombarded with electrons, was polished. The ABCD face and the face opposite to it formed the mirrors of the optical resonator and were either polished or split along natural cleavage planes. The other faces of the sample were not subjected to any special treatment. The resonator length L = AF was typically ~ 0.7 mm. The transverse dimension AD (2-5 mm) was considerably larger than the width ad of the active region, shown shaded in Fig. 4. The third dimension (AB) was 0.15-1.0 mm, which was greater than the depth of penetration of the electron beam.

Figure 4 also shows the coordinate axes which will be used later. The x axis coincides with the direction of the electron beam and the z axis is the axis of the optical resonator, i.e., it is perpendicular to the resonator mirrors.

The samples used in our study were indium-soldered to copper substrates. The whole operation was carried out in vacuum ($\sim 10^{-5}$ mm Hg) at 420-450°C. Each sample was exposed to 300-450°C for about 30 min.

§ 3. Experimental Method

The substrate and the sample were mechanically clamped to the heat sink in the cryostat (Fig. 5). The substrate was separated from the sink by a copper foil to which a thermocouple was attached. This thermocouple was used to determine the time-average value of the temperature. The systematic error in the measurement of the temperature did not exceed ± 5 deg and the random error was within ± 0.1 deg.

The temperature of the sample was varied (after evaporation of liquid nitrogen from the cryostat) by a special heater wound on the heat sink (this heater is not shown in Fig. 5). When the heater was switched off, the temperature rose at a rate of about ± 0.25 deg/min at $T_0 \approx$ 80°K, whereas the time needed for measuring the laser threshold or for recording the emission spectrum did not exceed 5 min. Thus, the temperature did not vary by more than 1 deg during a given measurement.

---

[*]This was due to the fact that the high-voltage pulses were not rectangular and the focusing properties of the magnetic lens depended on the electron energy: when this lens focused the electrons corresponding to the top of the pulse of the accelerating voltage, the electrons corresponding to the edges of the pulse were scattered.

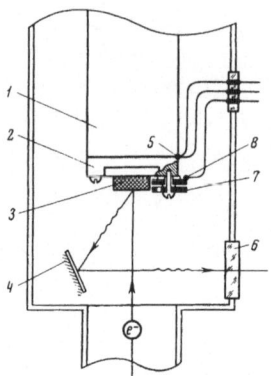

Fig. 5. Position of the sample
in the vacuum chanber and
some auxiliary parts: 1) heat
sink; 2) substrate; 3) sample;
4) mirror; 5) thermocouple;
6) optical window; 7) diaphragm; 8) collector.

The electron density was measured by deflecting the electron beam to a diaphragm 7
(Fig. 5) and by measuring the current passing through this diaphragm with an S1-11 oscillograph. In most cases, we used a diaphragm with an aperture 0.45 mm in diameter, i.e., we
measured the average current density over an area of 0.16 mm². The electrons were collected
by a flat copper plate 8 and the input resistance of the oscillograph (75 Ω) was used as the load
resistance. The total error in the measurement of the current density was about ± 10% (this included a random error of ±5%).

The electron energy was measured at a fixed value of the voltage provided by a source
which supplied the charging line (CL) of the high-voltage pulse generator (Fig. 3). The ampli-

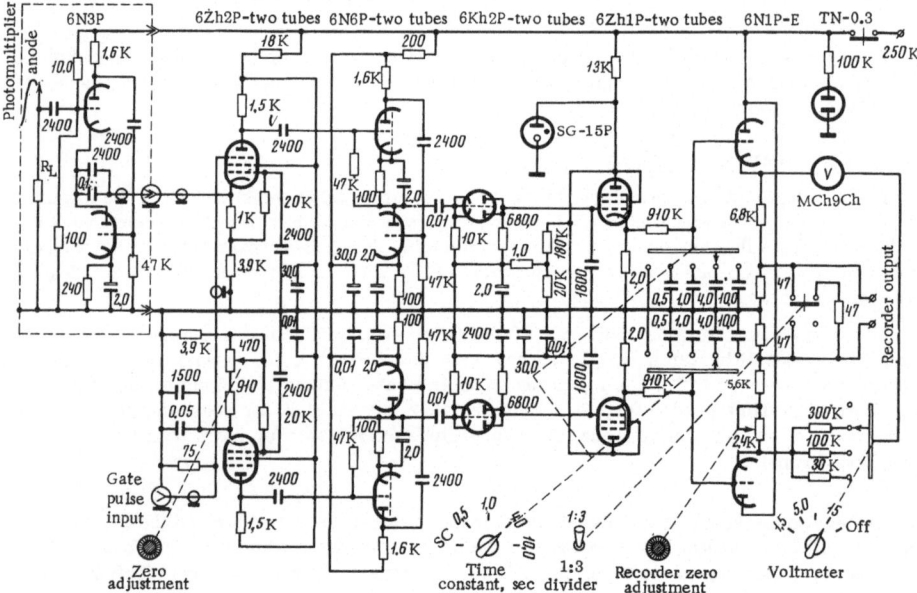

Fig. 6. Basic circuit of the synchronous pulse detector.

tude of the accelerating voltage pulses was measured with the aid of a divider (made of TVO resistors) and the S1-11 oscillograph. The electron energy was determined to within ±10% and the values of this energy were reporducible to within ±3%.

The optical radiation generated in a sample was observed along the z axis (Fig. 4). (In Fig. 2 this direction lies in the plane of the figure, whereas in Fig. 5 it is perpendicular to the figure.) The spontaneous radiation was recorded along a direction inclined at about 30° with respect to the electron beam and it was extracted from the vacuum chamber with the aid of a mirror 4, as shown in Fig. 5.

The output radiation spectrum was recorded with a DFS-12 spectrometer. The radiation detector was an FÉU-22 photomultiplier. The spectrum was recorded on the chart of an ÉPP-09 (or PS1-02) electronic potentiometer with the aid of a synchronous detector.

The synchronous detector transformed the voltage pulses reaching it into a constant voltage and acted as a time selector. The basic circuit of the detector is shown in Fig. 6, which consisted of two identical channels, one of which (the upper channel in Fig. 6) received the signal and a gate pulse and the other received just the gate pulse. The signal was taken from the photomultiplier (load resistance $R_L = 10$ kΩ), passed through a cathode follower, and applied to the cathode of a 6Zh2P tube, which acted as the time selector: this tube passed only that part of the input signal which was within the limits of the duration of the gate pulse. The signal then passed through a peak detector (6Kh2P) and a dc amplifier (6Zh1P + 6N1P). At the output the second-channel signal was subtrated from the first-channel voltage. The dc signal output of the synchronous detector was proportional to the amplitude of the input pulses and independent of their duration (for $\tau_p \geq 70$-80 nsec) and the repitition frequency (for $f \geq 5$-10 pulse/sec).

The distribution of the radiation field in the near zone was studied visually with the aid of an image converter and a microscope. When the current density was gradually increased, the laser action started in a very narrow region (a "point") and the brightness of this point increased strongly when the threshold was reached. Therefore, the threshold was deduced from the appearance of such a luminous point. The position of this point relative to the edges of the ±30 μ. The density of the current at the laser threshold was measured as described above. This method of measuring the threshold was very accurate and the values obtained agreed with those deduced from the narrowing of the emission spectrum.

The total output power of the emitted radiation was measured in a solid angle of about 1 sr. These measurements were performed with a coaxial photocell of the FÉK-09 or FÉK-14 type. The photocell was calibrated by comparing its sensitivity at various wavelengths with the known absolute sensitivity of a "standard" photocell.* The total calibration error did not exceed ±15% and was primarily due to the inaccuracy of the calibration of the "standard" photocell.

The distribution of the radiation field in the near zone was determined not only visually (as described above) but also more accurately by scanning with a photomultiplier which had a slit in front of its photocathode. The direction of the scan was along a magnified image of the face of the sample from which the radiation was emitted. This operation was performed with an FÉU-28 photomultiplier. The magnification was about 8.5. The width of the slit was 5-10 μ. The photomultiplier signal was recorded with the aid of a synchronous detector and an ÉPP-09 electronic potentiometer. (The scanning rate was selected so as to suit the rate of recording.) The spatial resolution in this procedure was at least 10 μ.

---

*The "standard" photocell (FÉK-09 No. 128-66) was supplied to us by Yu. V. Talanov of the All-Union Scientific Research Institute of Optical Measurements.

CHAPTER II

PRINCIPAL PROPERTIES OF GALLIUM ARSENIDE LASERS

§ 1. Laser Threshold, Output Power and Efficiency

As in the case of optical pumping and injection of nonequilibrium carriers across a p−n junction, the output power P of an electron-beam-pumped semiconductor laser is described by the following semiempirical relationship:

$$P = \beta\gamma f(\varkappa L)(P_p - P_{th}). \tag{15}$$

Equation (15) is derived from definitions of the coefficients which occur in this equation and from the experimental data on the nature of the dependence of the output power P on the pumping power P_p. Let us now consider the other quantities which occur in the above equation.

The coefficient β describes the energy lost by the nonequilibrium carriers during their thermalization and it is equal to the ratio of the average energy $\hbar\omega$ of a photon emitted as a result of recombination of an electron−hole pair to the average energy of formation ε_i of such a pair. Since $\hbar\omega \approx \Delta$ and $\varepsilon \approx 3\Delta$ (Introduction, § 3), it follows that $\beta \approx \frac{1}{3}$.

The quantum efficiency γ of radiative recombination is calculated making allowance for the losses resulting from nonradiative recombination. This efficiency is equal to the ratio of the number of electron−hole pairs recombining per unit time to form photons whose energy lies within the limits of the width of the spontaneous emission line* to the total number of electron−hole pairs excited by the pumping raidation per unit time. The quantum efficiency γ depends strongly on the quality of the semiconductor crystal and on the excitation level. However, the experimental data obtained in investigations of injection lasers [51, 68] show that the technology of growth of GaAs crystals is sufficiently advanced for the synthesis of crystals whose efficiency is $\gamma \approx 1$ above the laser threshold.

The factor in parentheses reflects the experimentally observed dependence of P on P_p and describes the losses due to spontaneous emission. In accordance with the experimental results, it is assumed that when $P_p > P_{th}$, the output power of the spontaneous radiation is independent of the pumping level and equal to $\beta\gamma f(\varkappa L)P_{th}$, where P_{th} is the threshold pumping power.

Some of the radiation generated in the active region of a semiconductor laser is absorbed in the laser resonator because of nonresonant losses (absorption by free carriers, scattering by optical inhomogeneities in a crystal, etc.). These losses are represented by a loss function $f(\varkappa L)$, which is defined as the ratio of the total power emerging from the laser resonator to the power generated inside this resonator. The loss function depends on $\varkappa L$ ($\varkappa$ is the nonresonant-absorption coefficient and L is the resonator length) and on the reflection coefficient of the mirrors R. The dependence of the output function on $\varkappa$ and L is governed by the distribution of the energy flux in the resonator. Two models have been used to derive the following formulas for the loss function [69, 70]:

$$f(\varkappa L) = \left\{ \frac{\varkappa L}{e^{\varkappa L} - 1} + \frac{\varkappa L}{1 - R} \right\}^{-1}, \tag{16}$$

$$f(\varkappa L) = \left\{ 1 - \frac{\varkappa L}{\ln R} \right\}^{-1}. \tag{17}$$

---

*We are talking here of the principal spontaneous emission line located near the fundamental absorption edge and representing radiative transitions between those states which are active in the stimulated emission above the laser threshold.

The formula (16) corresponds to the model which presumes a homogeneous volume distribution of the output power density and postulates that the medium filling the resonator can be represented by an absorption coefficient $\varkappa$. In this case, the energy flux in the resonator can be represented by a damped wave and the field at any point in the resonator is an incoherent sum of the fields contributed by all the other points. Therefore, strictly speaking, Eq. (16) represents the loss function of the spontaneous radiation at a pumping level such that the absorption coefficient of the semiconductor is equal to zero (this leaves only the nonresonant losses).

The principal assumption made in the derivation of the formula (17) is the uniform distribution of the gain along the resonator axis so that the energy flux can be represented by an exponentially growing wave. This assumption is valid for pumping levels just above the laser threshold when the intensity of the modes generated in the resonator is low compared with the spontaneous radiation field so that this intensity has no influence on the nonequlibrium carrier density which governs the gain.

The differences between the numerical values of the loss function calculated using Eqs. (16) and (17) is small. Therefore, we can use either of them in approximate calculations.

The overall and differential efficiencies of a laser are defined as follows:

$$\eta = \frac{P}{P_p} = \beta\gamma f(\varkappa L)\left(1 - \frac{P_{th}}{P_p}\right), \tag{18}$$

$$\eta_d = \frac{dP}{dP_p} = \beta\gamma f(\varkappa L). \tag{19}$$

Theoretical estimates of the capabilities of electron-beam-pumped semiconductor lasers based on the relationships mentioned above indicate that it is quite realistic to expect overall efficiencies of ~20-30%. If this efficiency can be achieved, the output of electron-beam-pumped semiconductor lasers should be considerably higher than the output of injection lasers because the volume of the active region and the area of the emitting faces should be considerably higher for electron-beam pumping than for carrier injection.

It would appear that the most promising are the purest undoped crystals because the nonresonant losses in these crystals should be smaller and the lifetimes longer than in doped crystals [65]. Therefore, the first electron-beam experiments were performed on undoped gallium arsenide. The early measurements of the output power and efficiency of electron-beam-pumped lasers were reported in [8]: the experimental values of these parameters were fairly low.[*] For example, it was reported in [8] that at liquid nitrogen temperature a gallium arsenide laser pumped by bombardment with 50 keV electrons (current density 10 A/cm^2) had an output of 10-20 W in a resonator L = 0.48 mm long and the efficiency of this laser was about 5%. Laser action could not be achieved at all in some of the undoped GaAs crystals available at that time (1964).

In view of these results, it was particularly important to study the selection of semiconducting materials suitable for electron-beam-pumped lasers and to investigate the possibility of achieving high efficiencies by electron-beam pumping.[†]

---

[*]Similar low values of the output power and efficiency were also reported for other semiconductor crystals [12, 16-18, 71, 72].

[†]The first successes were reported by Hurwitz [23, 73] for CdS and CdS$_x$Se$_{1-x}$ crystals grown by a special method. At T $\approx$ 110°K Hurwitz was able to achieve an output power of about 340 W and an efficiency of 26%. However, the parameters of these lasers deteriorated rapidly with increasing temperature and Hurwitz reported no measurements at room temperature.

TABLE 3. Parameters of Lasers Made of Undoped Gallium
Arsenide

Sample No.	L, mm	T ≈ 85° K				T ≈ 300° K			
		jth, A/cm²	P, W	D, mm	j, A/cm²	jth, A/cm²	P, W	D, mm	j, A/cm²
41—2	0.45P	4.4	48	0.31	11	9.4	—	—	—
1—6	0.60C	3.4	55	0.43	10	—*	—	—	—
40—1	0.44P	4.7	5,6	—	8.8	—*	—	—	—
64—1	0.54P	4.2	11	0.1	10	—*	—	—	—
10—3	0.50P	3.2	23	—	10.5	—*	—	—	—
51—5	0.50P	3.1	55	0.27	10	8.4	10	0.13	10
13—2	0.43P	2.5	30	0.17	8.4	6.7	10	0.14	8.4

*No laser action for j ≤ 10 A/cm².

TABLE 4. Parameters of Lasers Made of Tellurium-Doped
Gallium Arsenide

Sample No.	L, mm	T ≈ 85° K				T ≈ 300° K			
		jth, A/cm²	P, W	D, mm	j, A/cm²	jth, A/cm²	P, W	D, mm	j, A/cm²
30—8*	0.43p	1.9	120	0.39	10.4	—	31	0.29	10.4
30—8*	0.43P	—	—	—	—	6.0	44	0.41	10.4
28—1	0.46P	1.2	110	0.39	9.0	5	36	0.35	9.0
28—3*	0.46P	1.5	125	0.40	9.6	6.5	40	0.28	11
28—3*	0.46P	0.9	240	0.45	9.6	—	—	—	—
32—1	0.85C	1.3	430	0.69	11	9.7	8	0.13	11
32—4*	0.66C	—	370	0.46	11	8.8	12	0.19	11
49—1*	0.47P	2.5	230	0.44	8.7	7.5	24	0.22	9.6
49—1	0.47P	1.9	—	—	—	—	—	—	—
54—1	0.50P	1.7	75	0.35	8.4	7.6	7,4	0.12	8.4
55—1*	0.49P	2.6	200	0,58	8.4	—	83	0.45	8.4
55—1*	0.49P	—	—	—	—	5.9	—	—	—
61—4	0.50C	1.2	175	0.59	7.3	2.85	83	0.64	9.0
61—15	0.67C	0.84	370	0.61	10	2.8	120	0.43	10
34—1	0.67C	2.0	250	0.66	10	5.0	23	0.27	9.6
34—3	0.80C	1.0	400	0.45	9.5	4.3	120	0.51	12
34—5*	0.54C	0.6	280	0.55	10	3.6	114	0.43	11
34—5*	0.54C	0.6	325	0.55	10	—	—	—	—
31—2	0.49P	1.1	130	0,42	9.3	5.1	40	0.17	9.6
39—5	0.44C	1.6	84	0.33	7.6	3.6	52	0.27	9.2
39—14*	0.70C	—	260	0.65	8.0	—	126	—·	10.6
39—14*	0.70C	2.4	—	—	—	—	120	0.35	9.8
39—16	0.65C	1.0	340	0.74	11	5.2	80	0.37	11
39—17	0.65C	0.8	270	0.72	10.5	3.4	200	0.60	10.9
57—1	0.44P	4.0	100	0.37	9	4.2	61	0.43	9.0
57—2	0.44P	3.6	145	0,37	9.2	3.8	48	0,22	9.0
56—1	0.47P	2.2	150	0.36	8.8	2.8	50	0.24	8.8
43—1	0.45P	No laser action for j ≤ 10 A/cm²							

TABLE 5. Parameters of Lasers Made of Zinc-Doped Gallium
Arsenide

Sample No.	L, mm	T ≈ 85° K				T ≈ 300° K			
		jth, A/cm²	P, W	D, mm	j, A/cm²	jth, A/cm²	P, W	D, mm	j, A/cm²
26—1*	0.557P	0.9	213	0.47	9	—	—	—	—
26—1*	0.557P	0.9	184	0.72	8.9	9.2	3.5	—	10
66—1	0.724C	1.9	215	0,685	8.6	9	—	—	—
22—2	0.43P	1.3	103	0.44	10.6	8.0	—	—	—
23—3	0.54P	1.0	130	0.77	10	10	—	—	—
65—1	0.37C	0.9	120	0,60	10	10	3.2	—	10

The present author and his colleagues [74] measured the output power and efficiency of over 150 gallium arsenide electron-beam-pumped lasers prepared from n- and p-type single-crystal ingots which had different impurity concentrations and were grown by different methods.

The results obtained [74] are given in Tables 3-5, in which the samples are listed in increasing order of the carrier density. The first column in these tables gives the sample numbers (the parameters of the cyrstals from which they were prepared can be found in Table 2); the second column gives the resonator size L (the method of preparation is indicated by a letter: P stands for polishing and C for cleavage). The other columns give the following laser parameters, measured at 85 and 300°K: the threshold current density j_{th}, the total output power P, the size of the emission region D, and the current density j at which the last two parameters were measured (the asterisks in the first columns of Tables 4-6 indicate that measurements were made on the same sample but at different points).

The highest output power was achieved for doped n-type gallium arsenide crystals. For many of these crystals the output power was 200-300 W at liquid nitrogen temperature and 50-100 W at room temperature.

The average output for 47 samples with tellurium concentrations in the range 3×10^{17}-4×10^{18} cm^{-3} was 215 W at T $\approx$ 80°K. For some of the GaAs samples (No. 32-4, $n_e = 7 \times 10^{17}$ cm^{-3}, $\mu = 3000$ cm$^2 \cdot$ V$^{-1} \cdot$ sec^{-1}; No. 61-15, $n_e = 2 \times 10^{18}$ cm^{-3}, $\mu = 2950$ cm$^2 \cdot$ V$^{-1} \cdot$ sec^{-1}; No. 34-3, $n_e = 2 \times 10^{18}$ cm^{-3}, $\mu = 2800$ cm$^2 \cdot$ V$^{-1} \cdot$ sec^{-1}) the experimental values of the output power were about 400 W at T $\approx$ 80°K. At room temperature the output power could be as high as 200 W (GaAs sample No. 39-17, $n_e = 4 \times 10^{18}$ cm^{-3}, $\mu = 2600$ cm$^2 \cdot$ V$^{-1} \cdot$ sec^{-1}).

Fairly high values of the output power were also achieved for p-type samples at liquid nitrogen temperature. For example, this output power was about 200 W for GaAs sample No. 26-1 ($n_h = 1.5 \times 10^{18}$ cm^{-3}, $\mu = 160$ cm$^2 \cdot$ V$^{-1} \cdot$ sec^{-1}) and for sample No. 66-1 ($n_h = 2 \times 10^{18}$ cm^{-3}, $\mu = 100$ cm$^2 \cdot$ V$^{-1} \cdot$ sec^{-1}). However, at room temperature the laser threshold was not reached in p-type samples (and in undoped samples) at current densities up to 10 A/cm^2.

The output power of the lasers made from undoped crystals was generally much lower than that which could be achieved using doped crystals. The highest value of the output achieved for an undoped crystal at liquid nitrogen temperature was below 55 W. The only exceptions were the epitaxially grown crystals. We had at our disposal only a few samples made of epitaxial films of undoped fallium arsenide (samples Nos. 58, 59, and 60 in Table 2). The output power of these samples was 170 W at liquid nitrogen temperature. Moreover, the laser threshold of the epitaxial samples was considerably lower (for example, $j_{th} = 2$ A/cm^2 for sample No. 60-2) than for the other undoped crystals. However, these results simply confirmed the general rule that much higher powers could be achieved using doped crystals. Since no data were available for doped epitaxial samples and relatively little was known about undoped epitaxial films, one could reasonably expect further improvement in the laser parameters as a result of advances in the technology of growth of semiconductor crystals (crystals with the fewest defects are currently obtained by the epitaxial method).

An analysis of the data on the laser threshold (Tables 3-5) indicated a correlation between the threshold current density and the degree of doping. For example, a typical value of j_{th} of undoped crystals at liquid nitrogen temperature was 3-4 A/cm^2, whereas the corresponding threshold current density was usually 1-1.5 A/cm^2 and sometimes less for doped n- and p-type crystals with carrier densities in the range $(0.3-2) \times 10^{18}$ cm^{-3} and 10^{18}-10^{19} cm^{-3}, respectively.

A clearer idea on the dependence of the laser threshold of n-type crystals on the degree of doping can be gained from Fig. 7. This figure shows the results obtained at liquid nitrogen and room temperatures for samples with resonators of L = 0.43-0.54 mm in size.

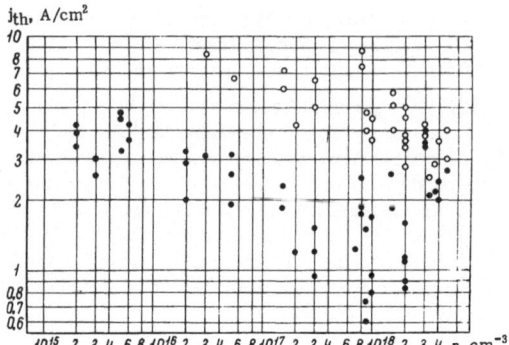

Fig. 7. Dependence of the threshold current
density j_{th} on the equilibrium electron density n in n-type lasers with resonators 0.43-0.54
mm long. The black dots represent the results obtained at T ≈ 80°K and the open circles
are the results obtained at T ≈ 300°K.

    The results plotted in Fig. 7 exhibit a considerable scatter of the experimental points due
to the many factors which can affect the quality of the semiconducting material and the laser
resonator (these factors include the presence of accidental impurities and defects, some differences in the resonator size, poor quality of the reflecting phases, etc.). However, in spite
of the scatter, we can say that the dependence of the laser threshold on the carrier density,
j_{th} (n), can be summarized as follows: at T ≈ 85°K the laser threshold is independent of n^* in
the range n < 10^{16} cm^{-3}, it decreases with increasing n in the range n > 10^{16} cm^{-3}, passing
through a minimum at n ≈ 10^{18} cm^{-3}, and rises in the range n > 10^{18} cm^{-3}. The room-temperature threshold also definitely decreases with increasing degree of doping. A minimum in the
dependence j_{th} (n) is also observed at room temperature but is shifted to n ≈ 3 × 10^{18} cm^{-3}.

    We shall now give a qualitative interpretation of the decrease in the laser threshold which
results from an increase in the degree of doping. The gain or absorption coefficient of a semiconductor[†] is governed by the density of electron and hole levels, their populations, and the
matrix elements of the transition between levels. Generally speaking, all these quantities vary
with the degree of doping. However, the strongest influence on the gain or absorption coefficient is exerted by the change in the population of electron states which results from the thermal ionization of some of the impurity atoms [33, 75].

    Therefore, we shall assume that doping with donors or acceptors produces a donor or
acceptor band and alters the population of levels in the conduction or valence band, but leaves
the other parameters unaffected.

    Let us now compare the gain of a donor-doped crystal with that of an undoped crystal at
some fixed pumping level. Obviously, the hole densities and, consequently, their quasi-Fermi
levels will be the same in both crystals. However, the electron density in the doped crystal
will be higher than in the undoped sample. This will also be true of the electron quasi-Fermi

---

[*]For n_e = 6 × 10^7 cm^{-3} (GaAs sample No. 41-2) the threshold current density is j_{th} = 4.4 A/cm^2
(Table 3).

[†]This applies also to any quantum system in which the initial and final states of an electron
form a continuous spectrum.

levels. Let us consider the spectral range in which the gains of both crystals are positive ($\hbar\omega - \Delta \leq F_e + F_h$). We can easily see that in this spectral range the gain at any wavelength will be higher for the doped crystal than for the undoped sample because the sum of the quasi-Fermi levels in the first case will be higher and the density of states not lower (because of the existence of an impurity band this density can only be higher) than in the second. Consequently, the laser threshold of a doped semiconductor crystal should be lower than the threshold of the corresponding undoped crystal provided the optical resonator losses are not raised by doping or any increase in these losses is small. The relative change in the laser threshold resulting from doping increases with increasing ratio of the equilibrium density of electrons in a doped crystal to the nonequlibrium density needed for the laser action in an undoped semiconductor. The difference between the threshold levels becomes significant when this ratio is of the order of unity. The order of magnitude of the electron density n_0 (T) required for the laser action in an undoped semiconductor is given by Eq. (8) and it is in good agreement with the experimental data plotted in Fig. 7.

The rise of the laser threshold at high doping levels may be due to an increase in the nonresonant losses in the optical resonator. In this case, a significant increase in the laser threshold should be observed at the doping level such that the nonresonant losses $\varkappa L$ (associated with, for example, an enhancement of the absorption by free carriers) become approximately equal to the losses due to the transmission of the resonator mirrors ln 1/R. If R = 0.3 and L = 0.5 mm, it follows from [45] that $\varkappa L = -\ln R$ when $n \approx (5-6) \times 10^{18}$ cm^{-3}, which is in satisfactory agreement with the experimental data given in Fig. 7.

A phenomenological theory of the dependence j_{th} (n) is given in [76]. This theory agrees only qualitatively with the experimental results plotted in Fig. 7.

The overall efficiency of a laser η is defined [8] as the ratio of the total output power P to the pumping power P_p which is acquired by the active region from the electron beam. The values of P are listed in Tables 3-5 and the pumping power P_p can be expressed in terms of the pumping current i_p which flows through the active region:

$$P_p = \frac{E_0 i_p}{e}.$$

Usually one measures no i_p but the current i passing through a diaphragm with an aperture of fixed diameter d (Chap. I). In our investigation this diameter (0.45 mm) was selected so that it was close to the average dimensions (L and D) of the active region. Therefore,

$$P_p = jLD\frac{E_0}{e}, \qquad (20)$$

where j represents the values of the current density listed in Tables 3-5. If the dimensions L L and D differ significantly* from the diameter d, we must replace Eq. (20) with the relationship

$$P_p = \frac{E_0 i}{e}\ \frac{4\Phi_0\left(\frac{L}{r_0}\right)\Phi_0\left(\frac{D}{r_0}\right)}{1 - \exp\left(-\frac{d^2}{2r_0^2}\right)}, \qquad (21)$$

which is derived on the assumption that the distribution of the electron-current density over the

---

*If the dimensions L and D lie between 0.3 and 0.56 mm, the values of P_p calculated using Eqs. (20) and (21) differ by not more than 8% from the true values.

TABLE 6. Efficiencies of Investigated Lasers

Sample No.	$T \approx 85°$ K $\frac{P_{th}}{P_p}$	η, %	$T \approx 300°$ K $\frac{P_{th}}{P_p}$	η, %	Sample No.	$T \approx 85°$ K $\frac{P_{th}}{P_p}$	η, %	$T \approx 300°$ K $\frac{P_{th}}{P_p}$	η, %
41—2	2.3	7.0	—	—	34—1	3.8	16	1.8	5.6
1.6	3.0	4.8	—	—	34—3	7.6	32	2.4	6.6
64—1	2.4	4.3	—	—	34—5*	14.5	24	2.9	10.6
51—5	3.2	9.1	1.2	3.4	34—5*	14.5	28	—	
13—2	3.4	10.7			31—2	8.0	16,5	1.8	10.5
			1.2	4.2	39—5	4.8	16	2.6	10.4
30—8*	5.6	15	—	4.8	39—14	—	21	—	12
30—8*	—	—	1.7	4.9	39—16	8.1	19	1.7	7.7
28—1	7.5	15	1.7	5.8	39—17	10	18	2.9	12
28—3*	6.4	15.5	1.6	6.4	57—1	2.2	15	2.0	8.4
28—3*	10.0	27	—	—	57—2	2.5	21	2.4	12
32—1	6.0	21.5	—	—	56—1	4.0	22	3.0	11.5
32—4	—	25	1.2	2.0	26—1*	10	20	—	—
49—1	3.2	29.5	1.2	5.4	26—1*	9.0	16,5	—	—
54—1	4.9	11	1.1	3.1	66—1	4.0	15	—	—
55—1	3.8	19	—	10	22—2	8.0	12	—	—
61—4*	4.0	19	2.9	7.2	65—1	11	14	—	—
61—4*	—	21	—	9.7					
61—15	9.4	25	2.9	11.0					

beam cross section is of the form[*]

$$j(r) = j_0 \exp\left(-\frac{2r^2}{r_0^2}\right). \tag{22}$$

Here, r is the distance from the center of the electron beam; r_0 is a parameter which depends on the beam focusing (we assumed that the value of r_0 corresponding to the maximum current density was 0.6 mm, which corresponded to an electron-beam diameter of 0.7 mm at the 0.5 level of the current density); Φ_0 (x) is the error function [77].

The overall efficiencies of lasers operated at 85 and 300°K were calculated in this way from the experimental data in Tables 3 and 5. These values are listed in Table 6.

We can see from Table 6 that the highest values of the efficiency are achieved for doped n-type samples and that these values can be as high as 30% at $T \approx 85°$K and 12% at $T \approx 300°$K. The data given in Table 6 should be supplemented by the observation that the average efficiency obtained at $T \approx 85°$K for 47 samples with tellurium concentrations in the range $3 \times 10^{17} - 4 \times 10^{18}$ cm^{-3} was 18.4%.

The differential efficiency obtained at T = 85°K is plotted in Fig. 8. The abscissa in this figure represents the equilibrium electron density n measured at room temperature and the ordinate is the differential efficiency defined by Eq. (19).

The experimental points in Fig. 8 are strongly scattered and the scatter is greater than the random experimental error. This means that the efficiency of a laser depends not only on the degree of doping but also on other parameters of a semiconductor crystal and many of these parameters cannot be controlled precisely.

In spite of the scatter of the experimental points, we can easily see from Fig. 8 that when the equilibrium electron density is increased from 10^{16} to 10^{18} cm^{-3} the differential efficiency

---

[*]This assumption is in agreement with the results of measurements of the electron-beam current passing through apertures of various diameters.

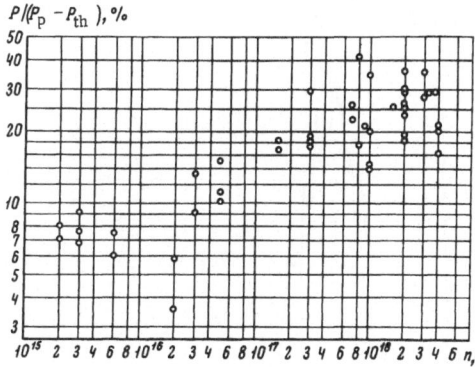

$P/(P_{\mathrm{p}} - P_{\mathrm{th}})$, %

Fig. 8. Differential efficiency of n-type lasers at $T \approx 85°K$, plotted against electron density n.

η_{d} rises. The average value of η_{d} for samples with carrier densities from 3×10^{17} to 4×10^{18} cm^{-3} is approximately 24%.

The maximum theoretically possible value of η_{d} of an electron-beam-pumped laser is about 1/3. An allowance for some of the additional losses (see, for example, [78] and § 2 in Chap. III) can reduce this value slightly. Therefore, we may assume that the experimental values of the efficiency of lasers made of n-type crystals differ from the maximum theoretical efficiency by an amount lying within the limits of the experimental error. The precision of the values obtained in this way is sufficient for practical applications of the results obtained.

The low values of the efficiency of some of the crystals (particularly undoped samples) are mainly due to the following two factors: 1) the quantum efficiency γ is less than unity; 2) the semiconductor crystal is inhomogeneous.

The relationship between the overall efficiency η and the quantum efficiency γ is self-evident. However, it is worth making some comments about the influence of the inhomogeneity of a semiconductor on the laser efficiency.

Let us assume that the active region of a semiconductor crystal is inhomogeneous along the y axis (Fig. 4) and the inhomogeneity gives rise to a dependence of the local laser threshold* on the coordinate y. We shall also assume that the differential efficiency is independent of y and equal to its maximum value, i.e., we shall assume that the quantum efficiency is $\gamma = 1$ at every point in the crystal under consideration. Let us now mentally divide the resonator into regions of width Δy, which can be regarded — in the first approximation — as independent lasers. In this situation, such independent lasers will be characterized by different values of the excess of the current density over the threshold and, consequently, different values of the efficiency η even when the distribution of the excitation is homogeneous. When this situation occurs in practice the overall and differential efficiencies of a laser are averages of the local values which vary throughout the active region. The laser threshold is defined as the current density at which stimulated radiation is observed at least at one point, i.e., where the local threshold is lowest. Clearly, when the threshold measured in this way is only slightly exceeded, the average value of the efficiency over the whole active region may be much lower than in the homogeneous case. It is also ob-

---

*The local laser threshold has the following meaning. Let us assume that the region excited by the electron beam can be described by $y_0 \leq y \leq y_0 + \Delta y$. Let us also assume that Δy is sufficiently large to allow us to ignore the diffraction losses due to the finite size of this interval but sufficiently small to permit us to assume that the semiconductor is completely homogeneous within this interval. Then, the laser threshold of this part of the resonator can be called the local laser threshold at the point y_0.

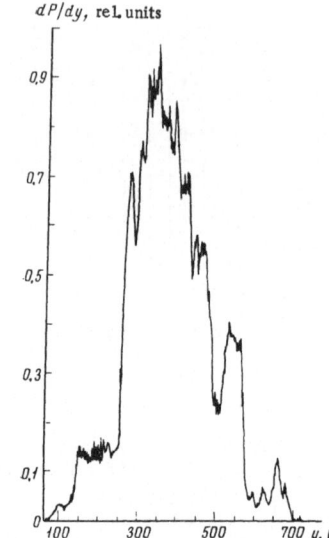

Fig. 9. Near-field distribution of
the output power along the y axis.

vious that this fall in the efficiency will become smaller when the excitation level is raised
well above the threshold and when the semiconductor crystal is made more homogeneous. The
inhomogeneity of semiconductor crystals is a well-established fact and it undoubtedly influences
the measured values of the efficiency. The following experimental observations support this
conclusion.

1. The later threshold of samples prepared from the same ingot varies from sample to
sample and even from one point to another within the same sample. For example, the laser
threshold measured at liquid nitrogen temperature for five samples prepared by polishing from
GaAs ingot No. 41 ($n_e = 7 \times 10^7$ cm^{-3}) ranged from 4 to 7 A/cm^2. The laser threshold of five
samples prepared by cleaving from the same ingot GaAs No. 39 ($n_e \approx 4 \times 10^{18}$ cm^{-3}) ranged
from 0.8 to 2 A/cm^2.

2. The near-zone distribution of the radiation field obtained with the aid of an image converter and a microscope revealed the following sequence of events which occurs in the majority of samples (particularly in the case of undoped crystals) when the current density is raised
gradually. The laser action always appears at a single point. When the current density is
raised, this point expands slightly and the laser emission is then observed at one or more neighboring points. At higher current densities such points may merge in a single continuous region
and the distribution of the emission intensity across this region is inhomogeneous. In our experiments the laser threshold of undoped samples was fairly high and there was an upper limit
to the current density which could be produced in our apparatus;* the luminous points usually
did not merge but overlapped slightly.

3. The highest values of the laser efficiency are obtained for those samples in which the
distribution of the radiation power along the y axis is relatively homogeneous. By way of example, Fig. 9 shows such a distribution for GaAs sample No. 28-3, obtained as described in
Chap. I. The spatial resolution of the results plotted in Fig. 9 is not less than 10 μ.

*Further rise in the current density was restricted not only by the capabilities of the apparatus
but also by "objective" factors associated with the danger of damage of the resonator mirrors
(Chap. III).

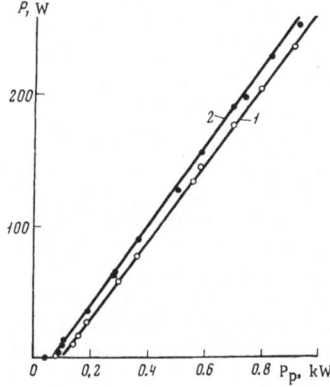

Fig. 10. Dependence of the total output power of GaAs lasers on the pumping power: 1) sample No. 61-6; 2) sample No. 34-5.

The values of the laser efficiency obtained in the present investigation did not indicate clearly which of the two possible factors was responsible for the low efficiency of some of the GaAs lasers. Moreover, it was not possible to determine definitely that these factors and no others were responsible for the reduced efficiency. Definite conclusions could only be drawn from additional experiments.

The cited values of the laser efficiency were calculated using the pumping powers deduced from the visually measured size of the active region. It was possible that some of the electrons incident on the semiconductor outside the active region made some contribution to the radiation emitted from this region or affected in some other way the measured total output power. More-over, other errors committed in the calculation of the total output power could reduce the precision of the experimental value of the efficiency.

In view of this, we carried out special measurements of the efficiency on the most homo-geneous samples. The excited region was defined by a diaphragm with a slit 0.48 mm wide. The pumping power was determined by measuring the total current passing through a rectan-gular aperture of the same dimensions as the region excited by the electron beam. The results of measurements carried out on GaAs samples No. 61-6 (L = 0.51 mm) and No. 34-5 (L = 0.54 mm) at liquid nitrogen temperature are plotted in Fig. 10. The following values were obtained for the overall and differential efficiencies of these samples:

$$\text{GaAs sample No. 61 — 6:} \quad \eta = (26 \pm 5)\%, \quad \eta_d = (29 \pm 6)\%;$$
$$\text{GaAs sample No. 34 — 5:} \quad \eta = (28 \pm 6)\%, \quad \eta_d = (30 \pm 6)\%.$$

All the experimental data on the laser threshold, the output power, and the efficiency re-ported in the present section were obtained using electron pulses of 50 Hz repetition frequency and 45-50 keV energy. Special experiments indicated that the laser parameters did not change when the repetition frequency was raised to 1000 Hz. However, when the electron energy was reduced, the output power and efficiency dropped sharply.[*]

§ 2. Temperature Dependence of the Laser Threshold

The gain necessary for the emission of coherent radiation from a semiconductor can be achieved by inverting the population of intrinsic, impurity, or exciton states. The population of

---

[*]The dependence of the laser threshold on the electron energy is discussed in § 6 in the present chapter.

these states and, consequently, the gain of a laser vary rapidly with temperature with the possible exception of liquid helium temperatures. Therefore, the laser threshold of a semiconductor laser should depend on the temperature irrespective of the pumping method. The nature of this temperature dependence is governed by the energy spectrum of those electron states which are involved in the transitions responsible for the gain. In particular, the temperature dependence of the laser threshold should be a function of the nature and concentration of the dopant.

The fullest information is available on the temperature dependence of the laser threshold of gallium arsenide injection lasers [79-82]. However, the injection pumping method is not very suitable for the determination of the relationship between the temperature dependence of the laser threshold and the degree of doping of the laser crystals because the active region in an injection laser must of necessity be heavily doped and strongly compensated. The electron-beam pumping method provides a better opportunity for investigating this relationship than do the injection and optical pumping methods because the excitation conditions (the rate of generation of electron—hole pairs) in an electron-beam-pumped laser are independent of the dopant concentration.

Some information on the temperature dependences of the laser threshold of electron-beam-pumped cadmium telluride and cadmium sulfide lasers is given in [71, 72]. However, the investigations reported in these papers were carried out on few samples and, therefore, the results obtained should be regarded as special cases, typical of undoped crystals of these compounds. The great variety of the temperature dependences of the laser threshold of gallium arsenide crystals with different degrees of doping was first demonstrated in [83] (see also [84]).

The present section is concerned with the results of an investigation of the temperature dependence of the laser threshold in the 80-300°K range, which was carried out on undoped and doped n- and p-type gallium arsenide. In view of the inhomogeneity of the laser threshold discussed in the preceding section, all the temperature dependences were recorded at a fixed point in a sample and the position of this point was known to within ± 30 μ.

The temperature dependences of the laser thresholds of undoped GaAs and of tellurium- and zinc-doped crystals are presented in Figs. 11 and 12. It is evident from these figures that the nature of the dependence varies with the degree of doping and there is a close correlation with the impurity concentration.

One of the most interesting features of the results obtained is exhibited by the samples with the intermediate degree of doping. The threshold current density for these samples is a nonmonotonic function of temperature T: the dependence j_{th} (T) has a falling region. Such a region is observed for GaAs sample No. 28-3 in the temperature range from 220 to 235°K. Similar falling regions are found in the temperature dependences of the laser threshold of p-type samples with an intermediate degree of doping (curve 1 in Fig. 12).

Fig. 11. Temperature of dependences of the laser threshold for undoped and tellurium-doped gallium arsenide samples: 1) No. 41-2 ($n_e \approx 6 \times 10^7$ cm^{-3}); 2) No. 1-5 ($n_e \approx 2 \times 20^{15}$ cm^{-3}); 3) No. 28-3 ($n_e \approx 3 \times 10^{17}$ cm^{-3}); 4) No. 49-1 ($n_e \approx 8 \times 10^{17}$ cm^{-3}); 5) No. 19-4 ($n_e \approx 1 \times 10^{18}$ cm^{-3}); 6) No. 34-5 ($n_e \approx 2 \times 10^{18}$ cm^{-3}).

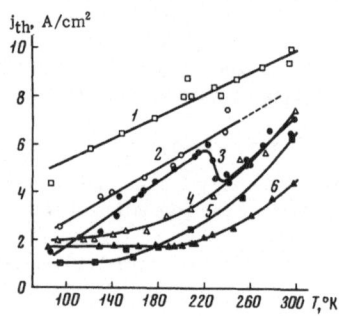

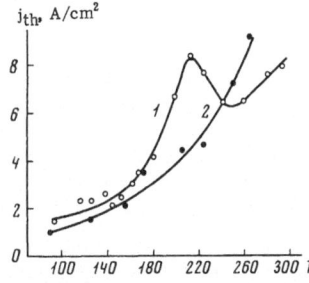

Fig. 12. Temperature dependences
of the laser threshold for zinc-
doped samples of gallium arsenide:
1) 22-3 ($n_h \approx 2 \times 10^{18}$ cm^{-3}); 2) No.
23-3 ($n_h \approx 2 \times 10^{19}$ cm^{-3}).

The existence of a falling region in the dependence j_{th} (T) can be explained qualitatively as follows. We shall consider a p-type semiconductor and assume that in the temperature range corresponding to the falling region the laser action is due to transitions with the final states in the valence band.*

We shall postulate that the stimulated emission resulting from these transitions can occur when the hole density reaches a certain value n_{th}, whose temperature dependence is plotted in Fig. 13a. In a doped semiconductor this density is the sum of two terms

$$n_{th} = n_0(T) + \Delta n(j), \qquad (23)$$

where n_0 (t) is the equilibrium density of holes in the valence band generated by the thermal ionization of the acceptors; Δn (j) is the nonequlibrium correction to this density due to pumping with a current of density j.

The dependence n_0 (T) is of the form shown in Fig. 13a: at low temperatures this density tends to zero whereas in the temperature range kT $\sim$ E_a it rises almost exponentially and then becomes saturated. Thus, it follows from Eq. (23) and from Fig. 13a that the nonequilibrium correction Δn (j), necessary for the attainment of the laser thesthold, depends on the temperature as shown in Fig. 13b. The nonequilibrium carrier density is proportional to the electron current density and, therefore, the experimentally determined dependence j_{th} (T) is in qualitative agreement with the dependence Δn (T) plotted in Fig. 13b.

According to this interpretation, the dependence j_{th} (T) should be monotonic for undoped and heavily doped samples because, in the case of undoped samples, the equilibrium density of electrons or holes n_0 (T) is considerably smaller than n_{th}, whereas, in the case of heavily doped

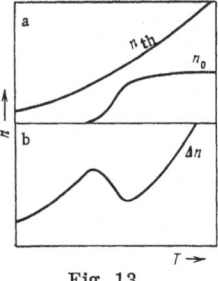

Fig. 13

*This will be proved in §§ 3 and 4.

samples, the intrinsic and impurity states of electrons overlap completely so that there are no bound electron or hole states.

A special feature of the temperature dependence of the laser threshold of heavily doped n-type samples is a fairly wide range of temperatures in which a plateau is observed. For example, such a plateau is found for GaAs sample No. 34-5 at temperatures below 200°K. This plateau in j_{th} (T) is evidently due to strong degeneracy of electrons. This is demonstrated in [43] for degenerate equilibrium holes. However, such degeneracy is lifted (for $N_a - N_d \approx 10^{19}$ cm^{-3}), because of the large effective mass, at temperatures as low at T $\approx$ 50°K.

The dependence j_{th} (T) obtained for heavily doped p-type samples is closest to that observed for injection lasers [79, 82]: j_{th} (T) $\propto$ exp (kT/E_0). The constant E_0 is about 0.3 meV for samples with acceptor concentrations $N_a \approx 2 \times 10^{19}$ cm^{-3}. This similarity is evidently due to the fact that in injection lasers the coherent radiation is generated in the p-type region, whose energy spectrum is approximately of the same form as that of a homogeneous crystal with a similar acceptor concentration.

The temperature dependence of the laser threshold of undoped crystals is nearly linear. Similar results are reported in [71, 72], which describe investigations of this temperature dependence for undoped crystals of CdS and CdTe. A nearly linear dependence follows also from the theory developed in [85] (see also [64]).

Thus, the experimental results obtained in the present investigation show that the data on gallium arsenide injection [79-82] and on electron-beam-pumped cadmium sulfide and cadmium telluride [71, 72] can be regarded as special cases of the great variety of temperature dependences of the laser threshold which can be observed in the same semiconductor at different doping levels.

§ 3. Emission Spectrum

We shall now consider the results of an investigation of the emission spectrum of gallium arsenide lasers made of crystals with different degrees of doping [83]. This investigation was carried out in the temperature range 80-300°K.

The emission spectrum of a laser represents eigenmodes in an optical resonator. In a Fabry–Perot resonator the difference between the wavelengths of adjacent longitudinal resonator modes is given by the following relationship [86], which makes an allowance for the disper-

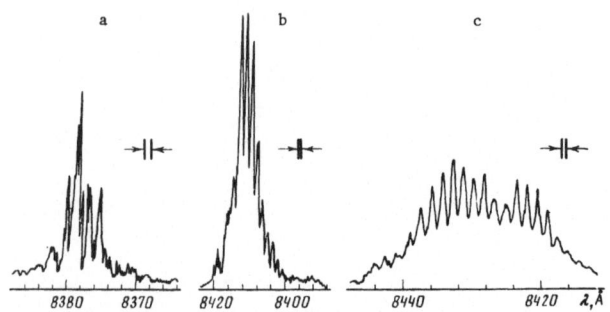

Fig. 14. Coherent emission spectra of GaAs sample No. 14-2 (L = 0.457 mm) obtained using different current densities (A/cm^2): a) 1.1; b) 4.0; c) 7.0. Threshold current density $j_{th} \approx 1$ A/cm^2.

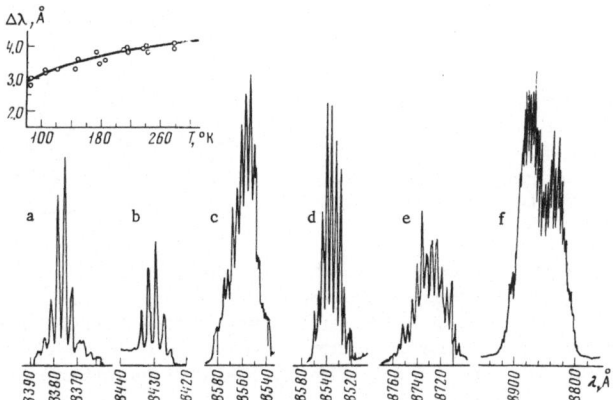

Fig. 15. Coherent emission spectra of GaAs sample No. 14-6
(L = 0.329 mm), recorded at various temperatures (°K): a)
83; b) 106; c) 177; d) 144; e) 240; f) 278. The temperature dependence of the mode separation $\Delta\lambda$ is plotted in the top left-
hand corner.

sion of the refractive index n:

$$\Delta\lambda = \frac{\lambda^2}{2L\left(n - \lambda \frac{dn}{d\lambda}\right)} . \qquad (24)$$

Stimulated emission appears in those modes which satisfy the self-excitation condition.

Usually several resonator modes are excited in a semiconductor laser* [5, 8-11] and the
number of these modes increases when the excitation level is raised.

Figure 14 shows the emission spectra of GaAs sample No. 14-2 ($n_h \approx 2 \times 10^{18}$ cm^{-3}) recorded at T ≈ 85°K under different excitation conditions. We can see that when the electron-
beam current density is increased not only the number of modes rises but also the spectral
line shifts in the direction of longer wavelengths. This shift may be due to the heating of the
active region by the electron beam. The results of calculations given in Appendix I can be used
to show that the heating at the end of a pumping pulse of 0.1 μsec duration is 5, 17, and 30 deg,
respectively for the cases represented in Figs. 14a, 14b, and 14c. If we use the data on the
temperature shift of the emission wavelength ($\Delta\lambda/\Delta T$ = 2.17 Å/deg), which are given below,
and if we assume that the modes with the longest wavelengths appear at the end of the pumping
pulse, we find that, in the cases represented by Figs. 14b and 14c, the shifts should be 30 and
55 Å, respectively, relative to the modes shown in Fig. 14a. This is in satisfactory agreement
with the observations. It also follows from Fig. 14 that the average separation between the
modes is independent (within the limits of the experimental error) of the pumping level and that
it amounts to 1.51 ± 0.04 and 1.52 ± 0.05 Å in the cases represented in Figs. 14a and 14c.

The influence of temperature on the emission spectrum is illustrated in Fig. 15. These
spectra were obtained for GaAs sample No. 14-6 with a resonator L = 0.329 ± 0.01 mm long at
a pumping level slightly in excess of the laser threshold (in all cases, the pumping level was

---

*It should be pointed out that not all the modes which are observed during a pumping pulse are
generated simultaneously. This is demonstrated for gallium arsenide injection lasers in [87].

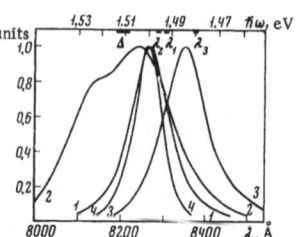

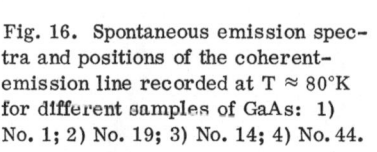

Fig. 16. Spontaneous emission spectra and positions of the coherent-
emission line recorded at T ≈ 80°K
for different samples of GaAs: 1)
No. 1; 2) No. 19; 3) No. 14; 4) No. 44.

$j/j_{th} \lesssim 1.5$). The instrumental width of the spectral line did not exceed 1 Å for any of the recorded spectra. We can see that the number of modes and the separation between them increase when the temperature is raised. The temperature dependence of the mode separation,
deduced from an analysis of the emission spectra of sample No. 14-6, is plotted in the top left-
hand corner of Fig. 15. This dependence is evidently due to the temperature dependence of the
refractive index [88].

The mode separation $\Delta\lambda$ depends also on the nature and concentration of the dopant. For
example, the average value of $\Delta\lambda$ for an undoped GaAs sample No. 6-1 with a resonator L =
0.48 mm long is 1.42 ± 0.06 Å at T ≈ 80°K, whereas the corresponding separation for a tellurium-doped GaAs sample No. 19-1 (L = 0.53 mm) is $\Delta\lambda$ = 2.00 ± 0.08 Å and that for a zinc-doped
GaAs sample No. 14-2 (L = 0.457 mm) is $\Delta\lambda$ = 1.51 ± 0.05 Å. When these values are compared
with Eq. (24), it is found that the quantity $\left(n - \lambda \frac{dn}{d\lambda}\right)$ is 4.9, 3.9, and 5.1, respectively. The first
two values are in poor agreement with the results reported in [88], whereas the last value is in
satisfactory agreement with an analysis of the spectra of gallium arsenide injection lasers giv-
en in [89].

The position of the stimulated emission line relative to the spontaneous radiation spectrum is plotted in Fig. 16. This figure gives the spontaneous radiation spectra and the posi-
tions of the stimulated line are indicated by λ_1, λ_2, and λ_3 along the photon energy scale $\hbar\omega$.
The same figure shows the position of the fundamental absorption edge Δ deduced from the re-
sults reported in [29]. Thick black lines are used to indicate indeterminacy of the positions of
λ_i and Δ because of a possible error in the measurements of the temperature of the sample
and probable influence of the heating of the active region during pumping. The spontaneous radiation spectrum was recorded at an angle of 30° with respect to the electron-bombarded face
of a plane-parallel disk ~1 mm thick and of ~0.5-0.8 cm diameter. In this sample the coherent
radiation could not be generated. All the spontaneous radiation spectra were recorded at a current density of 4 A/cm². The position of the stimulated emission line was determined using
samples of standatd form shown in Fig. 4. The symbol λ_1 and curve 1 of Fig. 16 represent an
undoped crystal (GaAs sample No. 1); the symbol λ_2 and curve 2 represent a GaAs crystal No.
19, etc.

Figure 16 also shows (for the sake of comparison) the spontaneous radiation spectrum of
an epitaxial film (GaAs sample No. 44) which could not be shaped to form an optical resonator
(the the thickness of the film was only 5-6 μ).

In all the cases represented in Fig. 16, the energy of the coherent photons was less than
the forbidden band width (measured for pure crystals) and it corresponded to the long-wavelength edge of the spontaneous radiation spectrum. The first of these observations will be discussed in § 4; the second observation reflects the difference between the wavelengths at which
maxima of the spontaneous radiation spectrum and the gain are observed [50, 85, 90]. The
spontaneous radiation spectrum has a maximum which corresponds approximately to a photon
energy $\Delta + F_e + F_h$ (the scale used in measurements of the Fermi levels F_e and F_h is discussed

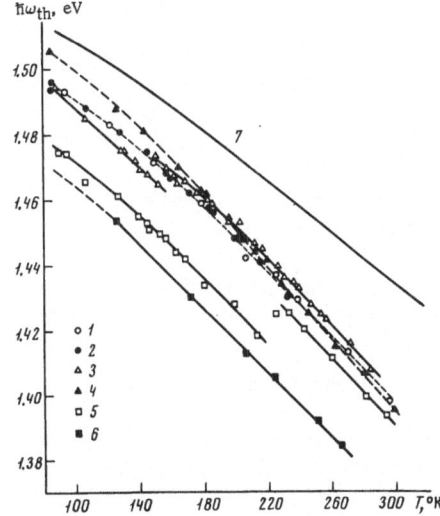

Fig. 17. Temperature dependences of the photon energy $\hbar\omega_{th}$ for GaAs samples: 1) No. 41-2 ($n_e \approx 6 \times 10^7$ cm^{-3}); 2) No. 40-1 ($n_e \approx 5 \times 10^{15}$ cm^{-3}); 3) No. 28-3 ($n_e \approx 3 \times 10^{17}$ cm^{-3}); 4) No. 34-5 ($n_e \approx 2 \times 10^{18}$ cm^{-3}); 5) No. 22-3 ($n_h \approx 2 \times 10^{18}$ cm^{-3}); 6) No. 23-3 ($n_h \approx 2 \times 10^{19}$ cm^{-3}); 7) dependence of Δ on T.

in the Introduction), whereas the gain at this photon energy is zero and positive at lower energies.

Figure 17 shows the temperature dependences of the photon energy $\hbar\omega_{th}$ at the center of the spontaneous emission line of n-type and p-type samples (these dependences were recorded at pumping levels slightly in excess of the laser threshold).

A special feature of the curves plotted in Fig. 17, which has not been observed before for semiconductor lasers, is the discontinuity of the temperature dependence of $\hbar\omega_{th}$ observed for n- and p-type samples with intermediate degrees of doping. In the case of GaAs sample No. 22-3, this discontinuity occurs at about 225°K, whereas in the case of GaAs sample No. 28-3 it occurs at 155°K (more precisely, between 147 and 160°K). In this range of temperatures the emission spectrum of GaAs sample No. 28-3 varies in the following manner (Fig. 18). At T < 145°K the stimulated emission spectrum has only one peak. When the temperature is raised to 147°K, a second peak corresponding to transitions with higher photon energies is observed. The intensity of the second peak is about two orders of magnitude less than that of the first peak

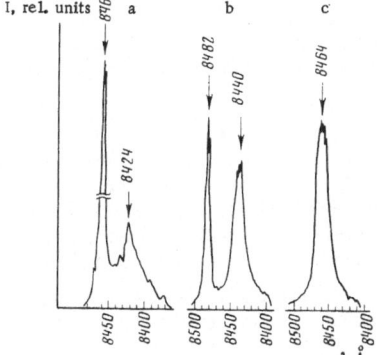

Fig. 18. Coherent emission spectra of GaAs sample No. 28-3 at various temperatures (°K): a) 147; b) 155; c) 163.

(Fig. 18a). When the temperature is raised still further, the intensity of the second (short-wavelength) peak increases rapidly and that of the first (long-wavelength) decreases. At T ≈ 154-156°K the intensities of the two peaks become equal (Fig. 18b) and at T ≈ 163°K the long-wavelength peak disappears (Fig. 18c). The difference between the photon energies of these two peaks is $(6-7.5) \times 10^{-3}$ eV for n-type samples and $(1.5-2.0) \times 10^{-2}$ eV for p-type samples. The first value corresponds to the ionization energy of donors and the second value is close to the ionization energy of acceptors [5].

In the case of undoped or heavily doped n- and p-type samples, the dependence $\hbar\omega_{th}$ (T) is continuous. In the case of heavily doped p-type samples, this dependence is closest to the corresponding dependence observed for gallium arsenide injection lasers [81].

It is worth noting that the photon energy $\hbar\omega_{th}$ is usually less than the forbidden band width (the upper curve in Fig. 17 shows the temperature dependence of the forbidden band width of gallium arsenide [29]).

The different nature of the dependence $\hbar\omega_{th}$ (T) for samples with different degrees of doping is due to certain features of the energy spectra.

We shall start by considering samples with intermediate doping levels. Semiconductors with a specific donor or acceptor impurity can emit stimulated radiation corresponding to 1) transitions between the conduction and valence bands or 2) transitions involving impurity states such as transitions from the conduction band to an acceptor band (in p-type samples) or from a donor band to the valence band (in n-type samples). The discontinuous nature of the dependence $\hbar\omega_{th}$ (T) is a consequence of a change in the laser action mechanism: at high temperatures this mechanism involves transitions of the first type and at low temperatures it is due to transitions of the second type. This interpretation is supported by the following qualitative considerations.

Let us first consider low temperatures, when practically all the impurity states are occupied by electrons in the case of donor impurities and by holes in the case of acceptor impurities. Obviously, beginning from some excitation level, the gain (regarded as a function of the photon energy $\hbar\omega$) should have two maxima corresponding to these two types of radiative transition. When the exictation level is raised, the absolute values of the gain at these maxima increase until one reaches the threshold value governed by the optical resonator losses. Above the threshold the gain is known to be practically independent of the pumping level. The absolute value of the gain at the maximum corresponding to transitions involving an impurity band increases with increasing impurity concentration. Therefore, at a given value of optical resonator losses, there is a minimum impurity concentration necessary for the achievement of the coherent emission as a result of the band−impurity transitions. If the impurity concentration is less than this value, the gain maximum corresponding to the band−band transitions is the first to reach the threshold level. We shall assume that the impurity concentration is sufficiently high so that the coherent emission at low temperatures is achieved first for band−impurity transitions.

We shall now discuss the case of fairly high temperatures $kT \gg E_{d,a}$, when the populations of the impurity and intrinsic states (near the bottom of the appropriate allowed band) differ only slightly. In this case, the gain at the maximum corresponding to the band−band transitions should have a higher value than at the maximum corresponding to the band−impurity transitions because the density of states in the former case is much higher than in the latter.*
Thus, if the impurity concentration is sufficiently high so that at low temperatures the stimu-

---

*According to Eq. (4), the higher the density of the electron and hole states between which transitions of energy $\hbar\omega$ take place and the higher the population of these states, the greater is the value of the gain at the frequency ω.

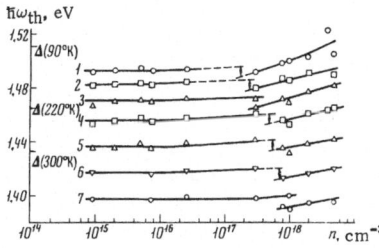

Fig. 19. Dependences of the photon energy $\hbar\omega_{th}$ on the equilibrium electron density n, recorded at various temperatures (°K): 1) 90; 2) 120; 3) 150; 4) 180; 5) 220; 6) 260; 7) 300. The electron energy was measured at T ≈ 300°K.

lated emission is due to the band—impurity transitions, the laser mechanism should change at sufficiently high temperatures to that primarily due to the band—band transitions. Obviously, there must be some intermediate temperature at which the threshold value of the gain is reached simultaneously (at the same pumping level) for both types of transition. The temperature at which this occurs obviously depends on the resonator losses and on the concentration and ionization energy of the impurities. The order of magnitude of this temperature is given by kT ~ $E_{d,a}$. However, if the impurity concentration is so low that even at low temperatures the stimulated emission is due to the band—band transitions, the laser mechanism remains the same at high temperatures.

Dependences of the photon energy $\hbar\omega_{th}$ on the carrier density in n-type samples are plotted in Fig. 19. The curves shown in this figure also have a discontinuity, indicating a change in the laser mechanism resulting from a change in the degree of doping: the higher the temperature, the higher is the impurity concentration at which this discontinuity is observed. This is easily explained on the basis of the proposed interpretation: the higher the density of states in the impurity band, the lower is the population (i.e., the higher the temperature) at which the gain reaches a value sufficient for stimulated transitions involving these states.

It also follows from Fig. 19 that, in the case of heavily doped n-type samples, the photon energy increases continuously with increasing equilibrium electron density and the rise is stronger at lower temperatures (this is, in effect, analogous to the Burstein effect [91]). At T ≈ 90°K the energy of photons $\hbar\omega_{th}$ emitted from samples with electron densities n_e ≈ (3-4) × 10^{18} cm^{-3} exceeds the forbidden band width Δ.

A comparison of the stimulated emission spectra of samples with different impurities and different impurity concentrations suggests the following scheme of radiative transitions which occur in stimulated emission at temperatures from 80 to 300°K.

In the case of undoped gallium arsenide, the main contribution to the stimulated emission line is made by the band—band transitions.

In the case of samples doped with donors (Te), we can have either the band—band transitions or the transitions from an impurity to the valence band (or its tail): which of these predominates depends on the temperature and degree of doping. The band—band transitions occur throughout the investigated range of temperatures in heavily doped samples and at high temperatures in samples with an intermediate degree of doping (the high-temperature range for GaAs sample No. 28-3 is defined by T ≥ 155°K and this range depends on the donor concentration). The transitions from an impurity to the valence band occur in samples with intermediate degrees of doping at low temperatures.

The situation is similar in samples doped with acceptors (Zn): we either have the band—band transitions (this occurs, for example, in GaAs sample No. 22-3 at T ≥ 225°K) or the transitions from the conduction band to an acceptor band.

Apart from the aforementioned observations relating to the correlation between the magnitude of the discontinuity in the $\hbar\omega_{th}$ (T) curve and the ionization energy of donors or acceptors, the proposed radiative transition scheme is also supported by the following results:

1. At high temperatures all the undoped and doped n- and p-type samples emit coherent photons of approximately the same energy. The only exception to this rule is represented by the samples with equilibrium hole densities $n_h \approx 2 \times 10^{19}$ cm^{-3}. However, since the temperature at which a change occurs in the laser mechanism is related to the impurity concentration (Fig. 19), we may conclude that at this hole density the temperature $T \approx 270-300°K$ is "insufficiently high" in the sense discussed above.
2. The difference between the energies $\hbar\omega_{th}$ of photons emitted from undoped and zinc-doped samples is 25-30 meV in the temperature range 80-300°K and this difference corresponds to the ionization energy of acceptors.
3. When the equilibrium electron density is raised from 3×10^{17} to 4×10^{18} cm^{-3}, the photon energy rises continuously from values typical of undoped samples to values exceeding (at $T \approx 90°K$) the forbidden band width. This means that the spectrum of electron states is continuous in this energy range and the density of states is a monotonic function of the energy.*

We thus find that, as in the case of the temperature dependence of the laser threshold, there is a great variety of the dependences $\hbar\omega_{th}$ (T) in the case of gallium arsenide samples with different degrees of doping.

§ 4. Nature of Radiative Transitions

We have already mentioned that the photon energy $\hbar\omega_{th}$ is usually (with the exception of heavily doped n-type samples) less than the forbidden band width deduced from the optical absorption in the purest crystals of gallium arsenide [29].

This problem has been encountered earlier in the physics of semiconductors. The concepts of the "forbidden band width" or the "fundamental absorption edge" are not very precise. The fundamental absorption edge of real crystals is always broad because photons of energies less than the forbidden band width Δ are absorbed quite strongly. The absorption coefficient α usually decreases exponentially: $\alpha (\hbar\omega) \propto \exp (\hbar\omega - \Delta/E_0)$, where E_0 is some constant [29, 45–47, 75]. If we use the criterion employed in the determination of the forbidden band width from the absorption spectra [29], we find that the absorption coefficient corresponding to $\hbar\omega = \Delta$ is approximately 10^4 cm^{-1}. It follows from Eq. (4) that at high excitation levels the absolute value of the gain at some frequency ω can reach the value of the absorption coefficient of an unexcited crystal absorbing radiation of the same frequency. On the other hand, the gain necessary for the laser action in optical resonators of the size used in the present investigation ($L \approx 0.5$ mm) does not exceed 100 cm^{-1}. Therefore, it is clear that the laser threshold conditions are satisfied at frequencies corresponding to absorption coefficients of unexcited crystals much lower than 10^4 cm^{-1}. In fact, a comparison of the experimental data plotted in Fig. 18 with the results given in [29, 45, 75] shows that, in those cases when radiative transitions in undoped and zinc-doped samples are attributed to the band–band mechanism,† the photon energy $\hbar\omega_{th}$ corresponds to the absorption coefficient of unexcited crystals ($\alpha \sim 100$ cm^{-1}).

---

*A qualitative analysis of Eq. (4) shows that the necessary condition for a change in the laser mechanism with increasing temperature can be formulated as follows. The density of states ρ (E) should be a nonmonotonic function of the energy, i.e., it should be a maximum at the center of the impurity band and a fairly deep minimum separating this band from the nearest allowed band (there may be some overlap between the two bands).

†The corresponding absorption coefficient of the tellurium–doped samples is less [45, 75] because of the Burstein effect.

Thus, the location of the stimulated emission spectrum within the forbidden band of a semiconductor and the broadening of the fundamental absorption edge are basically due to the same factor. This factor is related directly to the energy spectrum of a semiconductor near the valence and conduction band edges. It is mentioned in the Introduction that the nature of the spectrum of electron states of a semiconductor in the region of the allowed-band edges is governed by the interaction of electrons and holes with each other and with the impurity ions which are always present in real crystals.

The simplest manifestation of this interaction is the existence of exciton [92] and impurity states. Such bound states of carriers can be regarded as hydrogen-like atoms and the crystal lattice can be considered as a continuous medium. At low concentrations these hydrogen-like atoms are in the "gaseous" state and their energy spectrum is similar to that of the hydrogen atoms in vacuum (the only difference is in the energy scale). In this "gaseous" state of charge carriers in a semiconductor the forbidden band width, the energies of bound states, and other one-electron concepts have definite physical meaning. However, this state is hardly ever realized in gallium arsenide lasers.

At the excitation levels encountered in semiconductor lasers and at impurity concentrations typical of the majority of currently used gallium arsenide crystals, the system of interacting carriers is in the same state as strongly compressed matter. The scale and nature of changes in the energy spectrum of the carrier system, compared with the simplest "gaseous" state, can be understood quite easily on the basis of the following simple analogy with a system of monovalent atoms (H, Li, Na, etc.) in vacuum.

It is known [93] that changes in the energy spectra of such systems and their aggregate state resulting from an increase in the concentration are governed by the parameter

$$r_s = \left(\frac{3}{4\pi}\right)^{1/s} \frac{n^{-1/s}}{a_H}, \tag{25}$$

where n is the number of "particles" in the system considered; a_H is the Bohr radius of the corresponding bound state in a low-concentration system. In the case of a semiconductor, the parameter a_H represents the radius of a bound exciton or impurity state. In gallium arsenide the Bohr radius of an exciton or an electron localized at a donor impurity is equal to about 100 Å. Therefore, at electron densities $n_e \approx 10^{17}$ cm^{-3} we find that $r_s = 1.4$. At concentrations corresponding to this value of the parameter r_s a system of Li, Na, etc. atoms is in the solid state [94].

Hence, it is clear that in the case of gallium arsenide such one-electron concepts as the forbidden band width, excitons, impurity states, etc. lose their precise physical meaning at carrier densities of the order of 10^{16}-10^{17} cm^{-3} and the energy parameters become strongly dependent on the carrier density. Under these conditions, a quantitative interpretation of the experimental results must make an allowance for the basically many-electron nature of the problem.

The available theories [36-39, 95] are incapable of providing a basis for quantitative interpretation of the experimental results on the profile of the fundamental absorption edge and the stimulated emission spectrum of gallium arsenide although quite clearly the principal features of the absorption and emission spectra are governed by the existence of density-of-states tails and by consequent narrowing of the forbidden band. Estimates given in the Introduction show that the radiative transition scheme proposed in the present section is not in conflict with the results reported in [29]: in all the cases when radiative transitions are attributed to the band—band mechanism, the difference between the experimental values of $\hbar\omega_{th}$ (Fig. 17) and the forbidden band width given in [29] lie withing the range of the possible length of density-of-states tails and of the possible narrowing of the forbidden band.

§ 5. Influence of Excitation Inhomogeneity
in Electron-Beam-Pumped Lasers

A characteristic feature of the ionization of atoms in a solid by an electron beam is the inhomogeneous distribution of the volume excitation density with depth, which leads to an inhomogeneous distribution of the gain. This can affect strongly such important parameters as the threshold current density, the distribution of the radiation field in the far and near zones, and the overall efficiency. It is shown in [96, 97] that an inhomogeneous volume excitation of the active region of a laser may also have a strong influence on the dynamics of establishment of coherent oscillations.

In the case of semiconductor lasers, the experimental data are usually analyzed on the assumption that the resonator is an open Fabry–Perot structure and the distribution of the gain inside this resonator is homogeneous. In this approach the diffraction losses do not appear explicitly. Some of the experimental results, including the angular divergence of nondiffraction origin, show clearly that this approach is simply a very rough approximation to the real situation.

A realistic semiconductor laser scheme (Fig. 4) has the following properties: (1) in addition to reflection planes $z = \pm L/2$, there is an additional reflecting plane at $x = 0$; (2) the distribution of the excitation density is strongly inhomogeneous along the x axis.

The macroscopic approach to the effects associated with the excitation inhomogeneity [96] involves the solution of Maxwell's equations (subject to the appropriate boundary conditions) in conjunction with the constitutive equation. The latter equation describes the relationship between the gain (or some other parameter of matter associated with the amplification of the optical field in the resonator) at any given moment and the pumping density as well as the field in the resonator. A rigorous solution of this problem meets with serious difficulties of computational nature and of physical origin, associated with the derivation of the constitutive equation.

The nature of the expected effects can be determined by considering the following model of a laser which makes an allowance for the excitation inhomogeneity and the existence of a reflecting plane $x = 0$ in the optical resonator.

Let us assume that $z = \pm L/2$ planes form two totally reflecting resonator mirrors.* The medium which occupies the resonator consists of a semiconductor in the $x > 0$ half-space and vacuum in the $x < 0$ region. The complex permittivity $\varepsilon(x, t)$ at a frequency ω_0 is of the form:

$$\varepsilon(x, t) = \begin{cases} \varepsilon_0 - i \, [\varepsilon''(x, t) - \varepsilon_0''] \, \varepsilon_0 & \text{for } x \geqslant 0, \\ 1 & \text{for } x < 0. \end{cases} \tag{26}$$

In this definition the real part of the permittivity is assumed to be independent of the pumping level and equal to the permittivity ε_0 of the semiconductor crystal in its unexcited state, whereas the imaginary part has the following two components:

1) ε_0'' describes the attenuation of the field in the resonator as a result of transmission through the mirrors, the absorption by free carriers, and other processes which do not alter the total nonequilibrium carrier density:

$$\varepsilon_0'' = \frac{\lambda}{2\pi} \left(\varkappa + \frac{1}{2L} \ln \frac{1}{R_1 R_2} \right), \tag{27}$$

where R_1 and R_2 are the reflection coefficients of the mirrors, $\varkappa$ is the absorption coefficient

---

*The actual transmission losses of the mirrors are included in the volume losses in the resonator (this will be discussed later).

representing nonresonant processes (absorption by free carriers, scattering by inhomogeneities in the crystal, etc.), and λ is the wavelength in the medium considered;

2) ε'' (x, t) is due to the stimulated transitions and is related to the gain α at the frequency ω_0:

$$\varepsilon''(x, t) = \frac{\lambda}{2\pi} \alpha(x, t). \tag{28}$$

In this case, the permittivity of Eq. (26) is independent of the coordinates y and z and $|\varepsilon''(x, t) - \varepsilon_0''| \ll 1$, so that the field of any given resonator mode* can be found in the form

$$\mathscr{E}(x, z, t) = E(x, t)\exp\{i(kz + \omega_0 t)\}, \tag{29}$$

where the complex amplitude E (x, t) varies slowly compared with exp ($i\omega_0 t$), k is the wave vector, and $\omega_0 = \pm kc/\sqrt{\varepsilon_0}$. The amplitude E (x, t) satisfies the equation [96]

$$\frac{\partial E(x, t)}{\partial t} = \frac{\omega_0}{2}\left\{\frac{i}{k^2}\frac{\partial^2 E(x, t)}{\partial x^2} + [\varepsilon''(x, t) - \varepsilon_0''] E(x, t)\right\}. \tag{30}$$

It follows from Eqs. (28), (4), and (7) that ε'' (x, t) is a function of the nonequilibrium carrier density n (x, t): Eq. (4) defines the dependence of α on the quasi-Fermi levels F_e and F_h and the relationship with the electron and hole densities is given by Eq. (7). Consequently, a complete description of the resonator field must also include an equation for the nonequilibrium carrier density, which can be written in the form

$$\frac{\partial n(x, t)}{\partial t} = g(x) - \frac{n(x, t)}{\tau} - 2\omega_0\varepsilon''(x, t) u(x, t). \tag{31}$$

This is the equation of balance of the number of particles: the first term on the right-hand side describes the rate of generation of electron−hole pairs (9) and the other two terms describe the rates of recombination as a result of spontaneous and nonradiative transitions $|n(x, t/\tau|$ and as a result of stimulated transitions $|2\omega_0\varepsilon''(x, t) u(x, t)|$. The rest of the notation is as follows: τ is the lifetime of electron-hole pairs and u (x, t) is the density of the field energy in the resonator (photons/cm^3).

The dependence of ε'' on n (x, t), given by Eqs. (4), (7), and (28), is generally nonlinear.

However, numerical calculations carried out by Stern [43] show that the nonlinearity of the dependence of the gain α_{max} at the frequency corresponding to its maximum on the nonequilibrium carrier density is quite weak in the 3 cm$^{-1} \leq \alpha \leq$ 300 cm^{-1} range. The experimental data on injection lasers also indicate that in a certain range of excitation densities this dependence can be regarded as linear [51, 78, 80].

For these reasons we shall consider only a linear approximation for the dependence of ε'' on n (x, t):

$$\varepsilon''(x, t) = A[n(x, t) - n_0], \tag{32}$$

where A is a coefficient of proportionality; n_0 is defined by the condition $F_e + F_h = \hbar\omega - \Delta$, i.e.,

---

*Since we are considering the effects of an inhomogeneous distribution of the excitation along the x axis, it is natural to ignore the effects associated with the generation of many longitudinal modes because these can only give rise to an inhomogeneity in the longitudinal direction (along the z axis).

it represents that nonequilibrium carrier density at which the gain vanishes. Moreover, since we are interested not in a quantitative description of any given effect but in its existence, we shall assume that $n_0 = 0$. We thus suppose that $\varepsilon''(x, t) \to 0$ at $x \to \infty$ (i.e., in the unexcited part of the crystal). This is often justified. For example, the Burstein effect in heavily doped n-type crystals makes the absorption coefficient at the stimulated emission wavelength sufficiently small compared with the gain corresponding to the laser threshold (this applies to unexcited crystals).

When the foregoing assumption is made, Eq. (31) can be written in the form

$$\frac{\partial \varepsilon''(x, t)}{\partial t} = g(x) - \frac{\varepsilon''(x, t)}{\tau} - 2\omega_0 \varepsilon''(x, t) u(x, t). \tag{33}$$

The system (30) and (33) was integrated numerically on a computer subject to the following initial and boundary conditions:

$$\varepsilon''(x, 0) = 0, \qquad E(x, 0) = E_0$$

(E_0 is a fairly weak "initiating" field), $E(0, t) = 0$ (total internal reflection), and $E(\infty, t) = 0$.

The rate of generation of electron-hole pairs $g(x)$ was described by Eq. (9) and the ionization curve was approximated by Eq. (11). The following parameters were used in Eqs. (30) and (33): $\tau = 10^{-9}$ sec, $\hbar\omega_0 = 1.5$ eV, $\varepsilon'' = 10^{-4}$ (if $L = 0.5$ mm, this corresponds to an absorption coefficient $\alpha = 50$ cm^{-1}).

We shall now give the principal results of this calculation [98].

1. Laser Threshold. If $\partial^2 E/\partial x^2 = 0$ and $\varepsilon''(x, t)$ are independent of x, the laser threshold is defined by $\varepsilon'' = \varepsilon_0''$. If the excitation is inhomogeneous, it is insufficient to satisfy the condition $\varepsilon''(x_{max}) = \varepsilon_0''$ [x_{max} is the point corresponding to the maximum* of $\varepsilon''(x)$] in order to achieve stimulated emission. The threshold value of the gain $\varepsilon''(x)$ under inhomogeneous excitation conditions is higher than the corresponding value in the case of homogeneous excitation and the difference can be represented by a coefficient Λ which is numerically equal to the ratio $\varepsilon''(x_{max})/\varepsilon_0''$ of the threshold gains under inhomogeneous and homogeneous conditions. The dependence of Λ on the parameters x_0/λ_0 (x_0 is the depth of penetration and λ_0 is the wavelength in vacuum) is plotted in Fig. 20. The increase in the threshold value of ε'' is equivalent to the existence of additional "configurational" losses $\varepsilon_{conf}'' = (\Lambda - 1)\varepsilon_0''$. The physical meaning of these losses is as follows. If the distribution of the gain with depth is inhomogeneous and a reflecting plane is located at $x = 0$, some of the energy flows into the bulk of the crystal, i.e., the field "leaks" from the active to the loss region. The diffraction losses can be regarded as a component of the configurational losses.

2. Field Distributions in Near and Far Zones (Fig. 21). The asymmetry of the field at the end of a semiconductor crystal and the particularly strong asymmetry in the far zone can be seen quite clearly in Fig. 21. The laser beam is deflected away from the normal to the mirror by an angle φ_{defl}, which is approximately equal to half the angular divergence:

$$\varphi_{defl} \approx 0.26 \frac{\lambda_0}{x_0}, \qquad \varphi_{div} \approx 0.27 \frac{\lambda_0}{x_0}. \tag{34}$$

This deflection of the beam into the crystal has the following physical meaning. When only the

---

*Under steady-state conditions in the case of pumping levels slightly higher than the laser threshold, the maximum of $\varepsilon''(x)$ corresponds to the maximum of the ionization curve: $x_{max} = a$ (here, a is the parameter of the ionization curve defined in § 3 in the Introduction).

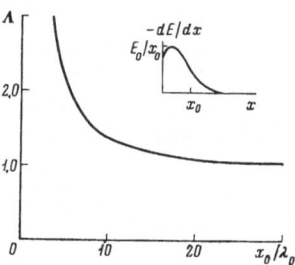

Fig. 20. Dependence of the configurational loss parameter Λ on
x_0/λ_0 calculated for the $(-dE/dx)$
distribution given by Eq. (11) and
shown in the top right-hand corner.

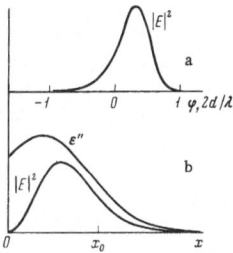

Fig. 21. a) Distribution of the
field $|E(\varphi)|^2$ in the far zone;
b) distribution of $\varepsilon''(x)$ and of
the field $|E(x)|^2$ on the mir-
rors (near zone).

surface layer of a crystal is excited, some of the energy flows into the bulk of the crystal, i.e.,
the field "leaks away" from the active to the loss region. Consequently, the wave vector of the
electromagnetic field acquires an x component which results in deflection of the far-zone field
by an angle $\varphi_{defl} \approx \langle k_x \rangle / k$, where the angular brackets represent averaging over x because the
far-zone field is formed by the whole end face of the laser crystal.

The distribution of the field in the near and far zones does not change significantly when
the laser threshold is exceeded quite considerably (for example, by a factor of up to 30).

3. Emission Intensity Pulsations. When Eqs. (30) and (33) are solved, it is
found that an electron-beam-pumped laser should be subject to pulsations similar to the spikes
observed in lasers made of luminescent crystals and glasses [99]. The pulsation period depends
on the excess of the pumping level over the threshold and, in the case of a laser with the parameters given above, this period lies within the range 0.1-0.3 nsec. These pulsations may be
damped or undamped. Numerical integration of the system (30) and (33) yields pulsations shown
in Fig. 22. The lower oscillatory curve represents the time dependence of the total energy of

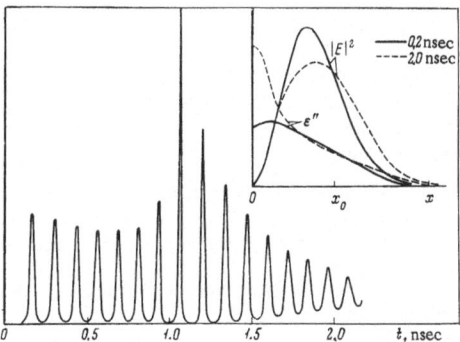

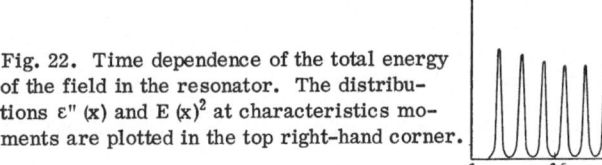

Fig. 22. Time dependence of the total energy
of the field in the resonator. The distributions $\varepsilon''(x)$ and $E(x)^2$ at characteristics moments are plotted in the top right-hand corner.

the field in the resonator, whereas the upper curves are the spatial distributions $\varepsilon''(x)$ and $|E(x)|^2$ at certain characteristic moments. We can see quite clearly that the transition to damped pulsations is associated with the displacement of the field away from the lateral surface and a shift of the gain maximum to the edge of the crystal.

The results of experimental investigations of the effects of excitation inhomogeneity are given in the next three sections.

§ 6. Dependence of Laser Threshold on Electron Energy

The dependence of the threshold current density on the electron energy is governed by the dependence, on the same energy, of the configurational losses and the volume density of the pumping power. According to Eq. (11), this density increases when the electron energy is reduced. Therefore, when we ignore the configurational losses, we find that the threshold current density should decrease with decreasing electron energy. An allowance for the configurational losses alters radically this dependence. If we assume that the imaginary part of the permittivity $\varepsilon''(x)$ is proportional to the volume density of the pumping power, we obtain

$$j_{th}(E_0) = B \frac{ex_0}{E_0} \Lambda(E_0),$$ (35)

where B is a coefficient of proportionality.

When comparisons are made with the experimental data, it is convenient to replace Eq. (35) with the following relationship:

$$\frac{j_{th}(E_0)}{j_{th}(E_0^*)} = \frac{x_0(E_0)}{x_0(E_0^*)} \frac{E_0^*}{E_0} \frac{\Lambda(E_0)}{\Lambda(E_0^*)},$$ (36)

which does not include the coefficient B. In the above equation the quantity E_0^* is some arbitrarily fixed value of the electron energy.

The dependence $j_{th}(E_0)$ given by Eq. (35) is of the following nature [98]. At high electron energies ($E_0 > 77$ keV) the threshold current density j_{th} depends very weakly on the electron energy E_0 but it rises rapidly with decreasing E_0 in the low-energy range ($E_0 < 50$ keV).

The dependence of the threshold current density of electron-beam-pumped semiconductor lasers on the electron energy E_0 was considered theoretically by Bogdankevich et al. [100]. They obtained the following analytic expression for the quantity Λ on the assumption that the distribution of the real part of the permittivity $\varepsilon''(x)$

$$\varepsilon''(x) = \frac{\varepsilon_m'' + \varepsilon_\infty''}{\cosh^2 px} - \varepsilon_\infty''.$$ (37)

In the above expression ε_m'' and ε_∞'' are the absolute values of the imaginary part of the permittivity at the maximum of the distribution given by Eq. (37), i.e., at x = 0 and in the unexcited part of the crystal, i.e., where $x \gg 1/p$; p is a parameter which is related to the depth of penetration of electrons.*

Experimental investigations of the dependence $j_{th}(E_0)$ were carried out in two different units. In the $E_0 \leq 50$ keV range we used the unit described in Chap. I, whereas in the $50\,\text{keV} \leq E_0 \leq 200$ keV range we employed a unit† with the following electron-beam parameters: elec-

---

*For $p = 0.88/x_0$ the distribution (37) is of the same width (at the midamplitude level) as the ionization curve of Eq. (11).
†A more detailed description of this unit can be found in [101].

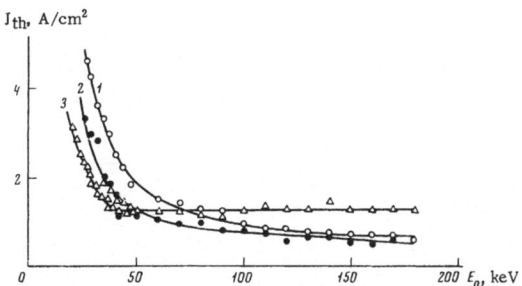

Fig. 23. Dependences of the threshold current density j_{th} at 80°K on the electron energy E_0, recorded for various GaAs samples: 1) No. 60-1; 2) No. 34-5; 3) No. 23-3.

tron energy 50 keV $\leq E_0 \leq$ 200 keV, current pulse duration 50 nsec, repetition frequency 10 Hz. The experiments were carried out on GaAs samples Nos. 60-1, 34-5, 39-11, 22-3, and 23-3 with resonators of lengths L = 0.51, 0.54, 0.55, 0.43, and 0.54 mm, respectively.

The experimental results obtained for undoped crystals GaAs sample No. 60-1 and for samples doped with tellurium (No. 34-5) and zinc (No. 23-3) are plotted in Fig. 23.

The experimental data of Fig. 23 are replotted in Fig. 24 alongside with theoretical results. The ordinate in Fig. 24 represents the ratio j_{th} (E)/j_{th} (200 keV), defined by Eq. (36). Curves 1-4 correspond to Eq. (36) in which the dependence Λ (E_0) is given by Eq. (17) in [100] with $\varkappa$ = 4, 10, 40, and 150 cm^{-1}, respectively. Curve 5 corresponds to the dependence Λ (E_0) plotted in Fig. 20, and curve 6 is deduced from the theory of Vainshtein [102]. The best agreement with the experimental results is obtained for the theory of Bogdankevich et al. [100] with the following values of the coefficient $\varkappa$:

Sample No. . . .	60—1	34—5	39—11	22—3	23—3
$\varkappa$, cm^{-1}	10	10	30	10	150

These values of $\varkappa$ are reasonable [45, 75] for the range of equilibrium carrier densities in the three samples (Table 2).

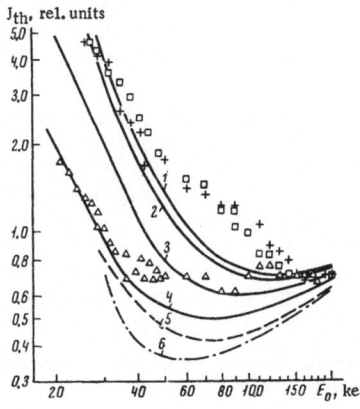

Fig. 24. Comparison of the theoretical and experimental dependences j_{th} (E_0): 1-4) theoretical results obtained in [100] for $\varkappa$ = 4, 10, 40, and 150 cm^{-1}, respectively; 5) calculated curve based on the dependence Λ (E_0) shown in Fig. 20; 6) theoretical curve calculated making allowance for diffraction losses in accordance with [102]. The experimental points are indicated by the same symbols as in Fig. 23.

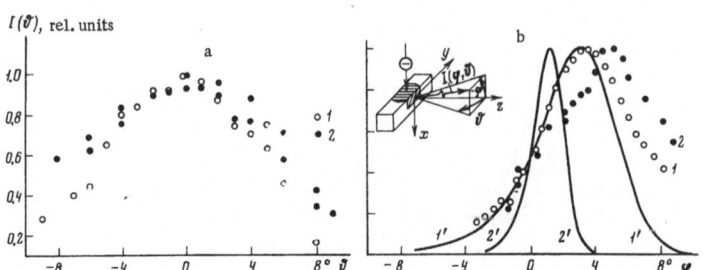

Fig. 25. Angular distributions of the output radiation. a) $I(\vartheta)$ for $\varphi =$ const: 1) $E_0 = 30$ keV ($\varphi = 3°$); 2) $E_0 = 48$ keV ($\varphi = 6°$). b) $I(\varphi)$ for $\vartheta = 0$: 1), 1') $E_0 = 30$ keV; 2), 2') $E_0 = 48$ keV; 1'), 2') theoretical curves corresponding to Fig. 21a.

§7.   Angular Distribution

An image converter study of the distribution of the radiation intensity in the far zone indicated that:

1) the divergence of the laser radiation is different in the horizontal and vertical planes[*] and is considerably larger than the diffraction limit;

2) the divergence in the horizontal direction is greater than in the vertical plane and it increases with increasing size of the active region (with increasing dimension ad in Fig. 4);

3) the distribution in the vertical plane is asymmetric relative to the laser axis (z axis in Fig. 4); in this plane the divergence is independent of the size of the active region and of the excess over the laser threshold (within a factor of 1.5-5);

4) the distribution in the far zone is practically independent of the type of material and is the same for gallium arsenide and for GaP_xAs_{1-x} solid solutions.

A quantitative description of the angular distribution for the employed laser geometry can be obtained in terms of two angles φ and ϑ, which are defined by

$$\tan \varphi = \frac{x}{\sqrt{y^2 + z^2}}, \quad \tan \vartheta = \frac{y}{z}. \tag{38}$$

Here, x, y, and z are the coordinates of the radius vector $\mathbf{r}$ (the selected coordinate axes are shown in Fig. 4). Thus, φ is the angle between the radius vector $\mathbf{r}$ and the (y, z) plane which is subjected to electron bombardment; ϑ is the angle between the projection of $\mathbf{r}$ onto the (y, z) plane and the z axis.

Quantitative measurements of the angular distribution were carried out for lasers made of GaP_xAs_{1-x} solid solutions because the emission spectrum of gallium arsenophosphides with x > 0.2 was located in the visible range. In these measurements an FÉU-28 photomultiplier was used to determine the radiation intensity at a distance of 1-2 m from the sample for various values of the angles φ and ϑ. The results obtained are plotted in Fig. 25. These results were obtained for a sample of GaP_xAs_{1-x} with a phosphorus concentration x ≈ 0.22 (this sample emitted stimulated radiation of wavelength $\lambda = 7070$ Å at liquid nitrogen temperature). The resonator length was L = 0.48 mm. The size of the active region was 0.3 mm in the case of bombardment with 48 keV electrons and 0.24 mm in the case of 30 keV electrons.

────────────

[*]These directions were selected in accordance with Fig. 4, in which the electron beam is directed vertically and the surface bombarded with electrons is horizontal.

The most interesting feature of the angular distribution was the observation that the direction corresponding to maximum emission intensity did not coincide with the resonator axis but was inclined at an angle φ_{defl} from this axis. This angle depended on the electron energy and, in the adopted coordinate system, was always positive. This was in full agreement with the theory put forward earlier. However, the experimental results indicated that the divergence of the radiation (for example, that measured at the midamplitude level) along the angle φ and the deflection φ_{defl} both increased when the electron energy was raised. The theory predicted the opposite effect. Moreover, the quantitative values of φ_{defl} and of the divergence along the angle φ also did not agree with the theory (the more or less satisfactory agreement at an electron energy of 30 keV was accidental).

The distribution along the angle ϑ was practically independent of the electron energy and, for the indicated dimensions D of the active region, the width of the distribution was approximately 12-16°. This should be compared with the diffraction limit of the divergence for the principal modes in the optical resonator, which was only several angular minutes.

The experimental results disagree with the theory for the following reasons. The theory is concerned with the single-mode case when the mode with the lowest transverse indices is excited. In this case, the resonator field is independent of the coordinate y and has a very simple distribution along the x axis (it does not vanish anywhere except at $x \to \infty$). On the other hand, it is evident that in semiconductor lasers the difference between the losses suffered by modes with different transverse indices can be small because of the strong loss background resulting from inadequate reflection coefficients of the mirrors, losses occurring in the bulk of the sample, etc. Therefore, higher modes may be excited in a semiconductor laser and these modes may increase the angular divergence of the output radiation.

Another possible reason for the strong divergence of the radiation emitted by semiconductor lasers is insufficient spatial coherence of this radiation. In other words, some parts of the resonator can behave as independent lasers. This is supported by the observation that a fairly sharp interference pattern is only rarely observed for semiconductor lasers.

Thus, the final determination of the cause of the strong angular divergence of semiconductor lasers must await additional experimental studies of the oscillation modes excited in these lasers. Moreover, a theoretical allowance would have to made for the possibility of excitation of higher modes in the optical resonator.

§ 8. Spiking Regime

Under pulsed conditions any laser with a finite time for the establishment of oscillations in the resonator T_1 ($T_1 \approx 1/\Delta f$, Δf is the mode width) and a finite lifetime of active particles in the excited state T_2 exhibits transient damped pulsations of the emission intensity, which precede the attainment of quasisteady-state conditions. The duration of this transient process and the pulsation period are functions of the excess of the excitation level over the threshold and of the absolute values of T_1 and T_2. Moreover, the quasisteady-state operation may become unstable and then the system goes over to the spiking regime [103]. This regime is characterized by regular or irregular pulsations of the energy of the field in the resonator and of the output power. Spiking in lasers made of luminescent crystals and glasses has been thoroughly investigated from the theoretical and experimental points of view. In particular, it has been shown that the main causes of the instability of the quasisteady-state operation are the inhomogeneity of the spatial distribution of the excitation and the interaction (via the active substance) of quasidegenerate resonator modes, for example, the modes with identical longitudinal but different transverse indices. In the first case, the appearance of spikes is due to a redistribution of the optical field energy in the active substance and, in the second, to a redistribution of the energy between oscillation modes [96, 97, 104].[*]

---

[*]Spiking is observed also in injection lasers [105, 106].

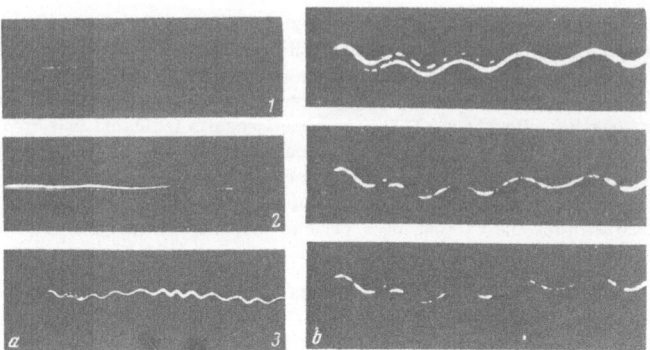

Fig. 26. Time dependences of the intensity of the radiation emitted
by a laser (photographs of the screen of a scanning image converter).
a) Frame 1 corresponds to $j = 4j_{th}$, frame 2 corresponds to $j = 10j_{th}$,
frame 3 corresponds to $j = 6j_{th}$ (the sinusoid gives the time scale: its
period is 1.6 nsec); b) photographs corresponding to frame 3 in Fig.
26a obtained using different exposures.

When a semiconductor laser is pumped by bombardment with electrons of energies of
several tens of kiloelectron volts, the distribution of the excitation is strongly inhomogeneous
and, therefore, we can expect to observe the spiking regime in such lasers. Solution of the system consisting of Eqs. (30) and (33) shows that pulsations of the output intensity should occur in
electron-beam-pumped lasers (Fig. 22).

Pulsations were not observed experimentally for a long time because of the lack of suitable apparatus with a sufficient time resolution. Such observations have been made recently
because of the availability of a suitable scanning image converter (streak camera), designated
by FÉR-2. This converter was developed and constructed at the All-Union Scientific Research
Institute of Optical Measurements.

The FÉR-2 image converter is intended for recording single short-duration events accompanied by the emission of radiation in the wavelength range from 4000 to 12,000 Å. The action of this converter can be described as fast-scanning image converter photography. The
image of a light signal is formed by an optical system with a slit. The image of the slit is projected onto the photocathode of the image converter where a scanner is employed to shift this
image at right-angles with respect to the axis of the slit. In this way, we can obtain a space
and time scan of the development of the process under investigation and can determine changes
in the intensity of the associated signal.

Our experiments were carried out on specimens prepared from GaAs sample No. 19 with a
resonator L = 0.528 mm long. The active region of the crystal (ad in Fig. 4) was projected by
an optical system (with a magnification close to unity) onto the photocathode of the scanning
image converter. The image was scanned and the luminescent screen was photographed in such
a way that each frame of the film corresponded to one pumping pulse.

Photographs of the time-scanned image of the emission region are shown in Fig. 26.
Here, the time scan is directed horizontally and the degree of blackening of the film is proportional to the instantaneous intensity of the light signal. Frame 1 in Fig. 26a corresponds to the
case when only one region in a crystal emits radiation. Frames 2 and 3 correspond to the case
when two regions separated in space are emitting. The scanning scale is shown in frame 3 of

Fig. 26a. The period of the sinusoid in this frame is 1.6 nsec ± 10%. In Fig. 26b the photograph corresponding to frame 3 in Fig. 26a is shown on an enlarged scale.

We have already mentioned that the time characteristics (pulsation period, regularity, etc.) depend strongly on the excess of the pumping level over the laser threshold. In view of this it should be mentioned that since the current pulse is bell-shaped, the excess over the threshold varies with the position in the pulse. At the leading and trailing edges of the pulse, the excess over the laser threshold is only slight.

The following conclusions can be drawn from the experiments described in the preceding paragraphs:

1) coherent emission of an electron-beam-pumped semiconductor laser exhibits pulsations of the intensity;
2) these pulsations can be regular or irregular, depending on the excess of the pumping level over the laser threshold;
3) in the case of regular pulsations, the period is 0.25-0.40 nsec, which is in satisfactory agreement with the theoretical values.

§ 9. Semiconductor Laser with Planar Resonator

In all the cases considered so far, the coherent radiation was generated in a narrow layer whose plane was perpendicular to the mirrors of the Fabry—Perot resonator. This is not the only possible geometry. Since high values of the gain ($\sim 10^3\,\mathrm{cm}^{-1}$) can be achieved quite easily in semicondcutors, it is possible to achieve laser action in a system which is known as the planar resonator [107]. In this case, the Fabry—Perot resonator is a plane-parallel plate made of a semiconductor crystal whose thickness is of the same order as the depth of penetration of electrons. The electron beam is directed at 90° or at some other angle with respect to the Fabry Perot mirror. This geometry is a simplified version of the arrangement suggested in [25] and known as the radiating mirror. The planar resonator has several advantages over the conventional laser geometry. In particular, the effects of excitation inhomogeneity can be eliminated practically completely.

However, the planar resonator or the radiating mirror system has certain disadvantages which present serious practical difficulties. These difficulties include, for example, a reduction in the carrier lifetime and a corresponding rise in the laser threshold (§ 2, Chap. III) because of the amplification of spontaneous radiation or stray emission of stimulated radiation along the semiconducting plate (if the self-excitation condition is satisfied because of reflection from lateral faces). These effects can be suppressed by introducing additional losses, which are anisotropic in respect of direction, into the active region. Such losses can be introduced

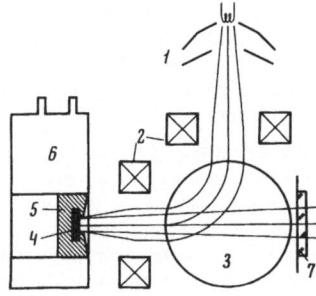

Fig. 27. Schematic representation of the apparatus used in the bombardment of a planar resonator: 1) electron source (Pierce gun); 2) magnetic focusing lenses; 3) magnetic deflection system; 4) semiconductor crystal; 5) heat sink; 6) cryostat; 7) optical window.

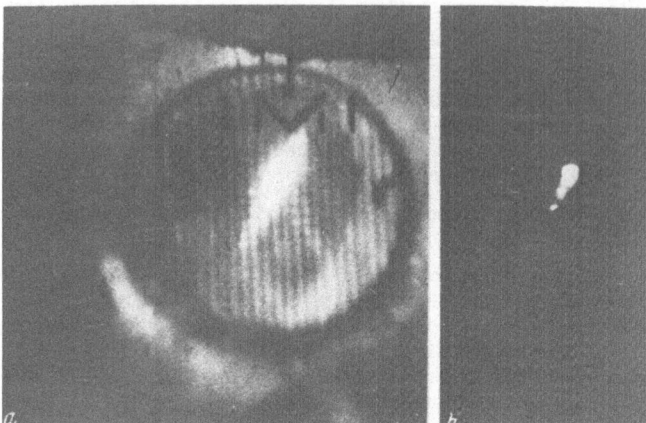

Fig. 28. Nature of the emission from a sample with a planar resonator at current densities below (a) and above (b) the laser threshold (photographs of the screen of an image converter).

by inserting a metal grid in the path of the electron beam in front of the semiconductor plate.
Then alternate active and passive regions will form in the plate. The passive regions will absorb the radiation traveling along the semiconductor plate, and at right angles to the plate the
laser action will occur in individual cells, formed by the metal grid, which can be regarded as
independent oscillators. If one of the mirrors of the Fabry—Perot resonator is external, one
can attain diffraction synchronization between such cells [108, 109].

The success in the suppression of amplification of spontaneous radiation along the active
layer, i.e., along the semiconductor plate, by means of a metal grid depends on the absorption in the unexcited part of the semiconductor crystal of the radiation generated in the excited part. The grid gives
the best results for p-type samples with a sufficiently high absorption coefficient (of the order of 100
cm^{-1}) and it is least effective in the case of n-type samples for which this coefficient can be
very small (~ 10 cm^{-1}). We used an undoped crystal (GaAs sample No. 10 with the parameters
listed in Table 2).

The parameters of the electron-beam pumping were as follows: the electron energy
ranged from 50 to 150 keV, the pulse duration was about 0.15 μsec, and the repetition frequency
was 10 Hz. The electron-beam pumping unit used in this study is shown schematically in Fig.
27. The electron beam was focused by magnetic lenses and rotated by 90° in a homogeneous
magnetic field. It reached a plane-parallel gallium arsenide plate at right angles to the face
forming a Fabry—Perot resonator mirror. The laser action along the active region was suppressed by means of a grid made of metal wires, 0.15 mm thick. The size of each cell was 0.3
mm and the thickness of the plane-parallel gallium arsenide plate was 0.1 mm. The emitted
radiation was recorded at right angles to the resonator mirror.

The distribution of the radiation emitted at current densities below (a) and above (b) the
laser threshold are shown in Fig. 28 (the distributions shown in that figure were photographed
directly from the screen of an image converter). Below the threshold the time average of the
intensity of the emitted radiation was comparable with the background illumination level so that
some details of the construction were visible (including the grid in front of the sample). When
the laser threshold was reached the intensity rose strongly and became much higher than the

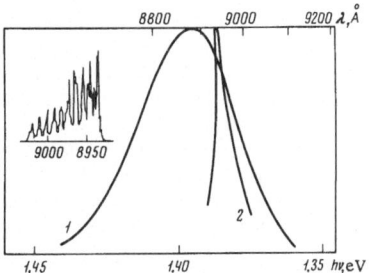

Fig. 29. Spectra of the spontaneous
(curve 1, j = 3 A/cm²) and stimulated (curve
2, j = 7 A/cm²) emission from a laser
with a planar resonator. Electron energy 150 keV, T = 300°K.

background level. Initially, the laser action was attained only in one cell but when the current
density was raised above the threshold the stimulated emission spread to several cells.

The spontaneous radiation spectrum and the position of the stimulated emission line, obtained at room temperature, are plotted in Fig. 29. The detailed structure of the stimulated
line is shown in the top-left-hand corner. The separate fine-structure maxima represent
the resonator modes. The results obtained show that the attainment of stimulated emission in a
system with a planar resonator is a practical proposition.

CHAPTER III

POSSIBILITIES OF INCREASING OUTPUT POWER OF ELECTRON-
BEAM-PUMPED SEMICONDUCTOR LASERS

§1.   Introductory Remarks

The output power of a laser is governed by its efficiency and by the pumping level. The
experimental values of the overall efficiency (§1, Chap. II) are close to the theoretical limit.
Therefore, the output power of electron-beam-pumped semiconductor lasers can only be increased by raising the total pumping power P_p, which can be done by increasing the electron
energy E_0, the current density j, and the geometrical dimensions of the resonator (its length L
and width D). The permissible values of these parameters are limited for the following reasons.

1. The electron energy cannot be raised indefinitely because at high values of E_0 there
is a danger that radiation defects will form in the active region of the laser. Therefore, the
highest electron energy that can be used to pump a semiconductor laser is represented by the
threshold of formation of radiation defects (in the case of gallium arsenide this value is about
250 keV [110]).*

2. When the current density is increased, the volume density of the pumping power rises
and, consequently, the density of the output radiation increases. At high laser radiation densities some irreversible changes may take place in the semiconductor crystal and these may
reduce strongly the output power and raise the laser threshold. The processes leading to these
irreversible changes are not yet fully understood but it has been established experimentally
that these processes result in damage to the crystal, which is concentrated mainly on the faces

---

*According to [111], the influence of high-energy electron bombardment of the parameters of a
laser is manifested by a reduction in its service life. Consequently, in those cases when the
service life is of little importance, it is permissible to use electrons of energies exceeding
the threshold of formation of radiation defects.

forming the resonator mirrors. This damage is of the threshold nature: a strong deterioration of the laser parameters and mechanical damage of the resonator faces are observed if the density of the laser radiation flux exceeds a certain critical value, which is usually called the damage threshold. For most of the investigated semiconductor crystals the damage threshold for output radiation pulses of $\sim 10^{-7}$ sec duration amounts to several megawatts per square centimeter [112, 113]. Thus, the permissible electron current density j_{max} is set by the condition that the density of the energy flux represented by the laser radiation passing through the resonator faces must not exceed the damage threshold. The current density limited by this condition is also a function of the resonator length L.

3. The maximum width D of a laser is limited by the effects associated with the amplification of spontaneous radiation. The dependence of the parameters of a laser on its transverse dimensions associated with these effects is considered in the next section.

The limitations discussed above set the upper limit to the output radiation power that can be emitted under pulsed conditions. In estimating the average value of this maximum power, we must also bear in mind the heating of the active region by electron bombardment. Such heating can be the principal factor which limits the highest permissible value of the electron-beam current density. This value can be estimated quite easily utilizing the data on the temperature dependence of the laser threshold (Chap. II) and calculating the temperature rise, as described in Appendix I. Heating of the active region does not restrict the output power under pulsed conditions but simply imposes a restriction on the maximum pulse duration.

§ 2. Dependences of Laser Parameters on Resonator Width

We shall consider the dependence of the laser parameters on the transverse dimensions of a resonator by analyzing the following specific case. We shall assume that the width D of the active region of a laser is increased but the density of the electron beam current is kept constant. Then, the total pumping power increases proportionally to the volume of the excited region. The question is, how does the total output power vary with the laser width D?

In this problem the basic point is this: when the active region of the laser increases in size, the losses in the form of spontaneous radiation do not rise proportionally to the volume of the active region but much faster because, in addition to the losses resulting from the generation of purely spontaneous (independent of the external field) radiation, there are also losses as a result of amplification of such radiation. This component of the losses increases rapidly with the size of the active region because it is proportional to the intensity of the spontaneous field which rises approximately exponentially in the active medium. Therefore, if the active region is sufficiently large, it may happen that the increase in the pumping power because of an increase in the active volume will be less than the corresponding increase in the losses. In this case, the total output power of the laser will decrease with increasing volume of the active region.

We shall now determine the actual form of the dependence of the total output power and of several other laser parameters on the transverse dimensions of the active region.

We shall approach this problem in the following way. We shall assume that the spontaneous radiation power p_{sp}^0 per unit volume of the active region and the gain α under coherent emission conditions are independent of the pumping level and of the coordinates. We shall calculate the losses P_{sp} resulting from the generation and amplification of spontaneous radiation. Next, we shall determine the output power from the law of conservation of energy for the whole coherent emission region. The internal losses experienced by the resonator modes will be allowed for by means of the loss function defined in § 1 of Chap. II. Other losses (for example, the configurational losses) will be allowed for indirectly (in those cases when they are important) by computing their contribution to the various coefficients in the initial equations.

We shall first consider the nature of the approximations which are implied in the assumption that p_{sp}^0 and α are independent of the pumping level and of the coordinates.

When the resonator width D is sufficiently small and the amplification of the spontaneous radiation can be ignored, the dependence of the output power of the pumping level is described by Eq. (15), which is basically the equation of energy balance in the laser. This energy balance implies (see § 1 in Chap. II) that the spontaneous radiation power is independent of the pumping level and equal to the value attained at the laser threshold: $P_{sp}^0 = (1/3)P_{th}^0$. Hence, it follows that above the threshold the total density of the nonequilibrium carriers and, therefore, the gain per pass through the resonator are independent of the pumping level. The second of these assumptions is a necessary condition for the generation of steady-state oscillations in the resonator. When the width D is increased, the gain per pass should remain constant* provided the laser action is maintained throughout the active region (this is the case of interest to us). The gain should still be independent of the pumping level because this independence is a condition for the establishment of steady-state oscillations in the resonator: if the gain were to change, these oscillations would grow (or decay) continuously with time. Hence, we may conclude that the power of the purely spontaneous radiation (per unit volume) should be independent of D and of the pumping level (the losses due to the amplification of spontaneous radiation will still increase with increasing D).

In general, the values of p_{sp}^0 and α depend on the coordinates. This dependence is due to the inhomogeneity of the pumping and of the distribution of the field of the resonator modes. The most important factor is the inhomogeneity of the depth of penetration of the electron beam. The inhomogeneity along other directions is usually ignored (see, for example, [100]). The inhomogeneity of the distributions of p_{sp}^0 and α cannot be allowed for in the approach adopted here because this can be done only by simultaneous solutions of the electromagnetic field and the constitutive equations. Therefore, we are forced to assume that p_{sp}^0 and α are constants, which is equivalent (to some extent) to averaging over the active region. The influence of the excitation inhomogeneity on the laser parameters can always be allowed for indirectly.

The losses due to the generation and amplification of spontaneous radiation can be calculated as follows.

We shall assume that an isotropic point source of power p_{sp}^0 is located at a point $\mathbf{r}'$. In a medium characterized by a gain α the energy flux (Poynting vector) of the field of this source is described by the expression

$$S_{sp}(\mathbf{r}) = \frac{p_{sp}^0}{4\pi} \frac{\mathbf{r} - \mathbf{r}'}{|\mathbf{r} - \mathbf{r}'|^3} \exp\{\alpha |\mathbf{r} - \mathbf{r}'|\}. \tag{39}$$

The energy flux defined by this expression is the solution of the equation

$$\operatorname{div} S_{sp} = \alpha |S_{sp}| + p_{sp}^0 \delta(\mathbf{r} - \mathbf{r}'), \tag{40}$$

which follows from the law of conservation of energy and differs from the steady-state Poynting theorem by the presence of a second term on the right-hand side (this term describes the point source of spontaneous radiation).

The field of Eq. (39) gives rise to stimulated transitions in the active medium and these transitions are responsible for the amplification of spontaneous radiation. The spontaneous

---

*Some reduction can occur because the diffraction losses increase. However, since the width D is finite, the contribution of this width to the diffraction losses is usually negligible compared with other types of loss.

radiation power $p_{sp}(\mathbf{r})$ generated per unit volume is given by

$$p_{sp}(\mathbf{r}) \equiv \operatorname{div} S_{sp}(\mathbf{r}) = p_{sp}^0 \delta(\mathbf{r} - \mathbf{r}') + \alpha \,|\, S_{sp}(\mathbf{r})\,|. \tag{41}$$

In the excited region of a semiconductor laser, the sources of spontaneous radiation are distributed continuously with a certain volume density p_{sp}^0. Apart from the amplification due to population inversion of the active levels, there is some absorption due to intraband transitions and other nonresonant processes. Moreover, in calculating the distribution of the spontaneous radiation flux, we must bear in mind the reflection from those faces which form the resonator mirrors. However, considerable mathematical difficulties are encountered in making such an allowance. It is clear that the greatest contribution to the laser losses should be made by that fraction of the spontaneous radiation which travels in the direction of the longest dimension of the active region. Therefore, we shall carry out calculations for two model cases, in which the distribution of the spontaneous radiation field at each point in the semiconductor is assumed to be simpler than that given by Eq. (39) so that the final results can be written in a relatively simple analytic form.

The first of these models is one-dimensional: the spontaneous radiation flux is assumed to be one-dimensional.

We shall postulate that the resonator geometry is of the type shown in Fig. 4 and that the active region is defined by the inequalities $|z| \leqslant L/2$, $|y| \leqslant D/2$, and $0 \leqslant x \leqslant x_0$. We shall assume that the layer $y' \ll y \ll y' + dy'$ in the active region emits a plane wave which travels along the y axis and which increases in amplitude because of amplification. Consequently, the energy flux along a cross section of the active region located at a point y can be written in the form

$$dS_{sp}(y, y') = \begin{cases} \dfrac{1}{2} p_{sp}^0 dy' \exp\left(\dfrac{y - y'}{y_0}\right) & \text{for} \quad y > y', \\[2mm] -\dfrac{1}{2} p_{sp}^0 dy' \exp\left(\dfrac{y' - y}{y_0}\right) & \text{for} \quad y \leqslant y', \end{cases} \tag{42}$$

where $p_{sp}^0 dy'$ is the spontaneous radiation power generated in the region $dV' = Lx_0 dy'$, whereas y_0 is a parameter which represents the growth of the field in space. This parameter is governed by the gain and the losses in the active region. The energy flux is assumed to be positive if it is directed along the y axis and negative if it is oppositely directed.

Integrating Eq. (42) throughout the active region, we obtain

$$S_{sp}^+(y) = \frac{1}{2} y_0 p_{sp}^0 \left\{ \exp\left(\frac{D}{2y_0} + \frac{y}{y_0}\right) - 1 \right\},$$
$$S_{sp}^-(y) = -\frac{1}{2} y_0 p_{sp}^0 \left\{ \exp\left(\frac{D}{2y_0} - \frac{y}{y_0}\right) - 1 \right\}, \tag{43}$$

where $S_{sp}(y)$ is the energy flux due to the spontaneous emission throughout the active region (the sign of the right-hand side represents the direction of this flux).

The losses due to spontaneous emission at some point consist of two terms representing the spontaneous emission at that point and the emission induced by the spontaneous radiation field throughout the rest of the active region. Thus, it follows from Eq. (43) that

$$\frac{dP_{sp}}{dy} = p_{sp}^0 + \alpha\,|\,S_{sp}^+\,| + \alpha\,|\,S_{sp}^-\,| = p_{sp}^0 + \alpha y_0 p_{sp}^0 \left\{ \exp\left(\frac{D}{2y_0}\right) \cosh\left(\frac{y}{y_0}\right) - 1 \right\}. \tag{44}$$

$$P_{sp}(D) = p_{sp}^0 D + \alpha y_0 p_{sp}^0 D \left\{ \frac{\exp\left(\dfrac{D}{y_0}\right) - 1}{\dfrac{D}{y_0}} - 1 \right\}. \tag{45}$$

where α is the gain ($\alpha \geq 1/y_0$), and P_{sp} (y) is the spontaneous radiation power within the volume defined by $Lx_0 (y-D/2)$ in the active region.*

The total output power P of a laser and its distribution along the y axis can be determined from the law of conservation of energy throughout the active region and in a layer of width dy, respectively:

$$\frac{1}{f (\varkappa L)} P + P_{sp} = \frac{1}{3} P_p \tag{46}$$

$$\frac{1}{f (\varkappa L)} \frac{dP}{dy} + \frac{dP_{sp}}{dy} = \frac{1}{3} \frac{dP_p}{dy} . \tag{47}$$

In these equations the nonresonant losses, as a result of which some of the power represented by the resonator modes is converted into heat, are allowed for by introducing the loss function $f (\varkappa L)$ defined in §1 of Chap. II. The quantum efficiency of the radiative recombination process is assumed to be unity (this is done for the sake of simplicity).

Utilizing the relationship[†] between p_{sp}^0 and the threshold pumping power P_{th}^0 and expressing the values of P_p and P_{th}^0 in terms of the electron beam parameters and the geometrical dimensions of the resonator, we obtain the following expressions from Eqs. (44)-(47):

$$P = \frac{1}{3} f (\varkappa L) \frac{E_0}{e} LD \{j - j_{th} (D)\}, \tag{48}$$

$$\frac{dP}{dy} = \frac{1}{3} f (\varkappa L) \frac{E_0}{e} L \left\{ j - j_{th}^0 - dy_0 j_{th}^0 \left[\exp \left(\frac{D}{2y_0} \right) \cosh \left(\frac{y}{y_0} \right) - 1 \right] \right\}, \tag{49}$$

where j_{th}^0 is the threshold current density in the limit $D \ll y_0$ and j_{th} (D) is defined by

$$j_{th} (D) = j_{th}^0 + \alpha y_0 j_{th}^0 \left\{ \frac{\exp \left(\frac{D}{y_0} \right) - 1}{D/y_0} - 1 \right\} . \tag{50}$$

Formulas (48) and (49) are valid at fairly high values of the current density j, when the whole of the excited region ($|y| \leq D/2$) is above the laser threshold (dP/dy > 0). In this case, the output power P is a linear function of the current density and, by definition, the value of j_{th} (D) is represented by the point of intersection of this function with the abscissa (j). Therefore, in a sense, j_{th} (D) can be regarded as the threshold current density. If the laser threshold j_{th}^* (y) is obtained in a similar manner for some part of an excited region, it is found from Eq. (49) that the value of this threshold is a function of the coordinate y and that it has a minimum at the center of the resonator, where y = 0.

Hence, it follows that when the current density is increased gradually, the laser threshold will be satisfied first at the center of the resonator and then the coherent emission region will gradually increase in size until it spreads over the whole of the excited region when $\frac{dP}{dy}\Big|_{y=D} = 0$. Below the value of the current density corresponding to this condition, the dependence of the output power of j is nonlinear, whereas at higher values of j it becomes linear and is described by Eq. (48). The nonlinear part of the dependence P (j) cannot be derived from the preceding

---

*The power emitted from this volume is less than that given by Eq. (44) and the difference is equal to the nonresonant losses within that volume.

†This relationship can also be derived from Eqs. (45) and (46) by going to the limit $D \ll y_0$ and comparing the results with Eq. (15).

equation because, in the case when only part of the excited region emits coherent radiation, an allowance must be made for the dependence of the gain on the coordinate y. For this reason, the above formulas cannot be used to calculate the absolute value of the current density at which the laser threshold is attained in some local part of the excited region.

The dependence P(D) described by Eqs. (48) and (50) is of the following nature: when D is increased, the output power P increases at first linearly (for $D \ll y_0$), then the rise slows down, the output power reaches its maximum at some value of the width D_{max} and then it begins to fall. The value of D_{max} is a function of the excess of the pumping level over the laser threshold and it is given by the formula

$$D_{max} = y_0 \ln \left\{ 1 + \frac{j - j_{th}^0}{\alpha y_0 j_{th}^0} \right\} . \tag{51}$$

It is worth pointing out that the distribution of the output power along the y axis is inhomogeneous. The energy flux, which is proportional to dP/dy, has its maximum value at the center of the resonator ($y = 0$).

Thus, formulas (48) and (50), which give the output power in terms of the constants j_{th}^0, α, and y_0, can be used to calculate the dependence of this power on the width of the coherent emission region.

The value of j_{th}^0 can be found either from experimental data (§6 in Chap. II) or from theoretical results obtained in [100]. The gain derived in [100] can be written in the form

$$\alpha = \varkappa + \frac{1}{L} \ln \frac{1}{R} + \frac{9\lambda}{4\pi x_0^2} \left\{ 1 + \sqrt{ 1 + 2 \left(\varkappa + \frac{1}{L} \ln \frac{1}{R} \right) \frac{4\pi x_0^2}{9\lambda} } \right\} , \tag{52}$$

where the last term on the right-hand side includes an allowance for the configurational losses. In the above formula R is the reflection coefficient of the mirrors; $\varkappa$ is the absorption coefficient associated with nonresonant losses in the active region (absorption by free carriers, scattering by inhomogeneities, etc.), and λ is the wavelength in the active medium. The parameter y_0 is governed by the difference between the gain and the losses at the spontaneous emission wavelength. Since the energy flux of the spontaneous radiation is assumed to be one-dimensional, the increase of this flux in space should be the same as that of the stimulated radiation field. Hence, we find that

$$\frac{1}{y_0} = \frac{1}{L} \ln \frac{1}{R} . \tag{53}$$

We shall now consider the case when the spontaneous radiation field can be represented by an assembly of cylindrical waves emitted from each point in the excited region and the coherent emission region is limited in the transverse direction by a cylindrical surface.

We shall introduce cylindrical coordinates (r, φ, z) in such a way that the z axis coincides with the resonator axis and the coherent emission region is defined by $0 \le z \le L$ and $r \le D/2$. The resonator mirrors are located in the planes $z = 0$ and $z = L$. The electron beam is directed along the z axis and the depth of penetration of the electrons is $x_0 = L$. This geometry corresponds to a laser with a planar resonator.

We find that, in the case of a laser with a planar resonator, there is no need to allow for the diffraction losses and the nonresonant absorption because these effects are negligible compared with the losses as a result of transmission of the resonator mirrors. Therefore, instead of Eqs. (52) and (53), we now have

$$\alpha = \frac{1}{y_0} = \frac{1}{L} \ln \frac{1}{R} . \tag{54}$$

The energy flux due to spontaneous emission in an element of volume $dV = Lr'dr'd\varphi'$ [at the point r with the coordinates (r', φ')] can be written in the form

$$dS_{sp}(\mathbf{r}, \mathbf{r}') = \frac{p_{sp}^0 \, dV' \exp\{\alpha \,|\, \mathbf{r} - \mathbf{r}'\,|\}}{2\pi L\,|\,\mathbf{r} - \mathbf{r}'\,|^2}(\mathbf{r} - \mathbf{r}'), \tag{55}$$

where p_{sp}^0 is the volume density of the spontaneous radiation power. Consequently, the volume density of the power at the point (r, φ) resulting from the spontaneous radiation emitted throughout the active region is given by

$$P_{sp}(r, \varphi) = p_{sp}^0 + \frac{\alpha p_{sp}^0}{2\pi} \int_0^{2\pi} d\varphi' \int_0^{D/2} \frac{\exp\{\alpha \sqrt{r^2 + r'^2 - 2rr'\cos(\varphi - \varphi')}\}}{\sqrt{r^2 + r'^2 - 2rr'\cos(\varphi - \varphi')}} r'dr'. \tag{56}$$

At the center of the active region $(r = 0)$ the density p_{sp} has its maximum given by

$$p_{sp}(0) = p_{sp}^0 \exp\left(\frac{\alpha D}{2}\right).$$

Therefore, the density of the radiation flux represented by the resonator modes has its highest value at the same point

$$|S(0)| = \left\{\frac{1}{3} P_p - p_{sp}^0 \exp\left(\frac{\alpha D}{2}\right)\right\} L, \tag{57}$$

where $p_p L = (E_0/e)j$ is the pumping energy flux per unit area.

We shall now calculate the total losses in a laser due to spontaneous radiation (including amplification of this radiation) throughout the active region. This will be done by calculating the radiative energy flux (55) generated in the volume dV' and emitted through the lateral surface $r = D/2$. The results of this calculation must be integrated over the whole active region. This procedure gives

$$P_{sp}(\alpha D) = p_{sp}^0 \frac{LD^2}{2} \int_0^1 x \, dx \int_0^\pi \frac{\exp\left\{\frac{\alpha D}{2} \sqrt{1 + x^2 + 2x\cos\psi}\right\}}{1 + x^2 + 2x\cos\psi}(1 + x\cos\psi)\,d\psi, \tag{58}$$

where $x = 2r'/D$ and $\psi = \pi - (\varphi - \varphi')$.

The integral in Eq. (58) can be expressed in terms of Bessel and Struve functions. If the variable x is replaced with ϑ, in accordance with

$$\cot\vartheta = \frac{x + \cos\psi}{\sin\psi},$$

Eq. (58) can be written in the form

$$P_{sp}(\alpha D) = \frac{1}{4} p_{sp}^0 L D^2 \int_0^\pi \int_0^\vartheta \frac{\sin 2(\psi - \vartheta)}{\sin^2 \vartheta} \exp\left\{\frac{\alpha D}{2} \frac{\sin\psi}{\sin\vartheta}\right\} d\psi \, d\vartheta. \tag{59}$$

Next, expanding the exponential function as a power series and integrating term by term, we obtain

$$P_{sp}(\alpha D) = \frac{1}{2} p_{sp}^0 \pi D^2 L \frac{I_1(\alpha D) + L_1(\alpha D)}{\alpha D}, \tag{60}$$

where the Bessel function with an imaginary argument $I_1(x)$ and the first-order Struve function $L_1(x)$ are defined as follows [114]:

$$I_1(x) = \sum_{m=0}^{\infty} \frac{\left(\frac{x}{2}\right)^{2m+1}}{m!\,(m+1)!}, \qquad L_1(x) = \frac{2}{\pi} \sum_{m=0}^{\infty} \frac{x^{2m+2}}{(2m+1)!!\,(2m+3)!!}. \tag{61}$$

The output power of the coherent radiation can now be found from the law of conservation of energy throughout the coherent emission region. The application of this law gives

$$P(\alpha D) = \frac{1}{3} \frac{E_0}{e} \frac{\pi D^2}{4} \left\{ j - j_{th}^0 \frac{I_1(\alpha D) + L_1(\alpha D)}{\frac{\alpha D}{2}} \right\}, \tag{62}$$

where j_{th}^0 is the threshold current density in the limit $\alpha D \ll 1$. For a given current density j the output power P has its maximum value for D_{max}, where D_{max} is the solution of the equation

$$I_0(\alpha D_{max}) + L_0(\alpha D_{max}) = \frac{j}{j_{th}^0}. \tag{63}$$

Here, the zero-order Bessel $I_0(x)$ and Struve $L_0(x)$ functions are defined by the following relationships [114]:

$$I_0(x) = \sum_{m=0}^{\infty} \frac{1}{(m!)^2} \left(\frac{x}{2}\right)^{2m}, \quad L_0(x) = \frac{2}{\pi} \sum_{m=0}^{\infty} \frac{x^{2m+1}}{[(2m+1)!!]^2}. \tag{64}$$

Tables of the Struve functions $L_0(x)$ and $L_1(x)$ for the $0.1 \leq x \leq 10$ range are given in Appendix II.

When $D = D_{max}$ the overall efficiency of the laser η depends only on αD_{max} and is given by the equation

$$\eta(\alpha D_{max}) = \frac{1}{3} \left\{ 1 - \frac{2}{\alpha D_{max}} \frac{I_1(\alpha D_{max}) + L_1(\alpha D_{max})}{I_0(\alpha D_{max}) + L_0(\alpha D_{max})} \right\}. \tag{65}$$

§ 3.   Pulse Output Power

In the present section we shall consider a semiconductor laser by single pulses of such a nature that the time average of the pumping power is negligibly small compared with the pulse power. If the pulse duration is $\sim 10^{-7}$ sec, these conditions are attained if the pulse repetition frequency does not exceed a few hundreds of hertz. Obviously, in this case the heating of a crystal by the incident electrons does not restrict the pulse output of the laser but simply imposes certain restrictions on the pulse duration. Therefore, the maximum pulse output power is governed by the optimal transverse dimension of the laser and the highest permissible flux density of the radiation emerging through the resonator faces.

We shall now give the results of calculations of the maximum pulse output power and of several other laser parameters obtained for the two models discussed in the preceding section. The first of these models corresponds to the conventional laser geometry and the second to the planar resonator case.

It follows from Eqs. (48), (49), and (51) that, in the case of conventional geometry, the maximum current density j_{max} at which the energy flux of the coherent radiation reaches the damage threshold S_{dt}^{*} and the transverse dimension D_{max} corresponding to the maximum out-

---

*We shall assume that $S_{dt} = 5 \times 10^6$ W/cm².

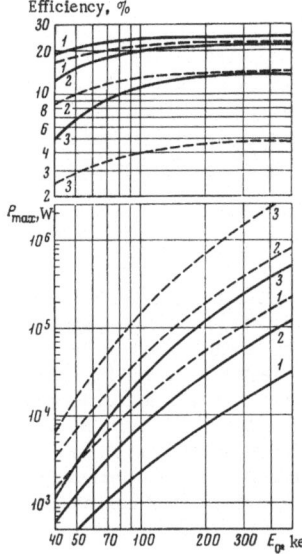

Fig. 30. Maximum output power
and the efficiency of lasers plotted against the electron energy.
The curves represent lasers with
different mirror reflection coefficients R and different resonator
lengths L. 1) L = 0.5 mm; 2) L =
2 mm; 3) L = 10 mm. The continuous curves correspond to R =
0.3 and the dashed curves to R =
0.8.

put power P_{max} can be found by the simultaneous solution of the following equations:

$$\frac{1}{3} f (\varkappa L) \frac{E_0 L}{ex_0} \left\{ j - j_{th}^0 + \frac{\alpha L}{\ln R} j_{th}^0 \left[\exp\left(- \frac{D_{max} \ln R}{2L} \right) - 1 \right] \right\} = 2S_{dt}$$

$$\frac{D_{max}}{L} \ln \frac{1}{R} = \ln \left\{ 1 + \frac{j - j_{th}^0}{\alpha L j_{th}^0} \ln \frac{1}{R} \right\},$$

(66)

$$P_{max} = \frac{1}{3} f (\varkappa L) \frac{E_0}{e} L D_{max} \left\{ j - j_{th}^0 + \frac{\alpha L}{\ln R} j_{th}^0 \left[\frac{\exp\left(\frac{D_{max}}{L} \ln \frac{1}{R} \right) - 1}{\frac{D_{max}}{L} \ln \frac{1}{R}} - 1 \right] \right\}.$$

Such calculations yield the dependences of the maximum power P_{max} and of the maximum
efficiency $\eta (D_{max})$ on the electron energy E_0, the resonator length L, and the reflection coefficient of the resonator mirrors R. These dependences are plotted in Figs. 30-32. In all cases,
the nonresonant absorption coefficient $\varkappa$,[*] which occurs in the loss function of Eq. (17), was assumed to be 1 cm^{-1}; the gain was determined using Eq. (52); the laser threshold j_{th}^0 was found
by assuming that the volume pumping power density $p_{th}^0 = (E_0/ex_0)j_{th}^0$ is proportional to the gain
and the coefficient of proportionality is such that $j_{th}^0 = 1$ A/cm^2 for $E_0 = 50$ keV, L = 0.5 mm,
and R = 0.3. The efficiency of the laser was calculated from the ratio P_{max} to the total pumping power $P_p = (E_0/e)j_{max} L D_{max}$.

The results obtained indicate that when the electron energy is increased to the threshold
of radiation-defect formation (250-300 keV for gallium arsenide) and the geometrical dimensions of the resonator are optimal, it is quite realistic to expect an output of the order of hundreds of kilowatts from the conventional resonator.

Similar calculations can be carried out for the planar resonator case on the basis of Eqs.
(57), (62), (63), and (65). The results obtained for R = 0.3 and $S_{dt} = 10^7$ W/cm^2 are given in
Table 7.

[*]It is shown experimentally in [112] that the contribution to $\varkappa$ representing the unavoidable
losses due to free-carrier absorption amounts to about 0.3 cm^{-1}.

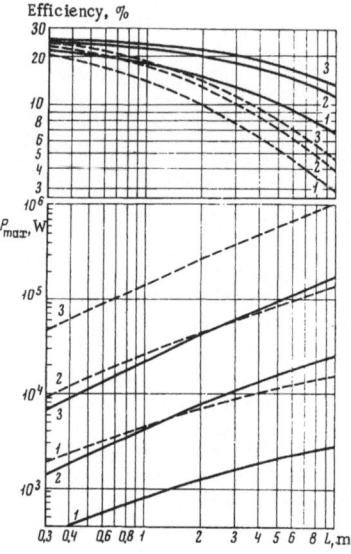

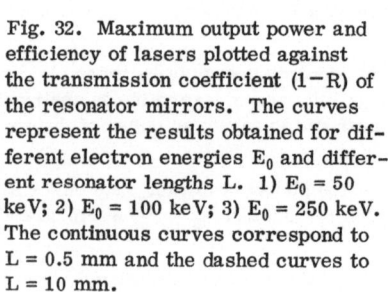

Fig. 31. Maximum output power and efficiency of lasers plotted against the resonator length. The curves represent the results obtained for different electron energies E_0 and different mirror reflection coefficients R. 1) $E_0 = 50$ keV; 2) $E_0 = 100$ keV; 3) $E_0 = 250$ keV. The continuous curves correspond to R = 0.3 and the dashed curves to R = 0.8.

In these calculations the threshold current density j_{th}^0 was again calculated on the assumption of proportionality between p_{th}^0 and α. We assumed that the gain α is given by Eq. (54) and we found the coefficient of proportionality from the experimental data reported in [101] for $e_0 \approx 200$ keV. This gave

$$j_{th}^0 = C \frac{e}{E_0} \ln \frac{1}{R}, \tag{67}$$

where C = 0.61 MW/cm^2. The efficiency in all the cases listed in Table 7 was about 22%.

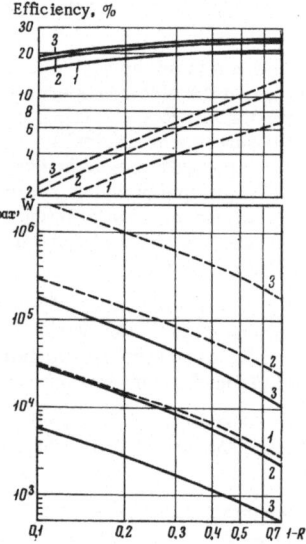

Fig. 32. Maximum output power and efficiency of lasers plotted against the transmission coefficient $(1-R)$ of the resonator mirrors. The curves represent the results obtained for different electron energies E_0 and different resonator lengths L. 1) $E_0 = 50$ keV; 2) $E_0 = 100$ keV; 3) $E_0 = 250$ keV. The continuous curves correspond to L = 0.5 mm and the dashed curves to L = 10 mm.

TABLE 7

E_0, keV	α, cm^{-1}	j_{th}^0, A/cm^2	j_{max}, A/cm^2	D_{max}, mm	P_{max}, kW
50	2410	29.4	1800	0.021	0.059
70	1360	21.0	1285	0.038	0.172
100	752	14.7	899	0.068	0.609
150	387	9.81	600	0.133	2.29
200	245	7.36	450	0.210	5.72
250	174	5.89	360	0.296	11.4
300	132	4.90	300	0.389	19.6
400	87	3.86	225	0.591	45.3
500	64	2.94	180	0.805	84.2

For $0.3 \leq R \leq 0.8$ and the maximum current density $j = j_{max}$ (governed by the damage threshold), we found that αD_{max} was within the range 5.2–6.0. The corresponding efficiency calculated from Eq. (65) was within the range 21.6–22%, i.e., it was practically independent of R. This led to the following dependence of the maximum output power on the reflection coefficient R:

$$\frac{P_{max}(R)}{P_{max}(R_0)} = \frac{P_p(R)}{P_p(R_0)} = \frac{\ln R\,[\alpha D_{max}(R)]^2}{\ln R_0\,[\alpha D_{max}(R_0)]^2}\,\frac{I_0\{\alpha D_{max}(R)\} + L_0\{\alpha D_{max}(R)\}}{I_0\{\alpha D_{max}(R_0)\} + L_0\{\alpha D_{max}(R_0)\}}, \tag{68}$$

where R_0 is some fixed value of R (for example, 0.3), which corresponds to the principal laser parameters listed in Table 7.

In particular, it follows from Eq. (68) that when $R = 0.8$ the maximum output power P_{max} should reach 200 kW for an electron energy of 250 keV.

The possibilities of increasing pulse output power are thus approximately the same for a laser with a planar resonator and one with a conventional geometry. However, in the former case, one would need current densities of the order of 100 A/cm^2, whereas in the latter case a current density one or two orders of magnitude lower would be sufficient.

§ 4. Average Output Power

We have already mentioned that the average output power is limited by heating of the active region.

We shall now consider the operation of a laser emitting pulses at a high repetition frequency and assume that the pulse duration is sufficiently short so that the pulse heating effect is much weaker than the average effect. In this case, the maximum average output P_{max} corresponds to the maximum pulse output power and the highest repetition frequency consistent with the limit set for average heating. It is convenient ot use the formula

$$\bar{P}_{max} = \eta(D_{max})\frac{E_0}{e}\bar{j}s, \tag{69}$$

where $\bar{j}$ is the average current density; $s = LD_{max}$ for the conventional geometry and $s = \pi D_{max}^2/4$ for the planar resonator case. The efficiency $\eta(D_{max})$ and the maximum transverse size D_{max} in Eq. (69) are given in § 3 (an allowance must be made for the fact that j_{th}^0 depends on the heating) and the highest permissible value of the pumping energy flux $(E_0/e)\bar{j}$ can be found using the results given in Appendix I.

Table 8 gives the results of calculations corresponding to $R = 0.3$ and $L = 0.5$ mm (conventional resonator geometry); the maximum permissible temperature of the active region was

B. M. LAVRUSHIN

TABLE 8

E_0, keV	$\frac{E_0}{e}\bar{j}$, kW/cm²	Conventional laser geometry			Laser with planar resonator		
		D_{max}, mm	$\eta(D_{max})$,%	P_{max}, W	D_{max}, μ	$\eta(D_{max})$, %	P_{max}, W
50	12.4	0.82	19.0	9.77	17,6	20	0,0059
70	11.8	0.93	20,4	11.7	31,0	20	0.0176
100	10.9	1.11	22.0	13.8	56,3	20	0.0537
150	9.48	1.30	23.5	15,1	109	20	0,176
200	8.34	1,40	24,1	14.7	172	20	0.385
250	7.44	1,45	24.4	13.7	243	20	0,685
300	6.72	1,48	24.6	12.7	320	20	1.07
400	5.58	1,54	24.9	11.1	485	20	2.05
500	4.77	1.55	24.9	9.59	662	20	3,26

assumed to be 300°K (for an initial temperature of 80°K) and the value of $(E_0/e)\bar{j}$ was calculated from the data plotted in Fig. 39; j^0_{th} (300°K) = 2.5j^0_{th} (80°K) (§ 2, Chap. II).

When the reflection coefficient of the mirrors R and the resonator length L are increased, the average laser power rises in the same way as the pulse output because such an increase in R and L does not affect the heat removal conditions.

The following results can be deduced from Table 8.

1. The average output power of a laser with the conventional resonator geometry depends weakly on the electron energy and is of the order of 10 W.

2. The average output power of a laser with a planar resonator is much lower than the output of a laser with the conventional geometry. This is basically due to the fact that, in the first case, the transverse dimensions are much smaller than in the second. The considerable difference between the dimensions has less effect on the pulse output power because, in the case of single short pulses, the current density in a laser with a planar resonator can be made much higher than in the conventional laser. However, in the case of the average output power, the difference between the transverse dimensions is of decisive influence because the average heating and, consequently, the time average value of the pumping power density are the same in both cases. It should also be mentioned that in the case of a laser with a planar resonator it would be difficult to attain the average powers listed in Table 8 because of the need to generate electron pulses of high current density (see § 3).

The possibility of increasing the pulse and average output powers by increasing the geometrical dimensions of the active region are considered in the present section and in § 3. One must mention the self-evident possibility of increasing the total output power by building a laser battery or a multielement laser. In this case, a semiconductor sample consists of optically isolated parts which operate as independent lasers with a common pumping source. The total output power of such a system is simply equal to the sum of the output powers of each of these parts. Therefore, the total output is limited only by the power of the pumping source and the efficiency of such a system is governed by the construction and the fabrication technology because of the unavoidable losses of the pumping power in the passive part of such a battery. The planar resonator geometry is more convenient from the point of view of fabrication technology and the synchronization of the separate elements in the multielement laser.

CONCLUSIONS

The main results of the present investigation can be summarized as follows.

1. An investigation was made of the dependences of the principal parameters of gallium arsenide lasers on the type of conduction and the impurity concentration. The nature of the

dependences of these parameters on the impurity concentration was determined and it was found that the use of doped gallium arsenide crystals in electron-beam-pumped lasers has definite advantages over the use of undoped samples. The best results were obtained for crystals with donor concentrations up to $(1-2) \times 10^{18}$ cm^{-3} for lasers operated at T $\approx$ 80°K and up to $(2-4) \times 10^{18}$ cm^{-3} in the case of lasers operated at T $\approx$ 300°K.

2. It was found that the output power of an electron-beam-pumped gallium arsenide laser was considerably higher than the output power of the best injection lasers and the efficiency was practically equal to the theoretical limit. This provided an experimental demonstration of the theoretically predicted advantages of the electron-beam-pumping method.

3. Detailed experimental temperature dependences of the laser threshold were obtained for n- and p-type gallium arsenide with different impurity concentrations. It was found that these dependences were governed by the type and concentration of the impurities. It was established that the threshold current density for samples with moderate impurity concentrations was a nonmonotonic function of the temperature.

4. The coherent emission spectra were investigated in the temperature range 80-300°K and it was shown that these spectra were related to the impurity concentration. A change in the laser action mechanism between 80 and 300°K was discovered for n- and p-type gallium arsenide with moderate impurity concentrations.

5. An analysis was made of the coherent emission spectra and of the way in which these spectra varied with the temperature and impurity concentration. The results of this analysis yielded a scheme of radiative transitions responsible for the laser action: the main contribution was made either by interband transitions or by transitions between impurity and intrinsic states (in those cases where division into these two types of state was physically meaningful). The actual transition mechanism depended not only on the impurity concentration but also on the temperature.

6. The following effects, associated with the inhomogeneity of the excitation of the active region by the electron beam, were discovered and investigated: (a) the threshold current density decreased with increasing electron energy E_0, in spite of the fact that the volume density of the pumping power corresponding to constant current density decreased when E_0 was raised; (b) the maximum in the angular distribution of the output radiation did not coincide with the normal to the optical resonator mirrors; and (c) under certain conditions the emission intensity pulsated with time (spiking).

7. A quantitative analysis was made of the importance of the losses associated with the amplification of spontaneous radiation. In particular, it was found that an allowance for these losses imposed restrictions on the transverse dimensions (width) of the active region. Estimates were made of the possibility of further increase of the pulse and average output power of semiconductor lasers by increasing their geometrical dimensions and the volume density of the pumping power.

The results obtained demonstrate the wide-ranging capabilities of electron excitation as a method for investigating stimulated emission from semiconductors and as a method for pumping semiconductor lasers. The investigation of the properties of gallium arsenide lasers reported in the present paper has demonstrated the advantages of the electron-beam pumping method compared with the method of carrier injection across a p–n junction and with optical excitation. The results obtained may be useful in industrial production of electron-beam-pumped lasers.

The author is grateful to N. G. Basov and O. V. Bogdankevich for suggesting the problem, their continuous interest, and discussion of the results. The author is also very grateful to N. A. Borisov and I. V.Kryukova for active participation in many of the investigations reported above; to V. A. Goncharov, A. N. Mestvirishvili, L. I. Lebedeva, and A. G. Devyatkov, who helped in some of the experiments, and to V. V. Karataev for supplying a large number of gallium arsenide crystals.

APPENDICES

I. THERMAL CONDITIONS IN ELECTRON-BEAM-PUMPED LASERS

A.   Formulation of the Problem

It is mentioned in the Introduction that when a semiconductor is excited by bombardment with electrons, about two-thirds of the pumping power is lost as heat in the process of thermalization of "hot" carriers. The volume distribution of the thermal losses suffered by the incident electrons in a crystal is given by the ionization curve $(-dE/dx)$ and by the distribution of the current density $j(y, z)$ across the electron beam. In the system of coordinates shown in Fig. 4 the power of the thermal losses suffered by the incident electrons in a unit volume of a crystal is given by

$$w(x, y, z, t) = -\frac{2}{3} \frac{dE(x)}{dx} \frac{j(y, z, t)}{e}. \tag{A.1}$$

The volume distribution of the temperature $T(x, y, z, t)$ at any time t is found by solving a three-dimensional heat conduction equation subject to the appropriate boundary and initial conditions. In general, this equation is nonlinear because the thermal conductivity $\varkappa$ and the specific heat C of semiconductor crystals are functions of the temperature T.[*]

The problem can be simplified bearing in mind the following points.

1. Under the experimental conditions the inhomogeneity of the electron beam is fairly weak: the transverse dimensions of the beam are much greater than the depth of penetration x_0. Consequently, we can ignore the flow of heat along the y and z axes compared with that along the x axis. The heat conduction equation then becomes one-dimensional. For similar reasons, a semiconductor crystal can be regarded as infinite along the y and z axes. Therefore, the whole problem becomes one-dimensional.

2. When short pulses are employed, the temperature of a crystal does not rise greatly (this point is discussed in Sec. C). Therefore, we can ignore the temperature dependences of $\varkappa$ and C and assume that $\varkappa(T) = \varkappa(T_0)$ and $C(T) = C(T_0)$ (T_0 is the initial temperature of the crystal); the heat conduction equation can thus be reduced to its linear form.

In view of the points just mentioned the problem can be formulated as follows. A homogeneous crystal is bounded by planes $x = 0$ and $x = x_1$ (Fig. 33). The $x = 0$ plane is bombarded with an electron beam in which the current density is $j(y, z, t)$. The dependence of this density on time t is described by the function which is shown in Fig. 33 and is a periodic sequence of rectangular pulses (duration t_0 and period t_1):

$$j(y, z, t) = j(y, z, 0) \sum_{k=0}^{\infty} [u(t - kt_1) - u(t - kt_1 - t_0)], \tag{A.2}$$

---

[*]The values of the thermal constants of gallium arsenide obtained in a wide range of temperatures are listed, for example, in [115-117].

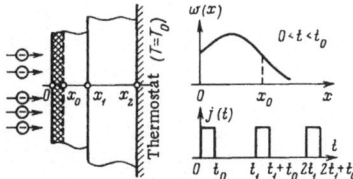

Fig. 33. Schematic representation of
the heating of a crystal by an electron
beam. The crystal occupies the region
$0 \le x \le x_1$. A heat sink is located in
the region $x_1 \le x \le x_2$ and a thermostat
occupies the region $x > x_2$. The distribution of the thermal losses suffered
by the electrons w (x) and the time dependence of the current j are plotted
on the right.

where u (t) is the unit function

$$u(t) = \begin{cases} 1 & \text{for} \quad t > 0, \\ 0 & \text{for} \quad t < 0. \end{cases} \tag{A.3}$$

The $x = x_1$ plane is in thermal contact with a heat sink whose $x = x_2$ plane is kept at a constant
temperature T_0.

The mathematical problem reduces to the solution of the one-dimensional heat conduction
equation

$$\frac{\partial T(x, t)}{\partial t} = \frac{\varkappa(T_0)}{\rho C(T_0)} \frac{\partial^2 T(x, t)}{\partial x^2} + \frac{w(x, y, z, t)}{\rho C(T_0)} \tag{A.4}$$

subject to the following boundary and initial conditions:

$$\frac{\partial T}{\partial x}\bigg|_{x=0} = 0,$$

$$T(x, t) \text{ and } \frac{\partial T(x, t)}{\partial x} \text{ continuous for } x = x_1 . \tag{A.5}$$

$$T(x_2 t) = T_0, \quad T(x, 0) = T_0.$$

The function w (x, y, z, t) in Eq. (A.4) is defined by Eq. (A.1). We shall omit the argument
(y, z) and assume that j (t) denotes the local current density in that part of a semiconductor crystal which is of interest to us. Consequently, (y, z) can be omitted from the arguments of the
function T (x, t) because this function depends parametrically on j.

B.   Solution of the Heat Conduction Equation

for a Semi-Infinite Sample

The temperature distribution in an infinite homogeneous body in which a heat ρC is evolved
instantaneously in a plane $x = x'$ at a time $t = t'$ is described by the Green's function of the heat
conduction equation:

$$G(x, x', t - t') = \frac{1}{\sqrt{4\pi a^2(t - t')}} \exp\left\{-\frac{(x - x')^2}{4a^2(t - t')}\right\}, \tag{A.6}$$

where $a^2 = \varkappa/\rho C$.

The Green's function (A.6) is the solution of Eq. (A.4) in the special case when $w(x, t) = \rho C \delta(x - x') \delta(t - t')$ and it satisfies the following initial and boundary conditions:

$$G(x, x', 0) = 0 \quad \text{for} \quad x \neq x',$$
$$G(\infty, x', t - t') = 0, \tag{A.7}$$
$$\left. \frac{\partial G}{\partial x} \right|_{x=x'} = 0.$$

It follows from Eq. (A.6) that in a time t the heat evolved travels a distance of the order of $\sqrt{2a^2 t}$ from the original source. Consequently, if $\sqrt{2a^2 t}$ is sufficiently short, namely, if

$$x_1 - x_0 \gg \sqrt{2a^2 t}, \tag{A.8}$$

we can ignore the presence of the boundary at $x = x_1$ and can replace the conditions of Eq. (A.5) with the following set:

$$\left. \frac{\partial T(x, t)}{\partial x} \right|_{x=0} = 0,$$
$$T(\infty, t) - T_0 = 0, \tag{A.9}$$
$$T(x, 0) - T_0 = 0.$$

The solution of Eq. (A.4) satisfying the above conditions can be expressed in terms of the Green's function of Eq. (A.6):

$$T(x, t) - T_0 = \int_0^t dt' \int_0^\infty dx' \frac{2}{3} \left[-\frac{dE(x')}{dx'} \right] \frac{j(t')}{e\rho C} \left[G(x, x', t - t') + G(x, -x', t - t') \right]. \tag{A.10}$$

Using the approximation of the ionization curve $-dE/dx$ by Eq. (13) and integrating the relevant part of Eq. (79), we obtain

$$T(x, t) - T_0 = \frac{2}{3} \frac{E_0' j}{e x_0 \rho C} \sum_{n=0}^\infty \left\{ u(t - nt_1)(t - nt_1) \left[F\left(\frac{x_0 + x}{\sqrt{2a^2(t - nt_1)}} \right) + F\left(\frac{x_0 - x}{\sqrt{2a^2(t - nt_1)}} \right) \right] - \right.$$
$$\left. - (t - nt_1 - t_0) u(t - nt_1 - t_0) \left[F\left(\frac{x_0 + x}{\sqrt{2a^2(t - nt_1 - t_0)}} \right) + F\left(\frac{x_0 - x}{\sqrt{2a^2(t - nt_1 - t_0)}} \right) \right] \right\}, \tag{A.11}$$

where

$$F(x) = (1 + x^2) \Phi_0(x) + \frac{x}{\sqrt{2\pi}} \exp\left(-\frac{x^2}{2} \right) - \frac{x|x|}{2}, \tag{A.12}$$

$\Phi_0(x) = \frac{1}{\sqrt{2\pi}} \int_0^x \exp\left(-\frac{t^2}{2} \right) dt$ is the error function [77], and $F(x)$ is an odd function of the argument x. The values of the function $F(x)$ in the range $x \geq 0$ are listed in Table 9.

Equation (A.11) is the exact solution of the problem formulated in Sec. A in the special case when the thickness of the crystal approaches the limit $x_1 \to \infty$. In the experiments involving electron bombardment of semiconductors, the conditions are usually such that Eq. (A.11) can be used in an approximate description of the temperature distribution in a crystal of finite thickness during and after a single pumping pulse. The formulas applicable to this case are given Appendix IC and the problem of the validity and precision of the formulas obtained is also discussed in that section.

TABLE 9. Function $F(x) = \dfrac{1+x^2}{\sqrt{2\pi}}\displaystyle\int_0^x \exp\left(-\dfrac{t^2}{2}\right)dt + \dfrac{x}{\sqrt{2\pi}}\exp\left(-\dfrac{x^2}{2}\right) - \dfrac{x^2}{2}$

x	$F(x)$	x	$F(x)$	x	$F(x)$	x	$F(x)$
0	0	1.0	0,424661	2.0	0,494232	3.0	0,499796
0.1	0.074921	1.1	0,439815	2.1	0,495719	3.1	0.499861
0.2	0.140639	1.2	0,452253	2.2	0,496848	3.2	0.499906
0.3	0.197940	1.3	0,462386	2.3	0,497698	3.3	0.499937
0,4	0.247597	1.4	0,470579	2.4	0,498332	3.4	0.499958
0.5	0.290361	1.5	0,477153	2.5	0,498800	3.5	0.499973
0.6	0.326951	1.6	0,482388	2.6	0,499145	3.6	0.499982
0.7	0.358051	1.7	0,486524	2.7	0,499395	3.7	0.499988
0.8	0.384310	1,8	0,489766	2.8	0,499576	3.8	0.499993
0.9	0,406328	1.9	0,492286	2.9	0.499705	∞	$1/2$

However, Eq. (A.11) cannot describe completely the thermal conditions in a semiconductor laser because this equation does not lead to the steady-state conditions when $t \to \infty$. In fact, as we approach the limit $t \to \infty$, the series in Eq. (A.11) diverges although it is physically self-evident that in a crystal of finite thickness the steady-state distribution T (x, t) should be a periodic function of time. This divergence is obviously due to the assumption that the thickness of the crystal is infinite. In fact, under steady-state conditions the average value (calculated for one period) of the flow of heat through any plane x = const > x_0 should be equal to the average flow across the x = 0 plane of that part of the pumping power which is lost in the form of heat. These conditions are established in the presence of a finite temperature gradient along the x axis throughout the crystal. If the crystal is infinitely thick, this gives rise to an infinitely high temperature at the x = 0 surface. Thus, in order to find the steady-state thermal conditions it is necessary to assume from the outset that the crystal is of finite thickness. The calculation of the period-average steady-state temperature distribution corresponding to this case is given in Appendix ID.

C. Heating of a Sample During One Pulse and Its Cooling During the Interval between Pumping Pulses

We shall consider the heating of a crystal during the first pumping pulse ($0 \le t \le t_0$). It follows from Eq. (A.11) that during this pulse

$$T(x, t) - T_0 = \frac{2}{3}\frac{E_0'j}{ex_0\rho C}\left[F\left(\frac{x_0 + x}{\sqrt{2a^2t}}\right) + F\left(\frac{x_0 - x}{\sqrt{2a^2t}}\right)\right]. \tag{A.13}$$

In the above formula the factor in front of the square brackets describes the adiabatic heating and the expression inside these brackets is due to heat conduction.

We shall consider two limiting cases for which we shall obtain approximate formulas representing the error function $\Phi_0(x)$ in the form of a series for small values of the argument and in the form of an asymptotic expansion for large values of the argument [114].

1. $2a^2t \ll (x_0 - x)^2$. In this case, we obtain the following formula from Eq. (A.13) if we limit ourselves to the lowest degree of the small parameter $\sqrt{2a^2t}/(x_0 \pm x)$:

$$T(x, t) - T_0 \approx \begin{cases} \dfrac{2}{3}\dfrac{E_0'jt}{ex_0\rho C} & \text{for } x < x_0, \\[2ex] \dfrac{1}{3}\dfrac{E_0'jt}{ex_0\rho C} & \text{for } x = x_0, \\[2ex] \dfrac{2}{3}\dfrac{E_0'jt}{ex_0\rho C}\dfrac{(2a^2 t)^{3/2}}{(x-x_0)^3}\dfrac{2}{\sqrt{2\pi}}\exp\left[-\dfrac{(x-x_0)^2}{4a^3t}\right]\left\{1-\left(\dfrac{x-x_0}{x+x_0}\right)^3 \times\right. \\[2ex] \left. \times \exp\left(-\dfrac{xx_0}{a^2t}\right)\right\} & \text{for } x > x_0. \end{cases} \tag{A.14}$$

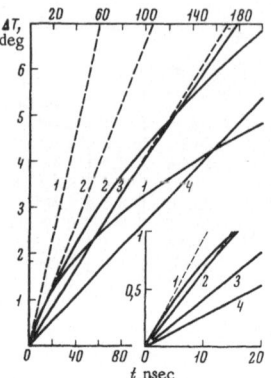

Fig. 34. Pulse heating of a crystal at x = 0 plotted as a function of time for a current density j = 1 A/cm² and different electron energies E_0 (keV): 1) 30; 2) 50; 3) 100; 4) 200. The continuous curves are plotted using Eq. (A.13) and the parameters (ρC = 0.859 J · cm⁻³·deg⁻¹ and $\varkappa$ = 2.95 W·cm⁻¹· deg⁻¹) corresponding to gallium arsenide at T = 80°K. The dashed curves represent the adiabatic heating of Eq. (A.14) for the same parameters.

2. $2a^2t \gg (x_0 \pm x)^2$. In this case, we find that Eq. (A.13) yields the following formula, accurate to within the square of the small parameter $(x_0 \pm x)/\sqrt{2a^2t}$:

$$T(x, t) - T_0 = \begin{cases} \dfrac{2}{3} \dfrac{E_0' j}{e} \dfrac{x_0}{\varkappa} \left(\dfrac{2}{\sqrt{\pi}} \dfrac{\sqrt{a^2 t}}{x_0} - \dfrac{1}{2} - \dfrac{x^2}{2x_0} \right) & \text{for } x \leqslant x_0, \\[4mm] \dfrac{2}{3} \dfrac{E_0' j}{e} \dfrac{x_0}{\varkappa} \left(\dfrac{2}{\sqrt{2\pi}} \dfrac{\sqrt{a^2 t}}{x_0} - \dfrac{x}{x_0} \right) & \text{for } x \geqslant x_0. \end{cases} \tag{A.15}$$

Thus, the time dependence of the temperature rise is of the following nature: the rise is proportional to t for sufficiently short time intervals and proportional to $\sqrt{t}$ for long time intervals.[*] The latter relationship is typical of all diffusion processes.

Figures 34 and 35 show the dependences of the pulse heating of Eq. (A.13) on the time t and the coordinate x.

Figure 34 shows the time dependence of the temperature rise ΔT = T (0, t)−T_0 at point x = 0 in the interval 0-200 nsec, calculated for different electron energies. We can see that heat conduction has a strong influence on the heating of a crystal at low electron energies. For example, when gallium arsenide is bombarded with electrons of 30 and 50 keV energies, the thermal conductivity of this material reduces the temperature rise after 100 nsec by a factor of 3 and 1.5, respectively (compared with the case when $\varkappa$ = 0). At higher electron energies, the effect of heat conduction is of little importance. For example, when E_0 = 100 keV the correction for heat conduction is only 7% for t ≤ 200 nsec, and when E_0 = 150 keV or higher the temperature rise in the interval t ≤ 200 nsec is practically unaffected by heat conduction and is governed solely by the specific heat, i.e., the heating is adiabatic.

The ordinate of Fig. 35 shows the ratio of the temperature rise at the point x to the corresponding rise at the point x = 0, whereas the abscissa represents $x/\sqrt{2a^2t}$, which is proportional to the distance x from the surface of the crystal in question. The parameter of the curves of the curves in Fig. 35 is $x_0/\sqrt{2a^2t}$. Curves 1-4 represent GaAs (T_0 = 80°K) bombarded with electrons of energies 33, 50, 75, and 100 keV, respectively (the curves are calculated for t = 0.1 μsec and a^2 = 3.44 cm²/sec). If we assume that x_0 is the same for all these curves, we find that

---

[*]The results obtained in [118] for a similar problem must be regarded as erroneous because it follows from Eq. (18) of that paper that at any moment $T(0, t) - T_0 = \dfrac{2}{3} \dfrac{E_0' t}{e x_0 \rho C}$, i.e., that the temperature rise is always proportional to t, which is clearly incorrect.

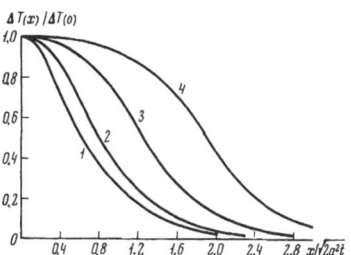

Fig. 35. Distribution of the pulse heating along the coordinate x. Curves 1-4 are plotted on the basis of Eq. (A.13) with $x_0/\sqrt{a^2 t} = 0.3, 0.6, 1.2$, and 1.9, respectively.

curves 1-4 represent heating at different moments. For example, if $x_0 = 5~\mu$ ($E_0 = 50$ keV) and $a^2 = 3.44$ cm²/sec, curves 1-4 correspond to t = 405, 101, 25, and 10 sec, respectively. We may also assume that x_0 and t are the same for all the curves and then we find that the curves correspond to different values of a^2, i.e., to different initial temperatures. Thus, if $x_0 = 16~\mu$ ($E_0 = 100$ keV) and t = 100 nsec, curves 1-4 describe the temperature distributions in gallium arsenide for initial temperatures $T_0 \approx 57, 90, 180$, and 320°K, respectively. It also follows from Fig. 35 that the temperature rise at the point $x = x_0 + 1.5\sqrt{2a^2 t}$ * does not exceed 5% of the rise at the point x = 0.

It follows from Eq. (A.13) that the temperature rise also depends on the initial temperature T_0 and on the density of the ionization losses suffered by electrons E_0/x_0. The dependences of the pulse heating on these parameters are plotted in Figs. 36 and 37.

Figure 36 shows the dependences of the temperature rise at the point x = 0 after a time t = 100 nsec; these dependences are plotted as a function of the electron energy E_0 for three values of the initial temperature $T_0 = 80, 120$, and 273°K. The adiabatic temperature rise (represented by the dashed curves) increases monotonically with decreasing E_0 because the thermal losses suffered by the electrons increase with decreasing electron energy. Heat conduction affects the temperature rise at electron energies which correspond to $x_0 \lesssim \sqrt{2a^2 t}$.† For t = 10^{-7} sec, this corresponds to electron energies $E_0 \lesssim 85$ keV ($T_0 = 80$°K), 65 keV ($T_0 = 120$°K), and 40 keV ($T_0 = 273$°K).

Figure 37 shows the dependence of the pulse heating on the initial temperature. The nature of this dependence is governed by the temperature dependences of ρC and $\varkappa$ [115, 116].

We shall now consider the limits of validity of Eq. (A.13). This equation was derived making the following assumptions:

1) the crystal heated by electron bombardment was assumed to be semi-infinite, i.e., unbounded in the x > 0 half-space;

2) the simplest possible approximation [Introduction, Eq. (13)] was used for the density of the thermal losses experienced by electrons;

3) use was made of the linearized heat conduction equation, i.e., it was assumed that the thermal constants of the crystal did not vary during pulse heating.

The first assumption limits the range of validity of Eq. (A.13) by imposing the following condition: the flow of heat across the $x = x_1$ boundary should be much less than the flow of heat

---

*The characteristic diffusion length of heat $\sqrt{2a^2 t}$ in gallium arsenide, corresponding to t = 10^{-7} sec, ranges from 8.3 μ for $T_0 = 180$°K to 2.4 μ for $T_0 = 273$°K.

†It follows directly from Eq. (A.13) that when $x_0 = 1.5 \sqrt{2a^2 t}$, the temperature rise decreases by about 5% because of heat conduction.

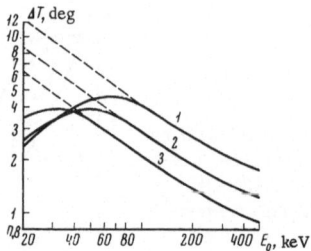

Fig. 36. Pulse heating plotted against the
electron energy for various initial temperatures T_0 (°K): 1) 80; 2) 120; 3) 273.
The continuous curves are plotted on the
basis of Eq. (A.13) with the following
parameters: $j = 1$ A/cm^2; $t = 0.1$ μsec;
$\rho C = 0.859$, 1.24, and 1.70 J $\cdot$ cm^{-3} $\cdot$ deg^{-1};
$\varkappa = 2.95$, 1.61, and 0.472 W $\cdot$ cm^{-1} $\cdot$ deg^{-1}
(GaAs at $T_0 = 80$, 120, and 273°K, respec-
tively). The dashed curves represent adiabatic heating.

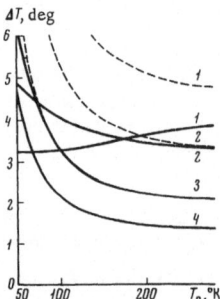

Fig. 37. Pulse heating of a crystal at
the point $x = 0$ plotted as a function of
the initial temperature T_0 for different
electron energies E_0 (keV): 1) 30; 2)
50; 3) 100; 4) 200. The continuous
curves are plotted on the basis of Eq.
(A.13) for $j = 1$ A/cm^2 and $t = 10^{-7}$ sec.
The experimental values of the specific
heat C were taken from [116]; the thermal conductivity was assumed to be $\varkappa =$
2.95 W $\cdot$ cm^{-1} $\cdot$ deg^{-1} at $T_0 = 80$°K and
its temperature dependence was as-
sumed to be given by Eq. (A.34). The
dashed curves correspond to adiabatic
heating ($\varkappa = 0$).

across the x = 0 boundary. Hence, we find

$$- \varkappa \frac{\partial T\,(x,\,t)}{\partial x}\bigg|_{x=x_1} \leqslant \frac{2}{3}\,\frac{E_0'j}{e}\,.$$ (A.16)

We can easily show that Eq. (A.13) obeys the above condition if

$$x_1 \gg x_0 + 1.5\,\sqrt{2a^2t} \quad \text{and} \quad x_0 \gg 0.1\,\sqrt{2a^2t}.$$ (A.17)

These relationships* are a refinement of the less accurate condition given by Eq. (A.8). Since the value of $\sqrt{2a^2t}$ does not exceed 10 μ for most semiconductor crystals if $T_0 \approx 50\text{--}70°K$ and the duration of the current pulses is typical of that employed in the electron–beam pumping of semiconductor lasers, it follows that the first of the assumptions mentioned above imposes only slight restrictions on the validity of Eq. (A.13).

The second assumption may lead to incorrect temperature distributions along the coordinate x, as calculated by means of Eq. (A.13). However, we can show [119] that when the electron energy is $E_0 = 50$ keV and the pulse duration is 50 to 150 nsec, the discrepancies between the true and calculated distributions should not exceed 10%. At lower values of the electron energy the discrepancy should be even smaller because, in this case, the distribution of the temperature rise is governed primarily by heat conduction and depends weakly on the actual distribution of the thermal losses suffered by the incident electrons. At high electron energies, when the heating depends weakly on heat conduction, the discrepancy can be larger but, in this case, fairly accurate results are given by the adiabatic approach which can be applied to any distribution of thermal losses. A more accurate result with allowance for heat conduction can be obtained from Eq. (A.10) if the ionization curve (−dE/dx) is described by a more accurate approximation such as that given by Eq. (11) in the Introduction. However, in this case, the final result can only be obtained by numerical integration.

The third assumption restricts the validity of Eq. (A.13) at high current densities (~100 A/cm² or higher for $T_0 \approx 80°K$). Calculations of the temperature rise in the active region of a crystal at high current densities are particularly important in assessing the possibility of increasing the output power of semiconductor lasers. In this case, the interest lies in the range of fairly high electron energies (Chap. III). It has been shown that, in this case, the heating can be regarded as adiabatic, which simplifies the problem considerably and makes it possible to apply the calculation procedure outlined below, making allowance for the dependence of the specific heat on the temperature rise.

The specific heat of gallium arsenide [115], like the specific heat of most other crystals, is described quite accurately by the well-known Debye law and this also applies to the temperature dependence of the specific heat in the 80–400°K range [120, 121]. The thermal energy per unit volume of heat is described by

$$u\,(T) = \frac{3RT}{A}\,\rho\,\frac{3}{x^3}\int_0^x \frac{t^3\,dt}{e^t - 1}\,,$$ (A.18)

where R is the universal gas constant; A and ρ are the gram-atom weight and the density of the substance, respectively; $x = \Theta/T$; Θ is the Debye temperature.

---

*At $x_1 = x_0 + 1.5\,\sqrt{2a^2t}$ the ratio of the left- to the right-hand side of the inequality (A.16) depends on x_0 (i.e., on the electron energy E_0) and it decreases from 10 to 1% when x_0 is increased from 0.1 $\sqrt{2a^2t}$ to 3 $\sqrt{2a^2t}$.

190   B. M. LAVRUSHIN

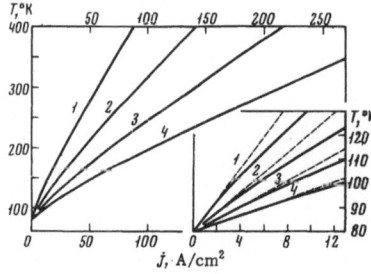

Fig. 38. Temperature of a crystal plotted against the current density for different electron energies E_0 (keV): 1) 50: 2) 100; 3) 200; 4) 500. The continuous curves are plotted on the basis of Eq. (A.18) with the following parameters: $T_0 = 80°K$, $t = 0.1$ μsec, $x = a$ (E_0), $\Theta = 335°K$ [116], $A = 72.3$ g, $\rho = 5.37$ g/cm³ [116]. The dashed curves are based on Eq. (A.14) with $\rho C = 0.859$ J·cm⁻³·deg⁻¹.

When a crystal is bombarded by an electron beam the temperature rise at a point x after a time t can be found from

$$u\,(T) - u\,(T_0) = \frac{2}{3}\,\frac{E_0' j t}{e x_0}\,\exp\left\{-\frac{(x-a)^2}{b^2}\right\},\tag{A.19}$$

where T_0 is the initial temperature; a and b are the parameters of the ionization curve (Introduction, § 3). The right-hand side of the above equation represents the amount of heat evolved per unit volume of the crystal during thermalization of the electron-beam-excited carriers. Since, in this case, the heating is assumed to be adiabatic, it follows that the temperature of the crystal is highest at the point $x = a$. The dependence of the temperature on the current density, given by Eq. (A.18), is plotted in Fig. 38 for four values of the electron energy $E_0 = 50$, 100, 200, and 500 keV. The dashed curves in Fig. 38 are plotted ignoring the temperature dependence of the specific heat and they show clearly the limits of validity of the adiabatic approximation.

We shall now consider the cooling of a crystal after the end of the first pumping pulse $(t_0 \leq t \leq t_1)$. It follows from Eq. (A.11) that the temperature is now described by

$$T\,(x, t) - T_0 = \frac{2}{3}\,\frac{E_0' j t}{e x_0 \rho C}\left\{F\left(\frac{x_0 + x}{\sqrt{2a^2 t}}\right) + F\left(\frac{x_0 - x}{\sqrt{2a^2 t}}\right) - \frac{t - t_0}{t}\left[F\left(\frac{x_0 + x}{\sqrt{2a^2\,(t - t_0)}}\right) + F\left(\frac{x_0 - x}{\sqrt{2a^2\,(t - t_0)}}\right)\right]\right\}.\tag{A.20}$$

However, the thermal conditions in a semiconductor laser can be studied without the very detailed information on the temperature distribution given by Eq. (A.20). All we need to determine is the time t_c during which a crystal cools sufficiently so that the temperature of the active region is close to the initial temperature. This cooling time is fairly long so that the inequalities $t_0 \ll t_c$ and $x_0 \ll \sqrt{2a^2 t}$ are applicable.

We shall consider the limiting case of long time intervals between pulses. We shall assume that $\frac{|x_0 \pm x|}{\sqrt{2a^2 t}} \ll 1$ and $\frac{t_0}{t} \ll 1$ in Eq. (A.20), expand the function F (x) as a series of powers of the argument, and retain terms up to the second power inclusive. If we restrict out treatment to the lowest power of the ratio t_0/t, we find that in the region where $x \leq x_0$, the temperature is

$$T\,(x, t) - T_0 \approx \frac{2}{3}\,\frac{E_0' j t_0}{\rho C e \sqrt{\pi a^2 t}}\tag{A.21}$$

[in this approximation T (x, t) is independent of the coordinate in the region $x \leq x_0$].

We shall define the cooling time t_c as the time during which the temperature of the active region $(x \leq x_0)$ decreases by a factor of 10. We then obtain from Eqs. (A.13) and (A.21) the relationship

$$t_c \approx \frac{100 x_0^2}{4\pi a^2} F^{-2}\left(\frac{x_0}{\sqrt{2a^2 t}}\right). \tag{A.22}$$

If $x_0 = 5\ \mu$ ($E_0 = 50$ keV), $t_0 = 0.1\ \mu sec$, and $a = 3.4$ cm^2/sec (GaAs at $T_0 = 80°K$), we find that $t_c \approx 5.4\ \mu sec$. If the initial temperature T_0 rises to $270°K$ ($a^2 = 0.28$ cm^2/sec), the cooling time of gallium arsenide increases to $29\ \mu sec$. It also follows from Eq. (A.22) that t_c decreases rapidly when the electron energy is raised. For example, if the electron energy is 100 keV ($x_0 = 16\ \mu$), we find that $t_c \approx 25$ and $290\ \mu sec$ for $t_0 \approx 80$ and $270°K$, respectively. We note that the thermal diffusion length corresponding to $t = t_c$ is always about $8x_0$ (and never less than this value.

Equation (A.11) and the formulas (A.20)-(A.22) which follow from it are, strictly speaking, valid only if the thickness of the crystal x_1 is infinite. In the case of crystals of finite thickness, Eq. (A.20) and the limiting cases represented by Eq. (A.21) and (A.22) are sufficiently accurate for practical purposes if the following inequality is satisfied:

$$x_1 \geqslant 2\sqrt{2a^2 t}. \tag{A.23}$$

In fact, it follows from Eq. (A.11) that the amount of heat which crosses the boundary $x = x_1$ in a time interval t is approximately 20 times less than the pumping energy passing during the same time across the boundary $x = 0$.

If the condition (A.23) is not satisfied, the cooling will be more rapid than that predicted by the formula (A.21) or (A.22). In this case, the cooling of a crystal in the interval between the pumping pulses can be described by a time-average value of the temperature. The calculation of such an average temperature for a crystal of arbitrary dimensions is given in Appendix ID.

In this and the preceding sections we have obtained formulas for the temperature of a crystal during and after the first pumping pulse. In fact, these formulas can be used also for time intervals corresponding to the n-th pumping pulse if T_0 is replaced by the temperature at the beginning of this n-th pulse ($t = nt_1$). This temperature can be estimated as described in Sec. D. Under normal conditions, this temperature is low and it depends weakly on the coordinates, compared with the pulse heating condition. This is why the formulas in the present section can be applied to any pumping pulse.

D. Time-Average Value of the Temperature Rise in a Crystal under Steady-State Conditions

In this section we shall consider the time-average value of the temperature

$$\overline{T}(x, t) = \frac{1}{t_1} \int_{t}^{t+t_1} T(x, t')\, dt' \tag{A.24}$$

under steady-state conditions when $\overline{T}(x, t)$ is independent of t. We shall made an allowance for the finite thermal conductivity of the heat sink and for the temperature dependence of the conductivity of the semiconductor crystal (GaAs). These allowances are important in the case of large temperature changes encountered in estimates of the maximum average output power of a semiconductor laser. When the temperature dependence of the thermal conductivity $\varkappa$ is taken into account, the heat conduction equation becomes

$$\frac{\partial Q(x, t)}{\partial x} = w(x, t) + \rho C \frac{\partial T(x, t)}{\partial t}, \tag{A.25}$$

where $Q(x, t) = -\varkappa(T)\frac{\partial T}{\partial x}$ is the heat flux and w(x, t) is given by Ea. (A.1).

Averaging over the period t_1, we obtain the following equations which are applicable to steady-state conditions:

$$\frac{d\overline{Q}(x)}{\partial x} = \overline{w}(x), \tag{A.26}$$

$$-\frac{1}{t_1}\int_t^{t+t_1} \varkappa[T(x, t')]\frac{\partial T(x, t')}{\partial x}\,dt' = \overline{Q}(x). \tag{A.27}$$

The solution of Eq. (A.27) satisfying the condition $\overline{Q}(0) = 0$ is

$$\overline{Q}(x) = \int_0^x \overline{w}(x')\,dx'. \tag{A.28}$$

We shall also assume that the change in the thermal conductivity during a single pulse is quite small, i.e., that

$$\varkappa[T(x, t_0)] - \varkappa[T(x, 0)] \ll \varkappa[\overline{T}(x)]. \tag{A.29}$$

Then, ignoring the dependence of $\varkappa$ on the integration variable t', we find that Eq. (A.28) yields

$$-\varkappa(\overline{T})\frac{\partial \overline{T}}{\partial x} = \int_0^x \overline{w}(x')\,dx'. \tag{A.30}$$

The temperature distributions in the semiconductor ($0 \leq x \leq x_1$) and in the heat sink ($x_1 \leq x \leq x_2$) can be found by solving Eq. (A.30) for each of these regions, subject to the following boundary conditions:

$$\begin{aligned}\overline{T}_1(x_1) &= \overline{T}_2(x_1),\\ \overline{T}_2(x_2) &= T_0,\end{aligned} \tag{A.31}$$

where $\overline{T}_1(x)$ and $\overline{T}_2(x)$ are the temperature distributions in the semiconductor and the heat sink, respectively.

The solution of this problem can be written in the form

$$\int_{T_0}^{\overline{T}_2(x)} \varkappa_2(T)\,dT = \int_x^{x_2} dx' \int_0^{x'} \overline{w}(x'')\,dx'', \tag{A.32}$$

$$\int_{\overline{T}_2(x_1)}^{\overline{T}_1(x)} \varkappa_1(T)\,dT = \int_x^{x_1} dx' \int_0^{x_1} \overline{w}(x'')\,dx'', \tag{A.33}$$

where $\varkappa_1(\overline{T})$ and $\varkappa_2(\overline{T})$ are the thermal conductivities of the semiconductor and the heat sink, respectively.

The relationships given by Eqs. (A.32) and (A.33) can be made more specific by postulating definite temperature dependences of $\varkappa_1$ and $\varkappa_2$ and by selecting a suitable approximation for the distribution $\overline{w}(x)$.

We shall consider the case when the temperature dependence of the thermal conductivity $\varkappa_1$ is of the form[*]

$$\varkappa_1(T) = \varkappa_1(T_0)\left(\frac{T_0}{T}\right)^{3/2}, \tag{A.34}$$

where $\varkappa_1(T_0)$ is independent of the temperature and the power density of the thermal losses experienced by the electron beam $\overline{w}(x)$ can be approximated by the expression

$$\overline{w}(x) = \frac{2}{3}\frac{E_0'\overline{j}}{ex_0}\exp\left\{-\frac{(x-a)^2}{b^2}\right\} \quad \text{for} \quad x \geqslant 0. \tag{A.35}$$

Here, $\overline{j}$ is the time average value of the current density; a and b are the parameters of the ionization curve.

Integrating Eqs. (A.32) and (A.33), we obtain

$$\overline{T}_2(x) - T_0 = \frac{2}{3}\frac{E_0'\overline{j}b\sqrt{\pi}}{ex_0\varkappa_2}\left\{(x_2-x)\,\Phi_0\left(\frac{a\sqrt{2}}{b}\right) + (x_2-a)\,\Phi_0\left(\frac{x_2-a}{b}\sqrt{2}\right) + \right.$$
$$\left. + \frac{b}{2\sqrt{\pi}}\exp\left[-\frac{(x_2-a)^2}{b^2}\right] - (x-a)\,\Phi_0\left(\frac{x-a}{b}\sqrt{2}\right) - \frac{b}{2\sqrt{\pi}}\exp\left[-\frac{(x-a)^2}{b^2}\right]\right\} \tag{A.36}$$

if $x_1 \leq x \leq x_2$, and

$$\sqrt{\frac{\overline{T}_2(x_1)}{\overline{T}_1(x)}} = 1 - \frac{2E_0'\overline{j}b\sqrt{\pi}}{3ex_0\overline{T}_2(x_1)\cdot 2\varkappa_1[\overline{T}_2(x_1)]}\left\{(x_1-x)\,\Phi_0\left(\frac{a\sqrt{2}}{b}\right) + (x_1-a)\,\Phi_0\left(\frac{x_1-a}{b}\sqrt{2}\right) + \right.$$
$$\left. + \frac{b}{2\sqrt{\pi}}\exp\left[-\frac{(x_1-a)^2}{b^2}\right] - (x-a)\,\Phi_0\left(\frac{x-a}{b}\sqrt{2}\right) - \frac{b}{2\sqrt{\pi}}\exp\left[-\frac{(x-a)^2}{b^2}\right]\right\} \tag{A.37}$$

if $0 \leq x \leq x_1$.

The thickness x_1 of the semiconductor crystal, which appears in the above formulas, should be considerably greater than the depth of penetration of the electrons x_0 because, otherwise, a considerable fraction of the electron-beam energy will be lost without doing any useful work. Therefore, we shall assume that $x_1 \geq 2x_0$. In this case, we obtain

$$\Phi_0\left(\frac{x_1-a}{b}\sqrt{2}\right) \approx \frac{1}{2}, \qquad \exp\left[-\frac{(x_1-a)^2}{b^2}\right] \ll \exp\left[-\frac{(x-a)^2}{b^2}\right].$$

Using the relationship between the parameters a, b, and x_0, we find that Eqs. (A.36) and (A.37) yield the following approximate formulas for the average temperature in the $x = 0$ plane where the heating is strongest:

$$\sqrt{\frac{\overline{T}_2(x_1)}{\overline{T}_1(0)}} \approx 1 - \frac{2E_0'\overline{j}}{3e\overline{T}_2(x_1)\varkappa_1[\overline{T}_2(x_1)]}\left(\frac{x_1}{2} - 0.24x_0\right), \tag{A.38}$$

$$\overline{T}_2(x_1) \approx T_0 + \frac{2}{3}\frac{E_0'\overline{j}}{e\varkappa_2}(x_2-x_1). \tag{A.39}$$

Under the conditions encountered experimentally, the average temperature $\overline{T}_1(0)$ differs little

---

[*]The temperature dependence of the thermal conductivity of gallium arsenide can be approximated by Eq. (A.34) in the range $K \leq t \leq 300°K$.

Fig. 39. Pumping power density (kW/cm^2) at which the average temperature of the surface of a GaAs crystal reaches 300°K (T_0 = 80°K), plotted against the electron energy. The curve is based on Eqs. (A.38). (A.39). and (A.34) with x_1 = $2x_0$, x_2-x_1 = 1 mm, $\varkappa_2$ = 4 W·cm^{-1}· deg^{-1}.

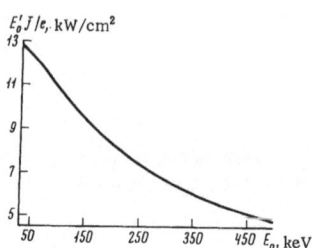

from $\overline{T}_2(x_1)$. Then,

$$\sqrt{\frac{\overline{T}_2(x_1)}{\overline{T}_1(0)}} \approx 1 - \frac{1}{2}\frac{\overline{T}_1(0)-\overline{T}_2(x_1)}{\overline{T}_2(x_1)}$$

and, instead of Eq. (A.37), we obtain*

$$\overline{T}_1(0) \approx \overline{T}_2(x_1) + \frac{2}{3}\frac{E_0'\overline{j}(x_1-0{,}48x_0)}{e\varkappa_1[\overline{T}_2(x_1)]}.\tag{A.40}$$

In particular, if E_0 = 50 keV (x_0 = 5 × 10^{-4} cm), j = 10 A/cm^2, t = 10^{-7} sec, f = 50 Hz, x_1 = 1 mm, x_2 = 5 mm, $\varkappa_1$ = 2.95 W·cm^{-1}·deg^{-1} (GaAs at T ≈ 80°K), and $\varkappa_2$ = 4 W·cm^{-1}·deg^{-1}, we find that $\overline{T}_2(x_1)-T_0 \approx 0.13$ deg and $\overline{T}_1(0)-T_0 \approx 0.19$ deg. The pulse repetition frequency f_0 at which the average temperature rise in the x = 0 plane becomes equal to the rise during a current pulse is given by

$$f_0 = \frac{2F\left(\dfrac{x_0}{\sqrt{2a^2t_0}}\right)}{x_0\rho C\left(\dfrac{x_2-x_1}{\varkappa_2}+\dfrac{x_1-0.48x_0}{\varkappa_1}\right)},\tag{A.41}$$

where F (x) is given by Eq. (A.12). For the values of the parameters just quoted, we find that $f_0 \approx 11$ kHz.

Figure 39 shows the effect of increasing the pulse repetition frequency for the semiconductor laser considered above. This figure is applicable to the case when the pumping pulses are short so that the temperature rise during a single pulse is considerably less than the average temperature rise in the case of a pulse train and the semiconductor (GaAs) crystal can be heated to 300°K from the initial temperature T_0 = 80°K. In this case, the dependence of the highest permissible pumping power density $E_0'\overline{j}/e$ (kW/cm^2) on the electron energy E_0 is given by the curve in Fig. 39. This dependence is plotted using Eqs. (A.38) and (A.39) on the assumption that the thickness of the semiconductor crystal (GaAs) is $2x_0$ (x_0 is the depth of penetration of the electrons), the thickness of the heat sink is (x_2-x_1) = 1 mm, and the thermal conductivity of the heat sink is $\varkappa_2$ = 4 W·cm^{-1}·deg^{-1}.

If these data are used, we can show that, in particular, a laser can be operated at a pulse repetition frequency of ~1.3 MHz if the pulse duration is 10 nsec, the electron energy is E_0 = 100 keV, and the density of the electron-beam current is j = 10 A/cm^2.

It also follows from these data that in order to operate a laser continuously during bombardment with electrons of E_0 = 50 keV energy, the laser threshold of the semiconductor should not exceed 0.25 A/cm^2 at room temperature.

---

*Equation (A.40) can also be derived directly from Eq. (A.33) if the temperature dependence of $\varkappa_1$ is ignored.

II. TABLE OF STRUVE FUNCTIONS

$$L_0(x) = \frac{2}{\pi} \sum_{m=0}^{\infty} \frac{x^{2m+1}}{[(2m+1)!!]^2}, \qquad L_1(x) = \frac{2}{\pi} \sum_{m=0}^{\infty} \frac{x^{2m+2}}{(2m+1)!!\,(2m+2)!!}$$

x	$L_0(x)$	$L_1(x)$	x	$L_0(x)$	$L_1(x)$
0.1	0.063733	0.0021235	5.1	29.6578	26.0718
0.2	0.127891	0.0085110	5.2	32.4552	28.6445
0.3	0.192903	0.019213	5.3	35.5224	31.4690
0.4	0.259204	0.034317	5.4	38.8856	34.5702
0.5	0.327241	0.053942	5.5	42.5739	37.9753
0.6	0.397472	0.078247	5.6	46.6191	41.7145
0.7	0.470376	0.107426	5.7	51.0563	45.8209
0.8	0.546452	0.141714	5.8	55.9241	50.3308
0.9	0.626223	0.181387	5.9	61.2645	55.2841
1.0	0.710243	0.226764	6.0	67.1245	60.7250
1.1	0.799102	0.278213	6.1	73.5548	63.7018
1.2	0.893426	0.336148	6.2	80.6117	73.2676
1.3	0.993887	0.401039	6.3	88.3572	80.4811
1.4	1.101207	0.473412	6.4	96.8590	88.4066
1.5	1.216162	0.553857	6.5	106.1919	97.1149
1.6	1.339593	0.643028	6.6	116.4380	106.6841
1.7	1.472407	0.741652	6.7	127.6876	117.1996
1.8	0.615589	0.850539	6.8	140.0400	128.7560
1.9	1.770210	0.970579	6.9	153.6043	141.4569
2.0	1.937434	1.102760	7.0	168.5006	151.4166
2.1	2.11853	1.24817	7.1	184.861	170.760
2.2	2.31487	1.40801	7.2	202.831	187.827
2.3	2.52798	1.58361	7.3	222.570	206.168
2.4	2.75949	1.77641	7.4	244.253	226.551
2.5	3.01121	1.98803	7.5	268.075	248.960
2.6	3.28511	2.22023	7.6	294.247	273.598
2.7	3.58333	2.47494	7.7	323.003	300.687
2.8	3.90824	2.75431	7.8	354.601	330.474
2.9	4.26242	3.06067	7.9	389.324	363.228
3.0	4.64868	3.39661	8.0	427.483	399.247
3.1	5.07014	3.76496	8.1	469.421	438.858
3.2	5.53019	4.16883	8.2	515.514	482.421
3.3	6.03253	4.61164	8.3	566.177	530.333
3.4	6.58126	5.09715	8.4	621.867	583.030
3.5	7.18085	5.62950	8.5	683.086	640.993
3.6	7.83620	6.21322	8.6	750.386	704.750
3.7	8.55269	6.85330	8.7	824.376	774.884
3.8	9.33624	7.55523	8.8	905.724	852.035
3.9	10.19332	8.32501	8.9	995.167	936.911
4.0	11.13105	9.16928	9.0	1093.517	1030.286
4.1	12.1572	10.0953	9.1	1201.66	1133.02
4.2	13.2804	11.1111	9.2	1320.59	1246.05
4.3	14.5100	12.2254	9.3	1451.38	1370.41
4.4	15.8564	13.4478	9.4	1595.22	1507.25
4.5	17.3309	14.7891	9.5	1753.41	1657.82
4.6	18.9459	16.2608	9.6	1927.41	1823.52
4.7	20.7151	17.8757	9.7	2118.80	2005.85
4.8	22.6536	19.6481	9.8	2329.32	2206.59
4.9	24.7778	21.5932	9.9	2560.90	2427.33
5.0	27.1059	23.7282	10.0	2815.65	2670.36

LITERATURE CITED

1. N. G. Basov, Advances in Quantum Electronics (Proc. Second Intern. Conf., Berkeley, California, (1961), Columbia University Press, New York (1961), p. 506 (discussion).
2. N. G. Basov and O. V. Bogdankevich, Zh. Eksp. Teor. Fiz., 44:1115 (1963).
3. N. G. Basov, O. V. Bogdankevich, and A. G. Devyatkov, Dokl. Akad. Nauk SSSR, 155:783 (1964); Zh. Eksp. Teor. Fiz., 47:1588 (1964).
4. C. E. Hurwitz and R. J. Keyes, Appl. Phys. Lett., 5:139 (1964).
5. D. A. Cusano, Solid State Commun., 2:353 (1964).
6. D. A. Cusano and J. D. Kingsley, Appl. Phys. Lett., 6:91 (1965).
7. L. N. Kurbatov, A. N. Kabanov, V. V. Sigriyanskii, V. E. Mashchenko, N. N. Mochalkin, A. I. Sharin, and N. V. Soroko-Novitskii, Dokl. Akad. Nauk SSSR, 165:303 (1965).
8. N. G. Basov, O. V. Bogdankevich, and B. M. Lavrushin, Gallium Arsenide Laser Excited by Fast Electrons, Preprint No. A-85 [in Russian], P. N. Lebedev Physics Institute, Academy of Sciences of the USSR, Moscow (1965); Fiz. Tverd. Tela, 8:21 (1966).
9. C. Benoit à la Guillaume and J. M. Debever, Proc. Seventh Intern. Conf. on Physics of Semiconductors, Paris, 1964, Vol. 4, Radiative Recombination in Semiconductors, publ. by Dunod, Paris; Academic Press, New York (1965), p. 255.
10. C. Benoit à la Guillaume and J. M. Debever, Solid State Commun., 2:145 (1964).
11. C. Benoit à la Guillaume and J. M. Debever, C. R. Acad. Sci., 259:2200 (1964) [Russian translation in: Lasers (ed. by F. V. Bunkin), IL, Moscow (1966)].
12. V. S. Vavilov and É. L. Nolle, Dokl. Akad. Nauk SSSR, 164:73 (1965).
13. É. L. Nolle, Spontaneous and Stimulated Recombination Radiation Emitted from Electron-Bombarded Cadmium Telluride (Thesis) [in Russian], P. N. Lebedev Physics Institute, Academy of Sciences of the USSR, Moscow (1965).
14. C. E. Hurwitz, Appl. Phys. Lett., 8:121 (1966).
15. N. G. Basov, O. V. Bogdankevich, A. N. Pechenov, G. B. Abdullaev, G. A. Akhundov, and É. Yu. Salaev, Dokl. Akad. Nauk SSSR, 161:1059 (1965).
16. F. H. Nicoll, Appl. Phys. Lett., 9:13 (1966).
17. C. E. Hurwitz, Appl. Phys. Lett., 9:116 (1966).
18. C. E. Hurwitz, IEEE J. Quantum Electron., QE-3:333 (1967).
19. O. V. Bogdankevich, M. M. Zverev, A. I. Krasil'nikov, and A. N. Pechenov, Phys. Status Solidi, 19:K5 (1967); O. V. Bogdankevich and M. M. Zverev (Sverev), II—VI Semiconducting Compounds (Proc. Intern. Conf., Providence, R. I., 1967), Benjamin, New York (1967).
20. C. Benoit à la Guillaume and J. M. Debever, Solid State Commun., 3:19 (1965).
21. C. E. Hurwitz, A. R. Calawa, and R. H. Rediker, IEEE J. Quantum Electron., QE-1:102 (1965).
22. F. M. Berkovskii, N. A. Goryunova, V. M. Orlov, S. M. Ryvkin, V. I. Sokolova, E. V. Tsvetkova, and G. P. Shpen'kov, Fiz. Tekh. Poluprov., 2:1218 (1968).
23. C. E. Hurwitz, Appl. Phys. Lett., 8:243 (1966).
24. N. G. Basov, O. V. Bogdankevich, P. G. Eliseev, and B. M. Lavrushin, Fiz. Tverd. Tela, 8:1341 (1966).
25. N. G. Basov, O. V. Bogdankevich, and A. G. Devyatkov, Proc. Seventh Intern. Conf. on Physics of Semiconductors, Paris, 1964, Vol. 4, Radiative Recombination in Semiconductors, publ. by Dunov, Paris; Academic Press, New York (1965), p. 225.
26. S. P. Owen, Phil. Mag., 47:736 (1924).
27. F. H. Pollak, C. W. Higginbotham, and M. Cardona, Proc. Eighth Intern. Conf. on Physics of Semiconductors, Kyoto, 1966, in: J. Phys. Soc. Jap., 21(Supplement):20 (1966).
28. M. Cardona, "Optical absorption above the fundamental edge," in: Semiconductors and Semimetals (ed. by R. K. Willardson and A. C. Beer), Academic Press, New York (1967), p. 125.

29. M. D. Sturge, Phys. Rev., 127:768 (1962).
30. R. C. Casella, J. Appl. Phys., 34:1703 (1963).
31. F. Stern and R. M. Talley, Phys. Rev., 100:1638 (1955).
32. W. Baltensperger, Phil. Mag., 44:1355 (1953).
33. J. I. Pankove, Proc. Eighteh Intern. Conf. on Physics of Semiconductors, Kyoto, 1966, in: J. Phys. Soc. Jap., 21(Supplement):298 (1966).
34. O. V. Emel'yanenko, T. S. Lagunova, Ọ. N. Nasledov, and G. N. Talalakin, Fiz. Tverd. Tela, 7:1315 (1965).
35. G. Lucovsky and A. J. Varga, J. Appl. Phys., 35:3419 (1964).
36. L. V. Keldysh, Semiconductors in Strong Electric Fields (Thesis) [in Russian], P. N. Lebedev Physics Institute, Academy of Sciences of the USSR, Moscow (1965).
37. E. O. Kane, Phys. Rev., 131:79 (1963).
38. V. L. Bonch-Bruevich, "Problems in electron theory of heavily doped semiconductors," in: Solid State Physics [in Russian], Moscow (1965).
39. T. N. Morgan, Phys. Rev., 139:A343 (1965).
40. K. G. Hambleton, C. Hilsum, and B. R. Holeman, Proc. Phys. Soc., London 77:1147 (1961).
41. E. D. Palik, J. R. Stevenson, and R. F. Wallis, Phys. Rev., 124:701 (1961).
42. O. Madelung, Physics of III–V Compounds, Wiley, New York (1964).
43. F. Stern, Phys. Rev., 148:186 (1966).
44. T. S. Moss and T. D. F. Hawkins, Infrared Phys., 1:111 (1961).
45. D. E. Hill, Phys. Rev., 133:A866 (1964).
46. G. Lucovsky, Appl. Phys. Lett., 5:37 (1964).
47. J. I. Pankove, Phys. Rev., 140:A2059 (1965).
48. A. S. Davydov, Quantum Mechanics, Pergamon Press, Oxford (1965).
49. J. Bardeen, F. J. Blatt, and L. H. Hall, Proc. Conf. on Photoconductivity, Atlantic City, 1954, publ. by Wiley, New York (1956), p. 146.
50. G. Lasher and F. Stern, Phys. Rev., 133:A553 (1964).
51. M. H. Pilkukn, Phys. Status Solidi, 25:9 (1968).
52. J. S. Blakemore, Semiconductor Statistics, Pergamon Press, Oxford (1962).
53. I. N. Oraevskii, Yu. M. Popov, and G. M. Strakhovskii, Phys. Status Solidi, 32:55 (1969).
54. E. Sègre (ed.), Experimental Nuclear Physics, Vol. 1, Wiley, New York (1953).
55. Yu. M. Popov, Tr. Fiz. Inst. Akad. Nauk SSSR, 31:3 (1965).
56. V. S. Vavilov, Effects of Radiation on Semiconductors, Consultants Bureau, New York (1965).
57. C. A. Klein, J. Appl. Phys., 39:2029 (1968).
58. B. Ya. Yurkov, Zh. Tekh. Fiz., 28:1159 (1958).
59. L. V. Spencer, Phys. Rev. 98:1597 (1955).
60. W. Ehrenberg and D. E. N. King, Proc. Phys. Soc. London, 81:751 (1963).
61. J. E. Holliday and E. J. Sternglass, J. Appl. Phys., 28:1189 (1957).
62. V. L. Levshin, É. Ya. Arapova, A. I. Blazhevich, Yu. V. Voronov, I. G. Voronova, V. B. Gutan, A. V. Lavrov, Yu. M. Popov, S. A. Fridman, V. A. Chikhacheva, and V. V. Shchaenko, Tr. Fiz. Inst. Akad. Nauk SSSR, 23:64 (1963).
63. O. N. Krokhin and Yu. M. Popov, Zh. Eksp. Teor. Fiz., 38:1589 (1960).
64. O. V. Bogdankevich, Semiconductor Laser Excited by a Beam of Accelerated Electrons (Thesis) [In Russian], P. N. Lebedev Physics Institute, Academy of Sciences of the USSR, Moscow (1966).
65. V. P. Sushkov and Yu. A. Moma, Fiz. Tekh. Poluprov., 1:1531 (1967).
66. A. V. Dudenkova and V. V. Nikitin, Fiz. Tverd. Tela, 9:851 (1967).
67. A. S. Nasibov, Generators of High-Power Short High-Voltage Pulses (Thesis) [in Russian], P. N. Lebedev Physics Institute, Academy of Sciences of the USSR, Moscow (1966). A. S. Nasibov and V. L. Lomakin, Prib. Tekh. Eksp., No. 3, p. 123 (1965).

68. N. G. Basov, P. G. Eliseev, S. D. Zakharov, Yu. P. Zakharov, I. N. Oraevskii, I. Z. Pinsker, and V. P. Strakhov, Fiz. Tverd. Tela, 8:2616 (1966).
69. G. Cheroff, F. Stern, and S. Triebwasser, Appl. Phys. Lett., 2:173 (1963).
70. J. R. Biard, W. N. Carr, and B. S. Reed, Trans. AIME, 230:286 (1964).
71. N. G. Basov, O. V. Bogdankevich, and A. G. Devyatkov, Fiz. Tverd. Tela, 8:1536 (1966).
72. V. S. Vavilov, É. L. Nolle, G. P. Golubev, V. S. Mashtakov, and E. I. Tsarapaeva, Fiz. Tverd. Tela, 9:842 (1967).
73. C. E. Hurwitz, Appl. Phys. Lett., 9:420 (1966).
74. O. V. Bogdankevich, N. A. Borisov, I. V. Kryukova, and B. M. Lavrushin, Output Power and Efficiency of Electron-Beam-Pumped Gallium Arsenide Lasers, Preprint No. 144 [in Russian], P. N. Lebedev Physics Institute, Academy of Sciences of the USSR, Moscow (1967); Fiz. Tekh. Poluprov., 2:1017 (1968).
75. W. J. Turner and W. E. Reese, J. Appl. Phys., 35:350 (1964).
76. C. A. Klein, Appl. Phys. Lett., 7:200 (1965).
77. Tables of Normal Error Function, Normal Density, and Its Normalized Derivatives [in Russian], Moscow (1960).
78. C. A. Klein, IEEE J. Quantum Electron., QE-4:186 (1968).
79. G. C. Dousmanis, H. Nelson, and D. L. Staebler, Appl. Phys. Lett., 5:174 (1964).
80. M. Pilkuhn, H. Rupprecht, and S. Blum, Solid-State Electron., 7:905 (1964).
81. P. G. Eliseev, I. Ismailov, A. I. Krasil'nikov, and M. A. Man'ko, Fiz. Tekh. Poluprov., 1:951 (1967).
82. V. I. Leskovich, G. T. Pak, A. I. Petrov, N. P. Chernousov, and V. I. Shveikin, Fiz. Tekh. Poluprov., 1:1440 (1967).
83. O. V. Bogdankevich, N. A. Borisov, I. V. Kryukova (Krukova), and B. M. Lavrushin, Phys. Status Solidi, 29:715 (1968).
84. O. V. Bogdankevich, N. A. Borisov, I. V. Kryukova, and B. M. Lavrushin, Proc. Ninth Intern. Conf. on Physics of Semiconductors, Moscow, 1968, Vol. 1, publ. by Nauka, Leningrad (1968), p. 575.
85. O. N. Krokhin, Fiz. Tverd. Tela, 7:2612 (1965).
86. M. I. Nathan, A. B. Fowler, and G. Burns, Phys. Rev. Lett., 11:152 (1963).
87. T. F. Nikitina, Yu. M. Popov, G. M. Strakhovskii, and N. N. Shukin, Fiz. Tekh. Poluprov., 3:164 (1969).
88. D. F. T. Marple, J. Appl. Phys., 35:1241 (1964).
89. G. Burns and M. I. Nathan, Proc. IEEE, 52:770 (1964).
90. F. Stern. Solid State Phys., 15:299 (1963).
91. E. Burstein, Phys. Rev., 93:632 (1954).
92. R. S. Knox, Theory of Excitons, Suppl. No. 5 to Solid State Phys., Academic Press, New York (1963).
93. C. Kittel, Quantum Theory of Solids, Wiley, New York (1963).
94. J. Callaway, Solid State Phys., 7:99 (1958).
95. A. A. Rogachev, Investigation of Optical Properties of Heavily Doped Semiconductors (Thesis) [in Russian], A. F. Ioffe Physicotechnical Institute, Leningrad (1967).
96. A. F. Suchkov, Zh. Eksp. Teor. Fiz., 49:1495 (1965).
97. V. S. Letokhov and A. F. Suchkov, Zh. Eksp. Teor. Fiz., 50:1148 (1966).
98. O. V. Bogdankevich, V. A. Goncharov, B. M. Lavrushin, V. S. Letokhov, and A. F. Suchkov, Fiz. Tekh. Poluprov., 1:7 (1967).
99. R. J. Collins, D. F. Nelson, A. L. Sachawlow, W. Bond, C. G. B. Garrett, and W. Kaiser, Phys. Rev. Lett., 5:303 (1960).
100. O. V. Bogdankevich, V. S. Letokhov, and A. F. Suchkov, Fiz. Tekh. Poluprov., 3:665 (1969).
101. N. G. Basov, O. V. Bogdankevich, A. N. Pechenov, A. S. Nasibov, and K. P. Fedoseev, Zh. Eksp. Teor. Fiz., 55:1710 (1968).

102. L. A. Vainshtein, Open Resonators and Open Waveguides [in Russian], Sovetskoye Radio, Moscow (1966).
103. O. V. Bogdankevich, V. A. Goncharov, Yu. A. Drozhbin, B. M. Lavrushin, A. N. Mestvirishvili, and V. A. Yakovlev, Zh. Eksp. Teor. Fiz., 53:785 (1967).
104. N. G. Basov, V. N. Morozov, and A. N. Oraevskii, Dokl. Akad. Nauk SSSR, 162:781 (1965); Zh. Eksp. Teor. Fiz., 49:895 (1965).
105. V. D. Kurnovos, V. I. Magalyas, A. A. Pleshkov, L. A. Rivlin, V. G. Trukhan, and V. V. Tsvetkov, ZhETF Pis. Red., 4:449 (1966).
106. Yu. A. Drozhbin, Yu. P. Zakharov, V. V. Nikitin, A. S. Semenov, and V. A. Yakovlev, ZhETF Pis. Red., 5:180 (1967).
107. N. G. Basov, O. V. Bogdankevich, V. A. Goncharov, B. M. Lavrushin, and V. Yu. Sudzilovskii, Dokl. Akad. Nauk SSSR, 168:1283 (1966).
108. É. M. Belenov and V. S. Letokhov, Zh. Tekh. Fiz., 35:2126 (1965).
109. N. G. Basov, É. M. Belenov, and V. S. Letokhov, Zh. Tekh. Fiz., 35:1098 (1965).
110. R. Bäuerlein, Z. Phys., 176:498 (1963).
111. O. V. Bogdankevich, V. S. Vavilov, V. A. Danilychev, V. V. Kalendin, and I. V. Kryukova, Influence of Irradiation on Principal Characteristics of Electron-Beam-Pumped Gallium Arsenide Lasers, Preprint No. 127 [in Russian], P. N. Lebedev Physics Institute, Academy of Sciences of the USSR, Moscow (1969); Fiz. Tekh. Poluprov., 4:1209 (1970).
112. N. G. Basov, A. Z. Grasyuk, V. F. Efimkov, and V. A. Katulin, Fiz. Tverd. Tela, 9:88 (1967).
113. I. G. Zubarev, Two-Photon Excitation of Semiconductor Lasers (Thesis) [in Russian], P. N. Lebedev Physics Institute, Academy of Sciences of the USSR, Moscow (1968).
114. I. S. Gradshteyn and I. M. Ryzhik (eds.), Table of Integrals, Series, and Products, Academic Press, New York (1965).
115. M. G. Holland, Phys. Rev., 134:A471 (1964); M. G. Holland, "Thermal conductivity," in: Semiconductors and Semimetals (ed. by R. K. Willardson and A. C. Beer), Vol. 2, Academic Press, New York (1966), p. 3.
116. U. Piesbergen, Z. Naturforsch., 18a:141 (1963); U. Piesbergen, "Heat capacity and Debye temperatures," in: Semiconductors and Semimetals (ed. by R. K. Willardson and A. K. Beer), Vol. 2, Academic Press, New York (1966), p. 49.
117. P. D. Maycock, Solid-State Electron., 10:161 (1967).
118. M. M. Bredov, Zh. Tekh. Fiz., 25:2104 (1955).
119. N. A. Borisov, Investigation of Spectral and Temperature Characteristics of Electron-Beam-Pumped Gallium Arsenide Lasers (Diploma Thesis), [in Russian], Moscow Physicotechnical Institute and P. N. Lebedev Physics Institute, Academy of Sciences of the USSR, Moscow (1967).
120. P. Debye, Ann. Phys. (Leipzig), 39:789 (1912).
121. L. D. Landau and E. M. Lifshitz, Statistical Physics, 2nd ed., Pergamon Press, Oxford (1969).
122. J. A. Beattie, J. Math. Phys. (Cambridge, Mass.), 6:1 (1926).
123. H. H. Landolt and R. Börnstein, Physikalisch-chemische Tabellen, Vol. 1, 5th ed., Springer, Berlin (1927), p. 705.

INVESTIGATIONS OF SOME FOUR-FIELD
PARAMETRIC INTERACTIONS

B. P. Kirsanov

A theoretical investigation is made of the four-field nonlinear mirror effect, in which a nonlinear medium exhibits frequency-selective reflection of some waves in the presence of other waves. Various mechanisms of such selection are considered and it is shown that the reflection coefficient of a nonlinear mirror may reach 100%. A study is made also of mechanisms of the four-field parametric frequency selection within wide stimulated emission lines. The possibility of observation of the effects in question is discussed and their practical applications are considered.

Introduction

Nonlinear parametric interaction of four fields is closely related to such interesting phenomena as frequency tripling [1, 2], modulation of the refractive index by a strong field [3, 4], Raman scattering of light [1, 2], small-angle "scattering of light by light" [5], etc. Veduta and Kirsanov [6] have investigated theoretically and experimentally (special cases) the parametric conversion of two quanta $\hbar\omega_1$, $\hbar\omega_2$ into two other quanta $\hbar\omega_3$, $\hbar\omega_4$ in a transparent nonlinear medium. Kirsanov and Selivanenko [7] have studied such parametric conversion under resonance conditions [7]. The parametric interaction of four waves is particularly effective when the spatial and time phase-matching conditions are satisfied. These conditions represent the laws of conservation of energy and momentum, which can be expressed as follows:

$$\omega_1 + \omega_2 = \omega_3 + \omega_4,$$
$$\mathbf{k}_1(\omega_1) + \mathbf{k}_2(\omega_2) = \mathbf{k}_3(\omega_3) + \mathbf{k}_4(\omega_4).$$

The second phase-matching condition may be satisfied by various spatial combinations because of the large number (four) of interacting waves. The present paper will be concerned with those cases when $\omega_1 = \omega_3$, $\omega_2 = \omega_4$ and $\mathbf{k}_2(\omega_2) = -\mathbf{k}_4(\omega_4)$. The earlier papers [6, 7] have dealt with the parametric interaction of four electromagnetic waves in which the two original quanta $\hbar\omega_1$, $\hbar\omega_2$ are converted into two other quanta $\hbar\omega_3$, $\hbar\omega_4$ in a single event. We shall discuss here the case when the parametric interaction of four fields cannot be regarded as a single elementary event. For example, two waves can interfere and become absorbed producing a spatially periodic structure of excited molecules which interact with two different waves.

These two types of parametric interaction can conveniently be discussed by considering periodic structures or gratings which result from the modulation of various parameters of a medium. This approach clarifies the physical nature of the phenomena under discussion. Quantitative estimates of the "diffraction" of optical fields by such structures can conveniently be obtained from Maxwell's equations and nonlinear polarizations, which result from the parametric interaction of four fields. We shall discuss theoretically the nonlinear mirror effect in which a

201

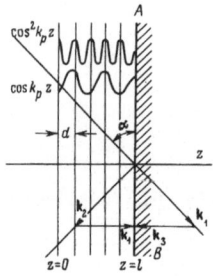

Fig. 1. Configuration of pumping
fields. AB represents a mirror
tuned to the frequence ω_1; α is the
glancing angle of the wave k_1;
$k_p = (k_1 - k_2)/2$; d is the period of
the standing wave with the $\cos^2 kr$
profile.

nonlinear medium reflects some waves in the presence of others. Moreover, we shall study
various mechanisms of the parametric frequency selection within wide stimulated emission
lines. We shall consider the possibility of observing these effects and applying them in practice.

§1. Nonlinear Mirror

Let us consider the situation pictured in Fig. 1. We shall assume that the pumping wave

$$E_1 = E(\omega_1) \exp(i\omega_1 t - i\mathbf{kr}) + \text{c.c.} \tag{1.1}$$

passes through a nonlinear medium and is totally reflected from a mirror AB. We shall postulate that the polarizations of all fields are collinear and directed at right-angles to the plane
of the figure. Interference produces a combined pumping wave (for the sake of simplicity we
shall assume the initial phase to be zero):

$$E_p = 2\,|\,E(\omega_1)\,|\exp\!\left(i\omega_1 t - \frac{\mathbf{k}_1 + \mathbf{k}_2}{2}\,\mathbf{r}\right)\cos^2 k_p\mathbf{r} + \text{c.c.} \; = 4\,|\,E(\omega_1)\,|\cos\!\left(\omega_1 t - \frac{\mathbf{k}_1 + \mathbf{k}_2}{2}\,\mathbf{r}\right)\cos(k_p\mathbf{r}), \tag{1.2}$$

whose amplitude is modulated by $\cos(k_p\mathbf{r})$, where

$$k_p = \frac{\mathbf{k}_1 - \mathbf{k}_2}{2}. \tag{1.3}$$

All the parameters of the medium are modulated along the direction k_p by $\cos^2(k_p\mathbf{r})$. The
planes of maxima of the grating produced in this way are separated by distances

$$d = \frac{\lambda_1}{2\sin\alpha}, \tag{1.4}$$

where λ_1 is the wavelength of the pump wave in the nonlinear medium and α is the glanding angle shown in Fig. 1. In particular, this situation leads to the formation of a grating of the correction to the refractive index which acts as an interference mirror and reflects a wave of
wavelength λ' whose glancing angle φ satisfies the Bragg law (the mirror AB reflects only the
λ_1 wave):

$$\lambda' = \frac{2d\sin\varphi}{m} = \frac{\lambda_1\sin\varphi}{m\sin\alpha}, \tag{1.5}$$

where m is an integer. If $\varphi = \pi/2$ and m = 1, we find that

$$\lambda' = 2d = \frac{\lambda_1}{\sin\alpha}. \tag{1.6}$$

We shall now consider some specific properties of gratings of this kind and we shall estimate the reflection coefficients of the corresponding nonlinear mirrors.

Grating Formed by Nonlinear Correction to the Refractive

Index of a Transparent Medium

If the fields shown in Fig. 1 are sufficiently strong, a nonlinear medium is modified by a periodic correction (period d) to the refractive index. This nonlinear correction may result from the electronic nonlinear polarizability, striction, orientation, etc. The value of this correction is given by

$$\delta n = \frac{12\pi}{n(\omega_3)} \chi_1 |E(\omega_1)|^2 \cos^2 k_p z. \tag{1.7}$$

Here, $n(\omega_3)$ is the refractive index. Following our earlier treatment [4], we have introduced χ_1 which is the component $\chi_{1111}^{(3)}$ of the third-rank tensor of the nonlinear susceptibility whose magnitude was determined in [4] for several liquids and glasses (the various mechanisms responsible for the correction to the refractive index are also considered in [4]). The reflection by the δ_n grating can be described in terms of parametric interaction of four waves which satisfy the phase-matching conditions [6]:

$$\begin{aligned} \mathbf{k}_1 + \mathbf{k}_3 &= \mathbf{k}_2 + \mathbf{k}_4, \\ \omega_1 + \omega_3 &= \omega_2 + \omega_4. \end{aligned} \tag{1.8}$$

In the case we are considering $\omega_1 = \omega_2$, $\omega_3 = \omega_4 = \omega'$, and $\mathbf{k}_3 = -\mathbf{k}_4 = \mathbf{k}_p$ (Fig. 1). We shall assume that the field E_3 is traveling along the positive direction of the z axis at right-angles to a plane-parallel sample of length l (Fig. 1) in which the necessary configuration of the fields E_1 and E_2 is established in accordance with Eq. (1.8) (the mirror AB is transparent for the ω_3 wave). The reflection coefficient of the nonlinear mirror which is obtained in the situation under consideration can be calculated by considering the fields in their complex form as given by Eq. (1.1). Then, substituting into Maxwell's equations the nonlinear polarizations

$$P^{nl}(\omega_{3,4}) = \chi_1 E(\omega_1) E^*(\omega_2) E(\omega_{4,3}) \exp[i(\omega_1 - \omega_2 + \omega_{4,3})t - i(\mathbf{k}_1 - \mathbf{k}_2 + \mathbf{k}_{1,3})\mathbf{r}], \tag{1.9}$$

and using the standard procedure [1], we obtain – in the approximation of strong fields $E(\omega_1)$, $E(\omega_2)$ – a system of "contracted" equations [6], which is applicable to steady-state conditions in the absence of losses if Eq. (1.8) is satisfied:

$$\begin{aligned} \frac{\partial E(\omega_3)}{\partial z} &= -\frac{2\pi i \omega_3}{c} \chi_1 E(\omega_2) E^*(\omega_1) E(\omega_4), \\ \frac{\partial E(\omega_4)}{\partial z} &= \frac{2\pi i \omega_3}{c} \chi_1 E(\omega_1) E^*(\omega_2) E(\omega_3) \end{aligned} \tag{1.10}$$

(c is the velocity of light), subject to the boundary conditions

$$E(\omega_3)|_{z=0} = E_0, \ E(\omega_4)|_{z=l} = 0. \tag{1.11}$$

If we use real amplitudes and phases $E(\omega_i) = A_i e^{i\varphi_i}$, we obtain the following system of equations

$$\begin{aligned} \frac{\partial A_3}{\partial z} &= \alpha_1 A_4 \sin \Phi, \\ \frac{\partial A_4}{\partial z} &= \alpha_1 A_3 \sin \Phi, \end{aligned} \tag{1.12}$$

$$\frac{\partial \Phi}{\partial z} = \alpha_1 \left(\frac{A_4}{A_3} + \frac{A_3}{A_4} \right) \cos \Phi, \tag{1.13}$$

where

$$\alpha_1 = \frac{2\pi\omega_3\chi_1}{c} A_1 A_2 = \frac{4\pi^2}{c^2} \chi_1 I_1 \omega_3, \tag{1.14}$$

$$\Phi = \varphi_4 - \varphi_3 + \varphi_2 - \varphi_1, \tag{1.15}$$

and the pumping flux density, obtained on the assumption that $A_1 = A_2$, is given by the expression

$$I_1 = c \frac{|E(\omega_1)|^2}{2\pi}. \tag{1.16}$$

It is evident from Eq. (1.13) that the steady-state value of the phase is $\Phi = \pi/2$ sign $[\chi_1]$. Using this value and the boundary conditions of Eq. (1.11), we obtain the following solutions for $A_{3,4}$ from Eq. (1.12):

$$\begin{aligned}
A_3 &= A_0 e^{-|\alpha_1| z}, \\
A_4 &= A_0 (e^{-|\alpha_1| z} - e^{-|\alpha_1| l}),
\end{aligned} \tag{1.17}$$

which shows that α_1 or, more precisely $|\alpha_1|$, is the distributed amplitude absorption coefficient.

It is evident from Eq. (1.17) that the nonlinear mirror under discussion does not increase the total energy of the fields of frequencies ω_3 and ω_4 but simply redistributes the energy between these fields. The energy reflection coefficient of the mirror is

$$R = \left[\frac{A_4^2}{A_3^2} \right]_{z=0} = (1 - e^{-|\alpha_1| l})^2. \tag{1.18}$$

Example. According to [4], nitrobenzene is characterized by $\chi_1 = 10^{-12}$ cgs esu; if $\omega = 2.7 \times 10^{15}$ sec^{-1}, $l = 1$ cm, $I_1 = 5 \times 10^8$ W/cm^2, we find from Eq. (1.14) that $\alpha_1 = 0.7$ cm^{-1} and Eq. (1.18) gives $R \approx 25\%$.

Grating Formed by Shifts of Energy Levels in a Strong Field

In a strong field the energy levels of a medium shift in proportion to $|E_p|^2$, as shown in [8]. It follows from Eq. (1.2) that a grating of energy level shifts established by such a strong wave produces an additional refractive index grating. We shall consider the case when $\Delta T_2 \lesssim 1$, where $\Delta = \omega_{10} - \omega_1$, ω_{10} is the frequency of the transition between levels 1 and 0, and T_2 is the transverse relaxation time of this transition. Following the method outlined in [8], we can show that a strong alternating pumping field of Eq. (1.2) shifts the energy levels by an amount

$$\Delta\omega = (\Delta\omega' - i\Delta\omega'') \cos^2 k_p \mathbf{r} \approx \left[\frac{4|d_{10}|^2|E(\omega_1)|^2}{\hbar^2\Delta} - i \frac{4|d_{10}|^2|E(\omega_1)|^2}{\hbar^2\Delta^2 T_2} \right] \cos^2 k_p \mathbf{r}. \tag{1.19}$$

Here, d_{10} is the dipole moment of the transition $1 \rightarrow 0$; $\Delta\omega_1 \cos^2 k_p \mathbf{r}$ corresponds to the shift of the level 1 and $\Delta\omega'' \cos^2 k_p \mathbf{r}$ to the broadening of this level (the corresponding level "broadening" grating is considered in Sec. 2). The resonant linear susceptibility at the frequency ω_3, where $|\omega_3 - \omega_{10}| \ll \omega_3$, is given by

$$\chi^{(1)}(\omega_3) \approx - \frac{|d_{01}|^2}{\hbar (\omega_{01} - \omega_3 + \Delta\omega - i/T_2)}. \tag{1.20}$$

If we assume that $\Delta\omega T_2 \ll 1$, $\Delta T_2 \ll 1$ and expand Eq. (1.20) in terms of $\Delta\omega$, we obtain

$$\chi^{(1)}(\omega_3) = \chi_0^{(1)}(\omega_3) + \delta\chi^{(1)} = \chi_0^{(1)'}(\omega_3) + i\chi_0^{(1)''}(\omega_3) + \chi_0^{(1)''}(\omega_3)(\Delta\omega' T_2) \cos^2 k_p \mathbf{r} - i\chi_0^{(1)''}(\omega_3)(\Delta\omega'' T_2) \cos^2 k_p \mathbf{r}, \tag{1.21}$$

where $\chi_0^{(1)}(\omega_3)$ is the linear susceptibility in the absence of the strong field, i.e., when $\Delta\omega = 0$. The corresponding change in the refractive index is

$$\delta n\,(\omega_3) = \frac{2\pi}{n\,(\omega_3)}\,\chi_0^{(1)''}(\omega_3)\,(\Delta\omega'T_2)\,\cos^2 k_p z. \tag{1.22}$$

Comparing Eqs. (1.7) and (1.22) and making an allowance for Eq. (1.9), we can introduce the equivalent nonlinear susceptibility

$$\chi_2 = \frac{\chi_0^{(1)''}(\omega_3)\,(\Delta\omega'T_2)}{6\,|E\,(\omega_1)|^2}$$

or we can employ Eq (1.14) to introduce the distributed absorption coefficient

$$\alpha_2 = \frac{2\pi\omega_3\chi_0^{(1)''}(\Delta\omega'T_2)}{6c} = \frac{\alpha_0\,(\Delta\omega'T_2)}{6}, \tag{1.23}$$

which includes the standard linear amplitude absorption coefficient

$$\alpha_0 = \frac{\sigma N}{2}. \tag{1.24}$$

Here, σ is the absorption cross section and N is the number of absorption centers.

Example. We shall consider a medium whose parameters (expressed in cgs esu) are close to the parameters of dye solutions considered in [9]: $\sigma = 10^{-16}$ ($d_{12} \approx 10^{-18}$), $\Delta N = 10^{-18}$, $T_2 = 10^{-12} \div 10^{-13}$, $\Delta = 10^{13}$, $|E\,(\omega_1)|^2 = 2 \cdot 10^5$ ($I_1 = 10^8$ W/cm^2). Then, it follows from Eqs. (1.19)–(1.23) that $\alpha_2 = 0.1 \div 1$ cm^{-1}.

Refractive Index Grating Due to Absorption of Light

Under the experimental conditions represented schematically by Fig. 1 the molecules which are excited by single-quantum absorption of a strong-field wave form a grating whose period d is the same as that given by Eq. (1.4).

These molecules can undergo transitions from an excited singlet state to a triplet state or they may undergo nonradiative transitions to the ground state in which heat is evolved. In this way a grating of excited molecules in singlet and triplet states is produced and a grating resulting from the heating of the medium is generated. These gratings give rise to corresponding refractive index gratings and nonlinear mirrors. During a pump pulse τ_{pl} of the type given by Eq. (1.2) the energy converted into heat in each element of volume is

$$W d^3 r = 8\mu\alpha_0\,\frac{c\,|E\,(\omega_1)|^2\,\tau_{pl}}{2\pi}\,\cos^2 k_p r d^3 r, \tag{1.25}$$

where $0 \le \mu \le 1$ is the coefficient of conversion of the absorbed energy into heat. The time needed for redistribution of this energy between the vibrational and translational degrees of freedom is short, of the order of the nonradiative transition time $T_1 \sim 10^{-10}$–10^{-13} sec. Bearing in mind the slowness of heat conduction, we find that the temperature at each point increases by

$$\Delta T \approx \frac{W}{C_v} = \frac{8\mu\alpha_0\,(I_1\tau_{pl})}{C_v}\,\cos^2 k_p r, \tag{1.26}$$

where C_v is the specific heat at constant volume. Such periodic inhomogeneous heating gives rise to a grating of regions where the pressure p is higher. These high-pressure regions expand and the regions where the pressure has not changed are compressed. The characteristic time needed for pressure equalization is $\tau \sim d/u \sim 5 \times 10^{-9}$ sec $\ll \tau_{pl}$, where $u \sim 10^5$ cm/sec is the velocity of sound in the medium under consideration. The total change in the refractive index is, therefore, [10]

$$\delta n \approx \left(\frac{\partial n}{\partial T}\right)_p \Delta T. \qquad (1.27)$$

This change is due to reduction in the density of the fluid because of heating [usually $(\partial n/\partial T)_p < 0$] and, according to [10], the value of $(\partial n/\partial T)_p$ is typically $(1-5) \times 10^{-4}$ deg^{-1}.

Comparing Eqs. (1.27) and (1.7) and making use of Eq. (1.26), we obtain the equivalent nonlinear susceptibility

$$\chi_3 = \frac{\mu \alpha_0 c \tau_{pl} \, n\,(\omega)}{3\pi^2 C_v} \left(\frac{\partial n}{\partial T}\right)_p. \qquad (1.28)$$

The corresponding distributed amplitude absorption coefficient of a nonlinear mirror follows from Eq. (1.14):

$$\alpha_3 = \frac{4\pi^2}{c^2} |\chi_3| I_1 \omega_3 = \frac{4}{3} \frac{\mu \alpha_0 \tau_{pl} \, n\,(\omega_3)\, \omega_3}{c C_v} \left(\frac{\partial n}{\partial T}\right)_p I_1. \qquad (1.29)$$

Example. If we ignore $(\partial n/\partial T)_v$ compared with $(\partial n/\partial T)_p$ and use the parameters $-(\partial n\ \partial T)_p = (1-5) \times 10^{-4}$, $I = 10^{15}$, $\tau_{pl} = 2 \times 10^{-8}$, $n(\omega_3) = 1.3$, $\omega_3 = 2.7 \cdot 10^{15}$, $C_v = 5 \cdot 10^7$, and $\mu = 1$ (all in cgs esu), we find that $\alpha_3 = -(6-30)\alpha_0$. We note that the temperature grating disperses quite slowly. If we use the heat conduction equation

$$\frac{\partial T}{\partial t} = \frac{k}{C_v \rho} \frac{\partial^2 T}{\partial z^2} \qquad (1.30)$$

(T is the temperature, k is the thermal conductivity, and ρ is the density), we find that the temperature drop ΔT in Eq. (1.26), established over a distance of the order d, disappears in a time of the order of

$$\tau_T = \frac{C_v \rho d^2}{k}, \qquad (1.31)$$

which gives $\tau_T \approx 10^{-3}$ sec for $\rho = 1$, $C_c = 5 \times 10^7$, $d = 5 \times 10^{-4}$, and $k = 10^4$ (all in cgs esu).

We shall now consider a grating formed by excited molecules. Under steady-state conditions the number of molecules in the excited (singlet) state is

$$\Delta N^* = \frac{4\alpha_0 c \, |E\,(\omega_1)|^2}{\pi \hbar \omega_1} T_1 \cos^2 k_p z, \qquad (1.32)$$

where the excited-state lifetime is $T_1 \ll \tau_{pl}$. The corresponding refractive index grating is

$$\delta n\,(z) = \Delta\beta \Delta N^* = \frac{4\alpha_0 \Delta\beta c \, |E\,(\omega_1)|^2 T_1}{\pi \hbar \omega_1} \cos^2 k_p z. \qquad (1.33)$$

Here, $\Delta\beta$ is the change in the susceptibility of an excited molecule relative to its unexcited state, which must be of the order of the susceptibility of a single molecule

$$\Delta\beta \approx \frac{n^2 - 1}{4\pi N} \sim \frac{0.1}{N} \sim 10^{-23} \text{ cm}^3. \qquad (1.34)$$

In general, the value of $\Delta\beta$ [11] can be larger (because the excited-state orbitals are larger) than the susceptibility of the ground state and the difference may amount to two orders of magnitude. A comparison of Eqs. (1.33) and (1.7) yields the nonlinear susceptibility

$$\chi_1 = \frac{1}{3}\frac{\alpha_0 c T_1 n\,(\omega_3)\,\Delta\beta}{\pi^2\hbar\omega_1} \tag{1.35}$$

and it follows from Eq. (1.14) that

$$\alpha_4 = \frac{4\pi^2}{c^2}\chi_1 I_1\omega_3 = \frac{4\Delta\beta T_1 n\,(\omega_3)\,\omega_3}{c\hbar\omega_1}I_1\alpha_0. \tag{1.36}$$

Example. If we use the parameters $\Delta\beta = 10^{-23}$, $T_1 = 10^{-10}$, $I_1 = 10^{15}$ $(10^8\ \text{W/cm}^2)$, $n(\omega_3) = 1.3$, and $\omega_1 \approx \omega_3$ (all in cgs esu), we obtain $\alpha_4 = 0.06\alpha_0$. We note that the length of a nonlinear absorbing mirror is limited (without allowance for bleaching) to the value $l_0 \sim 1/\alpha_0$ and, therefore, the nonlinear mirror discussed in the preceding paragraphs cannot have a large reflection coefficient R defined by Eq. (1.18).

We shall now discuss a grating formed by excited molecules in triplet states. We shall assume that the quantum yield of triplet excitations is η. Since the triplet-level lifetime is considerably longer than τ_{pl} (this lifetime may amount to seconds), it is necessary to replace T_1 in Eqs. (1.32), (1.33), (1.35), and (1.36) with $\eta\tau_{pl}$ so that the expression for the distributed absorption coefficient becomes

$$\alpha_5 = \alpha_0\left[\frac{4\Delta\beta\tau_{pl}\,n\,(\omega_3)\,\omega_3\eta}{3c\hbar\omega_1}\right]I_1. \tag{1.37}$$

Example. If we substitute $\Delta\beta = 10^{-23}$, $\tau_{pl} = 2 \times 10^{-8}$, $n(\omega_3) = 1.3$, $\omega_3 = \omega_1$, $\eta = 1$, and $I_1 = 10^{15}$ (all in cgs esu) into Eq. (1.37), we find that $\alpha_5 = 12\alpha_0$.

In this case the conditions may be less favorable because of the diffusion. However, an excited-triplet grating disperses slowly. The mean square displacement in one-dimensional diffusion is

$$\bar{z} = \sqrt{2D\tau_{pl}}\ . \tag{1.38}$$

Hence, the characteristic displacement for low-viscosity fluids is $\bar{z} = 65$ Å if the diffusion coefficient is $D = 10^{-5}$ cgs esu and $\tau_{pl} = 2 \times 10^{-8}$ sec. This displacement is much smaller than $d \sim 5000$ Å.

It follows that the largest reflection coefficient R can be obtained for mirrors based on triplet and temperature gratings. It is worth noting that all the coefficients derived above obey $\alpha_i \propto |E(\omega_1)|^2$.

§ 2. Parametric Frequency Selection within Wide

Stimulated Emission Lines

Frequency tuning within wide stimulated emission lines is usually attained by insertion of frequency-selective elements inside a lader resonator. We shall consider the possibility of frequency selection in which only the coherence of the pumping radiation is used and the pumping field configuration is of the type shown in Fig. 1. If such a pumping field is sufficiently strong, the spatial modulation of the parameters of the medium (susceptibility, refractive index, etc.) may give rise to an additional frequency-selective term in the expression for the gain or in the Q factor. This will occur at the wavelength λ' given by Eq. (1.6). Such frequency selection may be applied to the radiation emitted by dye solution lasers, to the stokes stimu-

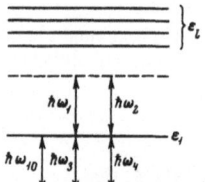

Fig. 2. Energy level scheme:
ε_k are the energy levels of the
active medium; ω_1 are the frequencies of the various fields.

lated Raman scattering (in the case of a wide line), or to the stimulated scattering in the
wing of the Rayleigh line. This frequency selection was achieved experimentally in [9] by
pumping dye solutions with coherent fields of configurations shown in Fig. 1. A narrow line
of wavelength λ' given by Eq. (1.6) was observed against the background of the wide stimulated
radiation spectrum. The effect observed in [9] was attributed to the resonant parametric scattering of four waves satisfying the phase-matching conditions of Eq. (1.8). An analysis of such
a resonant parametric interaction was given in our earlier paper [7]. However, the frequency
selection observed in [9] can be due to other mechanisms satisfying the phase-matching conditions of Eq. (1.8). The estimates given below show that the influence of the spatial modulation
of the population inversion can be two orders of magnitude stronger than the influence of the interaction considered in [7]. We shall now discuss various possible mechanisms of frequency
selection in the case of dye solutions.

Resonant Parametric Interaction of Four Fields

This resonant interaction was considered in [7]. The essence of the interaction is as follows. In a population-inverted medium (transition $1 \to 0$ in Fig. 2) we may encounter not only
single-quantum stimulated emission $\hbar\omega_3 = \hbar\omega_4 = \hbar\omega_{10}$ but also anti-Stokes double Raman scattering in which an excited molecule undergoes a virtual transition from state 1 to a state l as
a result of absorption of a quantum $\hbar\omega_1$ and then to a state k as a result of emission of $\hbar\omega_2$ and
finally to the ground state as a result of emission of $\hbar\omega_4$ (Fig. 2). If the phase-matching conditions (1.8) are satisfied, these incoherent processes are accompanied by coherent (parametric)
interference which depends on the phase relationships between the fields.

The first two incoherent processes reduce the population inversion ΔN whereas the resonant parametric interaction may increase or reduce the inversion, depending on the phases
of all four fields. The change in the population inversion resulting solely from the parametric
process in [7]

$$\left(\frac{d\Delta N}{dt}\right) = -\frac{4}{\hbar}\,\Delta N\,\mathrm{Re}\,\{\chi^{(3)''}E\,(\omega_1)\,E^*\,(\omega_2)\,E(\omega_3)\,E^*\,(\omega_4)\exp\left[i\,(\Delta\omega t - i\Delta\mathbf{k}\mathbf{r})\right]\}, \tag{2.1}$$

where $\chi^{(3)}$ is the imaginary component of the resonant susceptibility

$$\chi^{(3)} \sim -\frac{id_{10}T_2 d_{0r}d_{rs}d_{s1}}{\hbar^3\,(\omega_{1r}-\omega_3-\omega_4)\,(\omega_{1s}-\omega_4)}\ ; \tag{2.2}$$

$$\Delta\omega = \omega_1 - \omega_2 + \omega_3 - \omega_4,\quad \Delta\mathbf{k} = \mathbf{k}_1 - \mathbf{k}_2 + \mathbf{k}_3 - \mathbf{k}_4;$$

$1/T_2$ is the homogeneous broadening of the $1 \to 0$ line; d_{ik} is the matrix element of the transition $i \to k$. We can show [7] that the phase $\Phi = \varphi_1 - \varphi_2 + \varphi_3 - \varphi_4$ is such that the parametric process reduces the inversion. This gives rise to quanta $\hbar\omega_3$ and $\hbar\omega_4$ without any change in the
number of the quanta $\hbar\omega_1$ and $\hbar\omega_2$. Eqs. (2.1) and (2.2) can be used to estimate the additional
amplitude gain α_6 of fields of frequency ω_3 and ω_4 subject to the condition $|E\,(\omega_1)| = |E\,(\omega_2)|$:

$$\alpha_6 = \frac{2\omega_3 d_{1r}T_2 d_{0r}d_{rs}d_{s1}\,|\,E\,(\omega_1)\,|^2}{c\hbar^3\,(\omega_{1r}-\omega_3-\omega_4)\,(\omega_{1s}-\omega_4)}\,\Delta N. \tag{2.3}$$

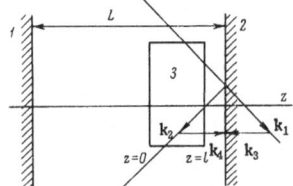

Fig. 3. Schematic representation of
a laser with frequency selection: 1)
mirror tuned to the frequency $\omega_3 = \omega_4$;
2) mirror tuned to the frequencies
$\omega_1 = \omega_2$ and $\omega_3 = \omega_4$; 3) active medium;
l is the length of the active medium;
L is the length of the resonator.

Example. If we use the parameters $d_{ik} \sim 10^{-18}$, $\Delta N \sim 10^{18}$, $T_2 = 10^{-12}$; $\omega_{lk} \sim \omega_{1,3} \sim 10^{15}$, $|E(\omega_1)|^2 = 2 \cdot 10^5$ (10^8 W/cm^2), all of which are given in cgs esu, we find that $\alpha_6 = 10^{-5}$ cm^{-1}. We note that since $\Delta N \propto |E(\omega_1)|^2$, it follows that $\alpha_6 \propto |E(\omega_1)|^4$.

Selection Resulting from the Presence of an Inversion Grating

A pump field of the configuration given by Eq. (1.2) gives rise to a population inversion which is modulated by $\cos^2(\mathbf{k_p r})$. We shall now consider the laser action in a dye solution excited by such a pumping field. The laser is shown schematically in Fig. 3. The field in the resonator can be represented by a superposition of standing waves of the same polarization:

$$E = \sum_n A_n u_n(z) \exp(i\omega_n t + i\varphi_n) + \text{c.c.} \tag{2.4}$$

Here, u_n is a real mode generated in the resonator shown in Fig. 3.

The method of slowly varying amplitudes yields the following expression for the values of A_n [12]:

$$\frac{\partial A_n}{\partial t} + \frac{1}{2}\frac{\omega_n}{Q_n} A_n = \alpha(\omega_n) A_n. \tag{2.5}$$

Here, Q_n is the Q factor at the frequency ω_n; $\alpha(\omega_n)$ is the average amplitude gain

$$\alpha(\omega_n) = \frac{1}{l} \int_0^l k(\omega_n, z) u_n^2(z) dz, \tag{2.6}$$

where $k(\omega_n, z)$ is the local amplitude gain which has the following form in the case of dye solutions [14]:

$$k(\omega_n, z) = \frac{\alpha_0}{2}\left\{\frac{n_2(z)}{n} - \exp\left[-\hbar(\omega_{el} - \omega_n)/kT\right]\left(1 - \frac{n_2(z)}{n}\right)\right\}. \tag{2.7}$$

Here, α_0 is the maximum gain; ω_{el} is the angular frequency of an electronic transition; k is the Boltzmann constant; T is the absolute temperature; n is the concentration of the dye molecules; $n_2(z)$ is the concentration of excited molecules, which is modulated along the z axis. The first term in the braces represents the gain and the second term represents the absorption of light of frequency ω_n. If the saturation can be ignored, the steady-state concentration of the excited dye molecules is given by [13]

$$n_2 = n\frac{\sigma_p c \overline{|E_p|^2}}{\hbar\omega_1 \cdot 4\pi} T_1, \tag{2.8}$$

where σ_p is the absorption coefficient for the pump radiation and T_1 is the excited-state lifetime. Substituting E_p into Eq. (1.2) and averaging over time, we obtain

$$\frac{n_2(z)}{n} = 4a \cos^2(k_p z), \tag{2.9}$$

$$a = \frac{\sigma_p c T_1 |E(\omega_1)|^2}{2\pi\hbar\omega_1} = \frac{\sigma_p T_1}{\hbar\omega_1} I_1. \tag{2.10}$$

Thus, in order to calculate α from Eq. (2.6) by means of Eqs. (2.7), (2.9), and (2.10), we must calculate the "overlap" integral

$$S = \frac{1}{l} \int_0^l \cos^2(k_p z) u_n^2(z) \, dz. \tag{2.11}$$

In the $0 \leq z \leq l$ region (Fig. 3) inside a plane-parallel resonator we may assume that the normalized form of the n-th mode is

$$u_n(z) = \sqrt{2} \sin(k_n z + \varphi_n). \tag{2.12}$$

Substituting Eq. (2.12) into Eq. (2.11) and bearing in mind that $k_n l$, $k_p l \gg 1$, we find that

$$S = \frac{1}{2} \left[1 - \frac{1}{2} \gamma(k_p, k_n) \cos 2p_n \right], \tag{2.13}$$

$$\gamma(k_p, k_n) = \frac{\sin[(k_p - k_n) l]}{(k_n - k_n) l}. \tag{2.14}$$

The function $\gamma(k_p, k_n)$ differs significantly from zero only for $|k_p - k_n| l \ll \pi/2$; when $k_p = k_n$ this function has its maximum value: $\gamma(k_p, k_p) = 1$. Since $L \gg l$ (Fig. 3), there is always some mode n for which $\varphi_n = \pi/2$ if $|k_p - k_n| l \ll \pi/2$, i.e., for which the overlap integral S has its maximum value

$$S = \frac{1}{2} \left[1 + \frac{1}{2} \gamma(k_p, k_n) \right]. \tag{2.15}$$

This corresponds to the case when the n-th mode fits best the inversion grating.

The amplitude gain is found from Eqs. (2.7)–(2.10) and (2.13)

$$\alpha(\omega_n) = \frac{\alpha_0}{2} \left\{ 2a \left[1 + \frac{1}{2} \gamma(k_p, k_n) \right] - \exp\{-\hbar(\omega_{e1} - \omega_n)/kT\} \left[1 - 2a \left(1 + \frac{1}{2} \gamma(k_p, k_n) \right) \right] \right\}. \tag{2.16}$$

It follows from Eq. (2.16) that the modes satisfying the condition $|k_n - k_p| l \ll \pi/2$, are subject to an additional selective gain

$$\alpha_\gamma(\omega_n) = \frac{\alpha_0 a}{2} \gamma(k_p, k_n) [1 + \exp\{-\hbar(\omega_{e1} - \omega_n)/kT\}] \approx \alpha_0' \gamma(k_p, k_n), \tag{2.17}$$

where $\alpha_0' = \alpha_p a/2$ is the standard amplitude gain obtained under the action of a pumping wave described by Eq. (1.1). The last term in the square brackets appears because the absorption at zero-inversion points will be minimal for the mode which is "coherent" with the inversion grating. If $k_n = k_p$ the gain $\alpha(\omega_p)$ is at least 1.5 times larger than the gain experienced by the waves satisfying $\alpha(\omega_n)$. The amplitude gain defined by Eq. (2.16) is plotted in Fig. 4.

The selective gain α_γ decreases when the inversion grating profile becomes smoother. Such smoothing may be due to the following factors: 1) diffusion, 2) inversion saturation, 3)

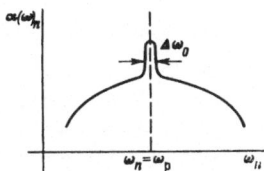

Fig. 4. Frequency dependence of the
gain $\alpha(\omega_n)$ in the presence of an inversion grating. Here, $\omega_p = k_p/c$ is
the frequency corresponding to
the gain maximum; $\Delta\omega_0$ is the width
of the curve representing the additional selective gain.

nonmonochromatic pumping, 4) angular divergence of the pumping wave, and 5) a large "tra-
veling-wave ratio" of the combined pumping wave.

We shall now consider the influence of all these factors on α_7.

1. As shown earlier, it follows from Eq. (1.38) that the diffusion which occurs during the
laser emission period $\tau \sim 10^{-8}$ sec cannot distort the inversion grating significantly.

2. It follows from Eqs. (2.8) and (2.10) that the saturation of inversion is unimportant if
[13]

$$4a = \frac{4\sigma p T_1}{\hbar\omega_1} I_1 \ll 1. \tag{2.18}$$

In the case of dyes with the parameters $\sigma = 10^{-17}$, $T_1 = 10^{-10}$, and $\omega_t = 2.7 \cdot 10^{15}$ (all in cgs esu),
the inequality (2.18) may be satisfied for $I_1 \ll 150$ mW/cm^2. However, even when saturation is
attained the grating does not disappear completely. In this case the overlap integral S can be
calculated from Eq. (2.11) by multiplying the integrand by the factor $1/[1 + 4a\cos^2(k_p z)]$.

3. Under nonmonochromatic pumping conditions represented by Eq. (1.2) the distribution
of inversion can be regarded as a superposition of various inversion gratings with different
values of k_p; this reduces the additional selective gain.

It is evident from Eqs. (2.17) and (2.14) that a monochromatic inversion grating with a
given value of k_p amplifies strongly a wave k_n which satisfies

$$|k_p - k_n| l < \frac{\pi}{2}. \tag{2.19}$$

Conversely, a given wave k_n will be amplified strongly by gratings for which the value of k_p
satisfies the inequality (2.19). Thus, when the width of the pumping spectrum obeys

$$\Delta\omega < \Delta\omega_0 = \frac{\pi c}{ln(\omega)\sin\alpha}, \tag{2.20}$$

the additional selective gain is given by Eq. (2.17).

Example. If $\alpha = 30°$, $l = 0.4$ cm, and $n(\omega) = 1.3$ (these parameters are similar to those
reported in [9]), we find that $\Delta\nu_0/c = \Delta\omega_0/2\pi c \approx 2$ cm^{-1}.

In the general case of an arbitrary spectral intensity of the pumping wave $\varphi(k_p)$, the selective gain can be described approximately by

$$\alpha_7(k_n) = \alpha_0' \frac{\int \gamma(k_p, k_n)\varphi(k_p)dk_p}{\int \varphi(k_p)dk_p}. \tag{2.21}$$

Hence, when the pumping spectrum consists of m modes $\left(\varphi(k) = \sum_{i=1}^{m} \delta(k-k_i)\right)$, which are sepa-

rated widely along the frequency scale $\left(\frac{\omega_i - \omega_k}{c} n(\omega_i) l \cos \alpha \gg \frac{\pi}{2}\right)$, the curve in Fig. 4 has m peaks near the appropriate values of ω_i and the amplitude of each peak is m times smaller than the amplitude in the case of single-mode pumping radiation. If the pumping spectrum is "rectangular" and its width is $\Delta\omega > \Delta\omega_0$ centered on ω_1, it follows from Eq. (2.21) that the curve of Fig. 4 will have a peak of width $\Delta\omega$ and amplitude equal to that obtained in the single-mode case but multiplied by the factor $\Delta\omega_0/\Delta\omega$.

4. In the preceding section we have assumed the absence of angular divergence of the pumping radiation. Such divergence gives rise to inversion gratings which are oriented differently in space. In this case we can use Eq. (2.7) provided that all the wave vectors of the gratings satisfy the condition

$$|k_{pz} - k_n| l < \frac{\pi}{2}, \tag{2.22}$$

where k_{pz} is the projection of k_p along the z axis.

5. The presence of the "traveling-wave ratio" in the combined pumping wave smoothes out the inversion grating. Let us assume that the energy reflection coefficient of the mirror AB in Fig. 1 is r. We can show that the formulas (2.16) and (2.17) are not affected if a is replaced with $a' = \frac{a}{4}(1 + \sqrt{r})^2$, and $\gamma(k_p, k_n)$ is replaced with $\gamma'(k_p, k_n) = \frac{4\sqrt{r}}{(1 + \sqrt{r})^2} \gamma(k_p, k_n)$. If $r = 0.8$ we find that $\gamma'(k_p, k_n) = 0.99 \gamma(k_p, k_n)$, i.e., the additional selective gain is not very sensitive to the presence of the "traveling-wave ratio" in the pumping wave. Thus, we can see that for reasonable values of the parameters all the five factors listed above have little influence on the efficiency of selection by means of an inversion grating. We note that $\alpha_7 \propto |E(\omega_1)|^2$ and, moreover, if $k_n = k_4$ and $-k_4 = k_3$, the phase-matching conditions (1.8) are satisfied.

Selection Resulting from the Presence of a Level-Broadening Grating

The resonant level broadening (1.19) in a strong alternating field alters the imaginary part of the linear susceptibility (1.21). If the pump field of Eq. (1.2) is periodic, a level broadening grating with a profile $\cos^2(k_p z)$ is generated and this gives rise to a gain (absorption) grating. If we use Eqs. (1.2), (1.19), (1.21) and (2.11), we find that the total gain $\alpha(\omega_n)$ of Eq. (2.5) is given by

$$\alpha(\omega_n) = \alpha'_0 - \alpha'_0(\Delta\omega'' T_2) S, \tag{2.23}$$

where α'_0 is the standard amplitude gain and

$$\Delta\omega'' = \frac{4|d_{12}|^2|E(\omega_1)|^2}{\hbar^2 \Delta^2 T_2}. \tag{2.24}$$

Since the correction (absorption) is negative, the strongest gain is experienced by that mode for which S is smallest. It follows from Eqs. (2.11) and (2.13) that the strongest gain corresponds $\varphi_n = 0$ and to

$$S = \frac{1}{2}\left[1 - \frac{1}{2}\gamma(k_p, k_n)\right]. \tag{2.25}$$

The modes whose wave vector is k_n are then subject to an additional selective gain

$$\alpha_8(\omega_n) = \frac{\alpha'_0 \Delta\omega'' T_2}{2} \gamma(k_p, k_n) = 2\alpha'_0 \left[\frac{|d_{12}|^2|E(\omega_1)|^2}{\hbar^2 \Delta^2}\right] \gamma(k_p, k_n). \tag{2.26}$$

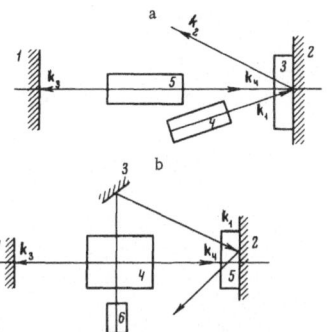

Fig. 5. a) Frequency selection achieved
with the aid of a nonlinear mirror: 1)
mirror tuned to the frequency $\omega_3 = \omega_4$;
2) mirror tuned to the frequencies $\omega_1 =
\omega_2$ and $\omega_3 = \omega_4$; 3) nonlinear mirror; 4)
triggering laser; 5) triggered laser. b)
Frequency selection by Q-factor switching: 1) mirror tuned to the frequency
$\omega_3 = \omega_4$; 2) mirror tuned to the frequencies $\omega_1 = \omega_2$ and $\omega_3 = \omega_4$; 3) mirror tuned
to the frequency ω_3; 4) dye solution; 5)
nonlinear mirror; 6) laser emitting at the
frequency ω_1.

Example. If we use the parameters $d_{12} = 10^{-18}$, $\Delta = 10^{13}$, and $|E(\omega_1)|^2 = 2 \cdot 10^5$ (all in cts
esu), we obtain $\alpha_8 (\omega_n = \omega_p) = 4 \cdot 10^{-3}\alpha_0$. Since $\alpha_0' \propto \Delta N \propto |E(\omega_1)|^2$ and $\Delta\omega'' \propto |E(\omega_1)|^2$, it fol-
lows that $\alpha_8 \propto |E(\omega_1)|^4$. The selection due to the presence of a level broadening grating can
be described in terms of the resonant parametric interaction of four fields satisfying Eq. (1.8)
and characterized by the resonant nonlinear susceptibilities of the type given by Eq. (2.2).

Frequency Selection Achieved by a Nonlinear Mirror

Let us consider the arrangement shown schematically in Fig. 5 where a dye laser is combined with a nonlinear mirror of one of the types described in § 1. The pumping source inverts
the population in a dye solution and switches on the nonlinear mirror. Since the reflection coefficient of such a mirror can reach even ~100%, a selective Q factor is established in the resonator for the wavelength λ' and since the Q factor for the other wavelengths is small, only the
narrow line λ' is amplified. Smooth frequency tuning within a wide dye-laser line can be
achieved by varying the angle α. In contrast to the case considered in [9] the emission of a
narrow line in the case of a generally low Q factor of the resonator need not be accompanied
by a strong background radiation.

We have considered so far the case of an external nonlinear mirror. When dye solutions
are pumped in the way shown in Fig. 1, the periodic correction (period d) to the refractive index (this may be due to a temperature grating) gives rise to an additional distributed nonlinear
mirror which couples the waves k_3 and k_4 emitted in opposite directions. We can show that in
this case we can, in principle, generate a narrow tunable line λ' even in the absence of a resonator.

Frequency Selection in Raman–Stokes Scattering

We shall now discuss the Raman-Stokes scattering in a resonator pumped by a wave of the type represented by Eq. (1.2). The amplitude gain of the Stokes wave is given by [2]

$$\alpha_S = \frac{2\pi\omega_3}{c} \Theta''(\omega_3) |E(\omega_p)|^2 N, \tag{2.27}$$

where ω_p is the pumping-wave frequency; n is the number of molecules $\Theta''(\omega_3)$ is the imaginary part of the Raman susceptibility, calculated per one molecule. In our case it follows from Eq. (1.2) that $E(\omega_p) = 2E(\omega_1)\cos(k_p r)$ and, therefore, a grating of the gain α_s with a profile $\cos^2 k_p r$ is established. The analysis is exactly the same as in the case of an inversion grating: in both cases the strongest gain is experienced by the waves with $k_3 = -k_4$ satisfying the phase-matching conditions of Eq. (1.8). The additional selective gain is then

$$\alpha_9(\omega_3) = \frac{2\pi\omega_3}{c} \Theta''(\omega_3) |E(\omega_1)|^2 \gamma(k_p, k_3) = \alpha_S \gamma(k_p, k_3). \tag{2.28}$$

Example. If the incident flux density is 10^8 W/cm^2 the gain for nitrobenzene α_s is 0.1 cm^{-1} [2], i.e., if $k_p = k_3$, $\alpha_9 \sim 0.1$ cm^{-1}.

An even stronger selective correction may appear in the case of resonant Raman–Stokes scattering ($\omega_1 \approx \omega_{01}$). In this case we may expect the Raman susceptibility to increase by a factor of $\omega_{01}^2 T_2^2$, i.e., the selective correction to the gain under resonance conditions should be

$$\alpha_{10} \sim \omega_{01}^2 T_2^2 \alpha_c. \tag{2.29}$$

Such selective gain may be observed also in dye solutions since the laser emission may be accompanied by the stimulated Raman scattering [13, 14]. In particular, the Raman scattering by the dye molecules may give rise to an additional selective gain $\alpha_9 \sim 0.1$ cm^{-1}. Moreover, a selective gain may result from the resonant Raman scattering by those dye molecules which have high values of the resonant Raman polarizability because of the presence of a large number of π bonds.

Example. In the latter case when $\omega_{01} \sim 10^{15}$ sec^{-1}, $T_2 = 10^{-13}$-10^{-12} sec, $\Delta N \sim 10^{18}$ cm^{-3}, we find that

$$\alpha_{10} \sim \alpha_9 (\omega_{01} T_2)^2 \frac{\Delta N}{N} \sim 10 - 10^{-1} \text{ cm}^{-1}, \tag{2.30}$$

where ΔN and N are, respectively, the numbers of the dye and the solvent molecules.

We have discussed various frequency selection mechanisms in the case of wide stimulated emission lines, subject to the conditions (1.8). The additional amplitude gains and the reflection coefficients of nonlinear mirrors, obtained for these mechanisms, are listed in Table 1.

§ 3. Discussion of Results

The nonlinear mirrors discussed in Sec. 1 have certain interesting practical applications. Frequency selection in accordance with the scheme shown in Fig. 5a is discussed in § 2. The advantage of this selection method compared with that described in [9] is the absence of a wide stimulated-emission background which accompanies the generation of a narrow line. Other cases worth experimental investigation are the frequency selection methods with the aid of a distributed nonlinear mirror inside the active medium (Fig. 3) and in the absence of an external resonator. In these cases one can also expect the wide background to be absent.

TABLE 1. Estimates of Frequency-Selective Absorption Coefficients
and of Corrections to Amplitude Gain

Mechanism	Value of α_i (cm^{-1}) for 10^8 W/cm^2 pumping and formula No.	Parameters of medium (cgs esu)	Dependence of $|E(\omega_1)|$
Nonlinear mirror formed in transparent medium because of "cubic" nonlinearity	$\alpha_1 = 0.1$ (1.14)	$\chi_1 = 10^{-12}$ $\omega_3 = 2.7 \cdot 10^{15}$ $n_{(\omega)} = 1.3$	$\propto |E(\omega_1)|^2$
Nonlinear mirror formed by level shift grating in resonant medium	$\alpha_2 = 2\,(10^{-2} - 10^{-3})\,\alpha_0 =$ $= 0.1 - 1$ (1.23)	$\Delta = 10^{13}$ $T_2 = 10^{-12} \div 10^{-13}$ $\alpha_0 = 50$	$\propto |E(\omega_1)|^2$
Nonlinear mirror formed by temperature grating	$\alpha_3 = (6-30)\,\alpha_0 = 300-1500$ (1.29)	$\left(\dfrac{\partial n}{\partial T}\right)_p = (1-5) \cdot$ $\cdot 10^{-4}$ $\alpha_0 = 50$	$\propto |E(\omega_1)|^2$
Nonlinear mirror formed by excited molecule (singlet) grating	$\alpha_4 = 0.1\,\alpha_0 = 5$ (1.36)	$\Delta\beta = 10^{-23}$ $T_1 = 10^{-10}$ $\alpha_0 = 50$	$\propto |E(\omega_1)|^2$
Nonlinear mirror formed by excited molecule (triplet) grating	$\alpha_5 = 12\,\alpha_0 = 600$ (1.37)	$\Delta\beta = 10^{-23}$ $T_1 \gg 2 \cdot 10^{-8}$ $\alpha_0 = 50$	$\propto |E(\omega_1)|^2$
Resonant parametric interaction of four fields	$\alpha_6 = 10^{-5}$ (2.3)	$d_{ik} \sim 10^{-18}$ $\Delta N \sim 10^{18}$ $T_2 \sim 10^{-12}$ $\omega_{ik} \sim \omega_{1,3} \sim 10^{15}$	$\propto |E(\omega_1)|^4$
Population inversion grating	$\alpha_7 = \dfrac{1}{2}\,\alpha_0' = 25$ (2.17)	$\alpha_0' = 50$	$\propto |E(\omega_1)|^2$
Level broadening grating	$\alpha_8 = 2 \cdot 10^{-3}\alpha_0' = 0.2$ (2.26)	$\alpha_{12} = 10^{-18}$ $\Delta = 10^{13}$ $\alpha_0' = 50$	$\propto |E(\omega_1)|^4$
Modulation of nonresonant Raman susceptibility	$\alpha_9 = \alpha_S = 0.1$ (2.28)	Nitrobenzene [2] $\alpha_S = 0.1$	$\propto |E(\omega_1)|^2$
Modulation of resonant Raman susceptibility	$\alpha_{10} = 10 - 10^{-1}$ (2.30)	$\omega_{0l} \simeq 10^{15}$ $T_2 = 10^{-13} \div 10^{-12}$ $\Delta N = 10^{18}$	$\propto |E(\omega_1)|^2$

A nonlinear mirror (Fig. 5b) can switch the Q factor: a pulse can be used to switch on and off the nonlinear mirror. It would be interesting to study the features of such switching in the case of a neodymium-glass laser. The nonlinear mirror in Fig. 5b can be used to synchronize the triggering of two lasers (suitable delay devices would be needed). Such synchronization is important in nonlinear optical experiments involving frequency shifts and other phenomena. Moreover, synchronized lasers can form a switch, which can be used in computer technology. If the delay time in the scheme of Fig. 5a is varied, the laser emission time can be reduced by varying amounts because of noncoincidence of the pumping period and the lifetime of the nonlinear mirror.

These applications can be realized if sufficiently efficient nonlinear mirrors are used. These efficiency requirements cannot be met by a refractive index grating in a transparent medium (α_1) or by a level shift grating (α_2). For example, in the case of nitrobenzene which has a relatively high value of the susceptibility $\chi_2 = 10^{-12}$ cgs esu, the reflection coefficient is only about 25% for $l = 1$ cm and $I_1 = 5 \times 10^8$ W/cm^2. Enhancement of the field by focusing does not help because it reduces the effective length l. The mirrors based on excited molecule (α_4, α_5) and temperature (α_3) gratings are more efficient. When molecules are in an excited state characterized by a short lifetime ($T_1 \sim 10^{-9} - 10^{-10}$ sec), it follows from Sec. 1 that the coefficient α_4 can be large but it is difficult to make it larger than the absorption coefficient α_0.

Therefore, if no bleaching takes place, the absorption limits the effective length l and the reflection coefficient corresponding to $\alpha_4 \leq \alpha_0$ does not exceed 40% [compare Eqs. (1.18) and (1.36)]. A triplet grating is more effective because the triplet-state lifetime may amount to several seconds. Consequently, the coefficient α_5 may be considerably larger than the absorption coefficient α_0. Thus, if $T_1 \gg \tau_{pl}$ we can achieve 100% reflection for a small value of l: for example, if $\alpha_0 = 50$ cm^{-1}, $I_1 = 10^8$ W/cm^2, $\tau_{pl} \sim 2 \times 10^{-8}$ sec, we find that $\alpha_4 = 1000$ cm^{-1} if $l = 10^{-3}$ cm^2. Morevoer, a mirror of this kind has a long lifetime ($T_1 \gg \tau_{pl}$). This is important in schemes of the type shown in Fig. 5b because a nonlinear mirror with a fairly long lifetime is needed to trigger the laser in Fig. 5b. A long-lived nonlinear mirror based on a triplet grating can be used to study the migration of energy from the triplet state. This can be done by investigating the reflection of gas-laser radiation of wavelength λ' from a grating of this kind with a period d given by Eq. (1.4). The deactivation of the grating and its spreading because of diffusion result in weakening of the reflected signal.

It is evident from Table 1 that a temperature grating can compete in efficiency with a triplet grating. A temperature grating is the easiest to establish experimentally and its lifetime is fairly long ($\sim 10^{-3}$ sec). We note that because of heating the refractive index decreases $[(\partial n/\partial T)_p < 0]$, whereas the excitation of molecules subject to the condition $\Delta \beta > 0$ results in an increase of the refractive index. Therefore, the variation of the reflection coefficient of a nonlinear mirror with time may be quite complex if two (triplet and temperature) gratings are present simultaneously.

It is evident from Table 1 that the strongest frequency selection within wide stimulated-emission lines can be achieved by forming an inversion grating (α_7) and by modulating the resonant Raman susceptibility (α_{10}). The estimates given in Table 1 are applicable to the dye solutions used in [9]. The results reported in [9] can be explained by the formation of an inversion grating (α_7). A resonant Raman susceptibility grating (α_{10}) may also form under these conditions. Unfortunately, there is little experimental information on the latter grating. A nonlinear mirror formed by a temperature grating (α_3) can play an important role in the frequency selection reported in [9]. It is quite difficult to determine which mechanism is responsible for the frequency selection because all the mechanisms obey formally the same phase-matching conditions (1.8) and a departure from these conditions always has the same effect. It follows from Table 1 that the selective corrections to the gain depend on the pumping field specific to a given mechanism. In particular, the resonat parametric correction is characterized by $\alpha_6 \propto |E(\omega_1)|^2 \Delta N \propto |E(\omega_1)|^4$ whereas α_7 and α_{10} are both proportional to $|E(\omega_1)|^2$. Hence, we can (at least in principle) determine which of these mechanisms is more important. This would require special investigations. The role of distributed mirrors (α_3, α_4) can be determined by repeating the experiments in the presence of a pump wave described by Eq. (1.2) but in the absence of external mirrors operating at the frequency ω_3.

Similar tunable selection can be achieved not only in the case of dye lasers but in the Raman scattering experiments. In particular, the substances containing the hydroxyl group OH (water, alcohols, formic acid, acetic acid, etc.) have wide spontaneous Raman scattering lines [15]. The stimulated Raman–Stokes scattering in water, characterized by a line ~ 1000 cm^{-1} wide, is reported in [16]. Moreover, a pumping wave of the type represented by Eq. (1.2) can be used in frequency selection of the stimulated scattering in the wing of the Rayleigh line, where the spontaneous width may reach 150 cm^{-1}.

Literature Cited

1. S. A. Akhmanov and R. V. Khokhlov, Problems in Nonlinear Optics [in Russian], VINITI, Moscow (1964).
2. N. Boembergen, Nonlinear Optics, Benjamin, New York (1965).

3. C. C. Wang, Phys. Rev., 152:149 (1966).
4. A. P. Veduta and B. P. Kirsanov, Zh. Eksp. Teor. Fiz., 54:1374 (1968).
5. R. Y. Chiao, P. L. Kelley, and E. Garmire, Phys. Rev. Lett., 17:1158 (1966); R. L. Carman, R. Y. Chiao, and P. L. Kelley, Phys. Rev. Lett., 17:1281 (1966).
6. A. P. Veduta and B. P. Kirsanov, ZhETF Pis. Red., 7:221 (1968); Zh. Eksp. Teor. Fiz., 56:1175 (1969).
7. B. P. Kirsanov and A. S. Selivanenko, Opt. Spektrosk., 26:986 (1969).
8. B. P. Kirsanov, Tr. Fiz. Inst. Akad. Nauk SSSR, 18:187 (1965).
9. L. D. Derkacheva and A. I. Krymova, ZhETF Pis. Red., 9:564 (1969).
10. I. L. Fabelinskii, Molecular Scattering of Light, Plenum Press, New York (1968).
11. G. A. Askar'yan, ZhETF Pis. Red., 4:400 (1966).
12. W. E. Lamb, Jr., Phys. Rev., 134:A1429 (1964).
13. B. I. Stepanov and A. N. Rubinov, Usp. Fiz. Nauk 95:45 (1968).
14. A. B. Tikhonov and N. T. Shpak, Abstracts of Papers presented at Fourth All-Union Symp. on Nonlinear Optics, Kiev, 1968 [in Russian].
15. M. V. Vol'kenshtein, Structure and Physical Properties of Molecules [in Russian], Izd. AN SSSR, Moscow–Leningrad (1955).
16. O. Rahn, M. Maier, and W. Kaiser, Opt. Commun., 1:109 (1969).

A CONTRIBUTION TO THE THEORY OF EXCITONS
IN MOLECULAR CRYSTALS

Yu. K. Khokhlov

The principal formulas in the theory of Frenkel excitons are derived including higher multipole interactions between molecules. These formulas are used in a discussion of the Ewald transformation, the structure of the exciton Hamiltonian for low values of **k**, and the effective mass of Coulomb excitons. The electrical and magnetic interactions are separated in the expression for the permittivity of a crystal and the macroscopic field is calculated.

The theory of excitons in molecular crystals (see, for example, [1, 2]) is usually restricted to the lowest (dipole−dipole) interaction between molecules. However, considerable advances in the experimental techniques have set problems which cannot be solved without extension of the theory to higher multipole interactions. It is found that the dipole−dipole interaction in some crystals is anomalously weak (naphthalene) or it may not account completely for the observations (anthracene, benzene). This has led to attempts to include the octupole−octupole interaction [3, 4] and to treatments in which the expansion of the interaction in terms of multipoles is completely avoided [5]. Similar problems are encountered in the calculation of the positions of multipole absorption lines of crystals [1].

The cited papers [1-5] leave open the question of generalization of important elements of the theory (such as the Ewald transformation and the elimination of the macroscopic field in calculation of the permittivity of a crystal) to higher multipole interactions. A considerable part of the present paper attempts to fill this gap. Moreover, we shall discuss the general form of the Hamiltonian of Coulomb excitons at low values of **k**, the effective mass of these excitons, and the separation of the electric and magnetic interactions in the expression for the permittivity of a crystal.

In this connection it is worth mentioning the work of Kornfeld [6] who generalized the Ewald transformation to include the quadrupole−quadrupole interaction or, more precisely, he extended it to the case when $l_1 + l_2 = 4$ [see Eq. (2.4)]. However, this generalization is insufficient for the problems encountered in practice. Moreover, Kornfeld [6] used the Cartesian multipole moments, which cannot be dealt with by the modern theory of angular momenta.

The macroscopic fields generated by higher multipole moments were calculated in [7] [see Eq. (4.5) in the present paper].

§ 1. Expansion of the Coulomb-Exciton Hamiltonian

in Terms of Multipole Interactions

The Hamiltonian of a molecular crystal can be written in the form

$$H = \sum_{n\alpha} H(n\alpha) + \frac{1}{2} \sum_{n\alpha} \sum_{m\beta}' V(n\alpha, m\beta), \qquad (1.1)$$

219

where H $(n\alpha)$ is the Hamiltonian of a single molecule located at a point $x_{n\alpha}$; V $(n\alpha, m\beta)$ is the interaction between molecules; n is the set of integral indices of a unit cell n_x, n_y, n_x); α is the index representing the position of the molecule in the unit cell; $x_{n\alpha} = x_n + x_\alpha$.

The wave functions of the ground and excited states of a crystal can be calculated in the form

$$\Psi_0 = \prod_{nx} \varphi_{nx}^0, \quad \Psi_f = N^{-1/2} \sum_{n\alpha} e^{ikx_{n\alpha}} a_\alpha(k) \varphi_{n\alpha}^f \prod_{m\beta}' \varphi_{m\beta}^0, \tag{1.2}$$

where $\varphi_{n\alpha}^0$, and $\varphi_{n\alpha}^f$ are the wave functions of the ground and excited states of a single molecule; k is the wave vector; N is the number of unit cells in a crystal. All the molecules are assumed to be identical and the excited state of these molecules is postulated to be nondegenerate.*

The coefficients $a_\alpha(k)$ are parameters of the variational problem for the Hamiltonian of Eq. (1.1). We thus arrive at the equation

$$\sum_\beta [\Gamma_{\alpha\beta}(k) - \mu(k)\delta_{\alpha\beta}] a_\beta(k) = 0, \tag{1.3}$$

in which

$$\Gamma_{\alpha\beta}(k) = \sum_n' \langle \varphi_{0\alpha}^f \varphi_{n\beta}^0, \ V(0\alpha, n\beta) \varphi_{0\alpha}^0 \varphi_{n\beta}^f \rangle \exp[ik(x_{n\beta} - x_\alpha)]. \tag{1.4}$$

Here, $\mu(k)$ is the required eigenvalue, which is related to the excitation energy of a crystal by

$$E_f - E_0 = \varepsilon_f - \varepsilon_0 + \mu(k); \tag{1.5}$$

$(\varepsilon_f - \varepsilon_0)$ is the excitation energy of a single molecules.†

We shall first consider Coulomb excitions for which the interaction V $(n\alpha, m\beta)$ represents a nonretarded Coulomb potential (see [1]). This potential can be expanded in terms of the multipole moment operators of the molecules $n\alpha$ and $m\beta$:

$$V(n\alpha, m\beta) = \sum_{l_1=1}^\infty \sum_{l_2=1}^\infty \hat{q}_{l_1}^{12\cdots}(n\alpha) \hat{q}_{l_2}^{1'2'\cdots}(m\beta)(-1)^{l_1} y_{l_1+l_2}^{12\cdots 1'2'\cdots}(\nabla_\alpha) \frac{1}{|x_{m\beta} - x_{n\alpha}|}. \tag{1.6}$$

This expansion is known as the two-center approximation.

In general, the operator of the Cartesian $2l$-pole moment is defined as

$$\hat{q}_l^{12\cdots l} = \frac{1}{l!} \sum_i e_i y_l^{12\cdots l}(r_i), \tag{1.7}$$

---

*In the presence of degeneracy we must assume that f can have several values which label the states and which belong to the same energy level. This increases the rank of the matrix equation (1.3). Then, the following substitutions should be made in Eq. (1.4): $\Gamma_{\alpha\beta}(k) \to \Gamma_{\alpha\beta}^{ff'}(k)$, $\varphi_{n\beta}^f \to \varphi_{n\beta}^{f'}$.
†The expression in square brackets in Eq. (1.3) was simplified by omitting the term $D_\alpha \delta_{\alpha\beta}$, where

$$D_\alpha = \sum_{n\beta}' [\langle \varphi_{0\alpha}^f \varphi_{n\beta}^0, \ V(0x, n\beta) \varphi_{0\alpha}^f \varphi_{n\beta}^0 \rangle - \langle \varphi_{0\alpha}^0 \varphi_{n\beta}^0, V(0\alpha, n\beta) \varphi_{0\alpha}^0 \varphi_{n\beta}^0 \rangle].$$

Usually this term vanishes or is unimportant. An allowance for this term would lead to trivial corrections.

where r_i is the coordinate of the i-th particle, measured from the center of mass of the particle system in question; e_i is the charge of this particle; $y_l^{12 \cdots l}(r)$ is the Cartesian harmonic polynomial, consisting of the following terms:

$$y_1^1(r) = r_1, \qquad y_2^{12}(r) = r_1 r_2 - \frac{1}{3} r^2 \delta_{12} \tag{1.8}$$

and so on [see also Eq. (2.5)]; the tensor indices $1, 2, \ldots l$, run through the values x, y, and z; the harmonic polynomial vanishes when contraction is applied to any pair of tensor indices. From this point onward we shall omit the sign of summation over the tensor indices $1, 2, \ldots, l$, and we shall also omit the indices themselves.

The first three moments (dipole, quadrupole, and octupole) will be given the special notation:

$$\hat{q}_1^1 = \hat{d}_1, \qquad \hat{q}_2^{12} = \hat{q}_{12}, \qquad \hat{q}_3^{123} = \hat{q}_{123}. \tag{1.9}$$

Substituting Eq. (1.6) into Eq. (1.4), we obtain

$$\Gamma_{\alpha\beta}(k) = \sum_n{}' \left[L_{\alpha\beta}(\nabla_\alpha) \frac{1}{|x_{n\beta} - x_\alpha|} \right] \exp\left[ik(x_{n\beta} - x_\alpha)\right] = L_{\alpha\beta}(\nabla_\alpha + ik) \sum_n{}' \frac{1}{|x_{n\beta} - x_\alpha|} \exp\left[ik(x_{n\beta} - x_\alpha)\right], \tag{1.10a}$$

where

$$L_{\alpha\beta}(\nabla) = \sum_{l_1=1}^{\infty} \sum_{l_2=1}^{\infty} q_{l_1} q'_{l_2} (-1)^{l_2} y_{l_1+l_2}(\nabla), \tag{1.11a}$$

$$q_l = \langle \hat{q}_l(\alpha) \rangle_{f0} \equiv \langle \varphi_{0\alpha}^f, \hat{q}_l(0\alpha) \varphi_{0\alpha}^0 \rangle, \tag{1.12}$$

$$q'_l = \langle \hat{q}_l(\beta) \rangle_{0f} \equiv \langle \varphi_{m\beta}^0, \hat{q}_l(m\beta) \varphi_{m\beta}^f \rangle.$$

The prime of the sum over n terms indicates that $n \neq 0$ for $\alpha = \beta$.

We shall use not only the Cartesian moments but also spherical multipole moments of the type

$$Q_{lm} = \sum_i e_i \mathscr{Y}_{lm}(r_i), \tag{1.13}$$

where $\mathscr{Y}_{lm}(r) = r^l Y_{lm}(\vartheta, \varphi)$ is the spherical harmonic polynomial. We shall convert the Cartesian to the spherical moments and back again using the formulas (A.4) and (A.6) derived in the Appendix. The action of the differential operator of Eq. (1.11b) is described by Eq. (A.7). These formulas are equivalent to those derived by Carlson and Rushbrooke [8].*

Conversion to the spherical multipole moments modifies Eqs. (1.10a) and (1.11a) to

$$\Gamma_{\alpha\beta}(k) = (4\pi)^{3/2} \sum_{l_1 m_1} \sum_{l_2 m_2} \sum_n{}' Q_{l_1 m_1} Q'_{l_2 m_2} (-1)^{l_2} B_{l_1 l_2}^{m_1 m_2} \frac{\mathscr{Y}_{LM}^*(x_{n\beta} - x_\alpha)}{|x_{n\beta} - x_\alpha|^{2L+1}} \exp\left[ik(x_{n\beta} - x_\alpha)\right], \tag{1.10b}$$

---

*The octupole–octupole interaction in naphthalene is considered in [4] on the basis of the two-center expansion of the Coulomb potential given in [9]. However, the expansion in [9] is applicable to the special case when both interacting molecules are located on a straight line parallel to the x axis of the coordinate system. In other words, this expansion does not include the angular dependence represented in Eq. (1.10b) by the function $\mathscr{Y}_{LM}^*(x_{n\beta} - x_\alpha)$. Since the number of molecule pairs located on lines parallel to the z axis is very small compared with the total number of interacting molecules in a crystal, the results obtained in [4] and in [3] are incorrect.

$$L_{\alpha\beta}(\nabla) = (4\pi)^{3/2} \sum_{l_1 m_1} \sum_{l_2 m_2} Q_{l_1 m_1} Q'_{l_2 m_2} (-1)^{l_1} B^{m_1 m_2}_{l_1 l_2} \frac{1}{(2L-1)!!} \mathscr{Y}^*_{LM}(\nabla). \tag{1.11b}$$

Here, the coefficient $B^{m_1 m_2}_{l_1 l_2}$ is given by Eq. (A.5); $Q_{l_1 m_1}$ and $Q'_{l_2 m_2}$ are defined in accordance with Eq. (1.12); $L = l_1 + l_2$, $M = m_1 + m_2$.

The matrix elements Q_{lm} are given in the laboratory system of coordinates S. Conversion to intrinsic multipole moments, defined in a coordinate system S' linked to a molecule, can be performed using the formula

$$Q_{lm}(S) = \sum_{m'} D^{(l)}_{m'm}(\alpha, \beta, \gamma) Q_{lm'}(S'),$$

where α, β, and γ are the Euler angles which describe the orientation of the molecule; $D^{(l)}_{m'm}$ is defined in [10].

The experimental data on the matrix elements of the intrinsic multipole moments usually apply to the real moments $\bar{Q}_{lm}$ rather than to the complex moments Q_{lm}. The real moments can conveniently be defined by means of the unitary transformation $\bar{Q}_{lm} = \Sigma_{m'} U_{mm'} Q_{lm'}$, where

$$U_{mm'} = \frac{(-1)^m \delta_{mm'} + \varepsilon_m \delta_{m,-m'}}{\sqrt{1+\varepsilon_m} + i\sqrt{\varepsilon^2_m - \varepsilon_m}}, \qquad \varepsilon_m = \begin{cases} +1, & m > 0, \\ 0, & m = 0, \\ -1, & m < 0. \end{cases}$$

The real moments $\bar{Q}_{lm}$ are associated with the real polynomials $\bar{\mathscr{Y}}_{lm}$, in particular the polynomials

$$\bar{\mathscr{Y}}_{10}(\mathbf{r}) = cz, \quad \bar{\mathscr{Y}}_{11}(\mathbf{r}) = cx, \quad \bar{\mathscr{Y}}_{1,-1}(\mathbf{r}) = cy, \quad c = \sqrt{4\pi/3}.$$

§ 2. Ewald Transformation for an Interaction of Arbitrary Multipole Order. Expansion of $\Gamma_{\alpha\beta}(\mathbf{k})$ in k. Effective Mass of Coulomb Excitons

According to Ewald [11] (see also § 30 in [12]), we can write

$$\sum_m' \frac{1}{|\mathbf{x}_{n\beta} - \mathbf{x}_\alpha|} \exp[ik(\mathbf{x}_{n\beta} - \mathbf{x}_\alpha)] = \frac{4\pi}{v_r k^2} e^{-k^2/4\rho^2} + \frac{\pi}{v_r \rho^2} \sum_{h\neq0}' F\left(\frac{|\mathbf{g}_h + \mathbf{k}|}{2\rho}\right) \times$$

$$\times \exp[i\mathbf{g}_h(\mathbf{x}_\alpha - \mathbf{x}_\beta)] + \rho \sum_n' H(\rho|\mathbf{x}_{n\beta} - \mathbf{x}_\alpha|) \exp[ik(\mathbf{x}_{n\beta} - \mathbf{x}_\alpha)] - \delta_{\alpha\beta}\delta_{n0} \frac{1}{|\mathbf{x}_\alpha - \mathbf{x}_\beta|} \Phi(\rho|\mathbf{x}_\alpha - \mathbf{x}_\beta|). \tag{2.1}$$

Here, h is the set of integral indices of a reciprocal lattice cell (h_x, h_y, h_z), $\mathbf{g}_h$ is the radius vector of a point in the reciprocal lattice; ρ is a parameter selected to ensure the best convergence of the sums over n and h; prime of the sum over n still indicates that $n \neq 0$ if $\alpha = \beta$; v_r is the volume of a unit cell;

$$F(x) = \frac{1}{x^2} e^{-x^2}, \quad H(x) = \frac{1}{x}(1 - \Phi(x)), \quad \Phi(x) = \frac{2}{\sqrt{\pi}} \int_0^x e^{-t^2} dt. \tag{2.2}$$

Substituting Eq. (2.1) into Eq. (1.10a) and performing the necessary computations, we obtain

$$\Gamma_{\alpha\beta}(\mathbf{k}) = \frac{4\pi}{v_r k^2} e^{-k^2/4\rho^2} L_{\alpha\beta}(i\mathbf{k}) + \frac{\pi}{v_r \rho^2} \sum_{h\neq0}' F\left(\frac{|\mathbf{g}_h + \mathbf{k}|}{2\rho}\right) \exp[i\mathbf{g}_h(\mathbf{x}_\alpha - \mathbf{x}_\beta)] L_{\alpha\beta}(i\mathbf{g}_h + i\mathbf{k}) +$$

$$+ \sum_{l_1}\sum_{l_2}\sum_n' q_{l_1} q'_{l_2} (-1)^{l_1} y_{l_1+l_2}(\mathbf{x}_{n\beta} - \mathbf{x}_\alpha) \frac{H_{l_1+l_2}(\rho|\mathbf{x}_{n\beta} - \mathbf{x}_\alpha|)}{|\mathbf{x}_{n\beta} - \mathbf{x}_\alpha|^{2(l_1+l_2)+1}} \exp[ik(\mathbf{x}_{n\beta} - \mathbf{x}_\alpha)]. \tag{2.3}$$

We note that the expression associated with the δ symbols in Eq. (2.1) is missing in Eq. (2.3).

In general, the functions $H_L(x)$ are given by

$$y_L(-\nabla)H(|\mathbf{r}|) = y_L(\mathbf{r})\frac{1}{|\mathbf{r}|^{2L+1}}H_L(|\mathbf{r}|),$$

where y_1 is given by Eq. (1.8). Direct calculations (differentiation) yield the following expressions for the first five functions

$$
\begin{aligned}
H_2 &= 3(1 - \Phi + x\Phi_1) - 2x^2\Phi_2, \\
H_3 &= 15(1 - \Phi + x\Phi_1) - 12x^2\Phi_2 + 4x^3\Phi_3, \\
H_4 &= 105(1 - \Phi + x\Phi_1) - 90x^2\Phi_2 + 40x^3\Phi_3 - 8x^4\Phi_4, \\
H_5 &= 945(1 - \Phi + x\Phi_1) - 840x^2\Phi_2 + 420x^3\Phi_3 - 120x^4\Phi_4 + 16x^5\Phi_5, \\
H_6 &= 10395(1 - \Phi + x\Phi_1) - 9450x^2\Phi_2 + 5040x^3\Phi_3 - 1680x^4\Phi_4 + 336x^5\Phi_5 - 32x^6\Phi_6.
\end{aligned}
\tag{2.4}
$$

The functions $\Phi(x)$ and $\Phi_n(x) = 2^{-n+1}d^n\Phi(x)/dx^n$ are tabulated in [13].

The functions H_2, H_3, and H_4 are identical with those deduced by Kornfeld [6].

The transformation performed as described above serves two quite different purposes. First, it reduces the initial sum of Eq. (1.10b) to two more rapidly converging sums. This is particularly important in the case of the dipole–dipole interaction, for which the sum of Eq. (1.10a) and Eq. (1.10b) converges quite slowly (the dipole–dipole interaction potential decreases with distance as R^{-3}, the quadrupole–quadrupole potential as R^{-5}, etc.). Secondly, and this is more important, the Ewald transformation allows us to separate from Eqs. (1.10a) and (1.10b) the part which is nonregular at the point $k = 0$, namely,

$$
\Gamma_{\alpha\beta}^{\text{nonr}}(\mathbf{k}) = \frac{4\pi}{v_r k^2}L_{\alpha\beta}(i\mathbf{k}) = \frac{4\pi}{v_r k^2}\Big\{ d_1 d_2'\Big(k_1 k_2 - \frac{1}{3}k^2\delta_{12}\Big) + i(q_{12}d_3' - d_1 q_{23}') \times
$$
$$
\times \Big[k_1 k_2 k_3 - \frac{1}{5}k^2(k_1\delta_{23} + k_2\delta_{31} + k_3\delta_{12})\Big] + \dots \Big\}.
\tag{2.5}
$$

The sum in Eqs. (1.10a) and (1.10b) cannot be calculated for low values of k, including $k = 0$, if the nonregular part of Eq. (2.5) is not separated first from Eqs. (1.10a) and (1.10b).

Let us consider the second aspect in greater detail. All the terms in Eqs. (1.10a) and (1.10b) considered as functions of three complex variables k_x, k_y, and k_z, are nonregular at the point $k = 0$ irrespective of the multipole-order interaction. [A regular function can be represented as a series of positive power of k_x, k_y, and k_z. In the case of Eq. (2.5) this cannot be done because of the presence of the factor k^{-2}.] Therefore, the sum in Eqs. (1.10a) and (1.10b) is nonregular as a whole. On the other hand, the expression under the summation sign in Eqs. (1.10a) and (1.10b) depends exponentially on k and, consequently, is a regular function of k. This means that if we wish to calculate the sum of Eqs. (1.10a) and (1.10b) for low values of k, we cannot expand Eqs. (1.10a) and (1.10b) in powers of k under the summation sign (in particular, we cannot assume that under that $k = 0$).[*]

For a fixed value of $\mathbf{s} = \mathbf{k}/k$, we can expand Eqs. (1.10a) and (1.10b) in powers of $k = |\mathbf{k}|$. Retaining terms up to k^2, we obtain

$$
\begin{aligned}
\Gamma_{\alpha\beta}(\mathbf{k}) &= \frac{4\pi}{v_r}\Big\{ d_1 d_2' y_2^{12}(\mathbf{s}) + ik(q_{12}d_3' - d_1 q_{23}')y_3^{123}(\mathbf{s}) + k^2\Big[(q_{12}q_{34}' - g_{123}d_4' - d_1 g_{234}')y_4^{1234}(\mathbf{s}) - \frac{1}{4\rho^2}d_1 d_2' y_2^{12}(\mathbf{s})\Big]\Big\} + \\
&+ \frac{\pi}{v_r\rho^2}\sum_{h\neq 0}' F\Big(\frac{|g_h|}{2\rho}\Big)\exp\left[ig_h(x_\alpha - x_\beta)\right]\Big\{1 + (\mathbf{k}\nabla_g) - 2\Big(1 + \frac{g_h^2}{4\rho^2}\Big)\frac{(\mathbf{k}g_h)}{g_h^2} + \\
&+ \frac{1}{2}(\mathbf{k}\nabla_g)^2 - \Big(1 + \frac{g_h^2}{4\rho^2}\Big)\Big[\frac{2}{g_h^2}(\mathbf{k}g_h)(\mathbf{k}\nabla_g) + \frac{k^2}{g_h^2}\Big] + 2\Big[2 + 2\frac{g_h^2}{4\rho^2} + \Big(\frac{g_h^2}{4\rho^2}\Big)^2\Big]\frac{1}{g_h^4}(\mathbf{k}g_h)^2\Big\}L_{\alpha\beta}(ig_h) + \\
&+ \sum_{l_i}\sum_{l_i}\sum_{n}' q_{l_i}q_{l_i}'(-1)^{l_i}y_{l_i+l_i}(x_{n\beta} - x_\alpha)\frac{H_{l_i+l_i}(\rho|x_{n\beta} - x_\alpha|)}{|x_{n\beta} - x_\alpha|^{2(l_i+l_i)+1}}\Big[1 + i\mathbf{k}(x_{n\beta} - x_\alpha) - \frac{1}{2}(\mathbf{k}, x_{n\beta} - x_\alpha)^2\Big].
\end{aligned}
\tag{2.6}
$$

---

[*]We must stress that this applies to Coulomb excitons in an infinite crystal. We shall not consider thin crystals or real excitons [1] for which the interaction is retarded.

(The effect of a gradient on a harmonic polynomial is discussed on p. 80 in [10]).

If we substitute Eq. (2.6) into Eq. (1.3), we obtain the value of μ (k) in the form of a similar expansion

$$\mu(k) = \mu^{(0)}(s) + ik\mu^{(1)}(s) + \frac{\hbar^2 k^2}{2m^*(s)}. \tag{2.7}$$

Here, m^* (s) is the effective mass of an exciton traveling in the direction s.

We shall now estimate the value of m^* (s) for a crystal with one molecule in each unit cell. (In this case the indices α and β vanish and we have μ (k) = Γ (k), $x_\alpha - x_\beta = 0$, and $q_i' = q_i^*$.)

We shall first assume that the expressions in Eq. (2.3) under the summation sign can be averaged out over the directions of the vectors g_h and x_n without significant changes in the values of the sums. The terms proportional to the spherical polynomials $\mathscr{Y}_{lm}$ (g_h) and $\mathscr{Y}_{lm}$ (x_n) with $l \neq 0$ vanish. The remaining terms yield

$$\frac{\hbar^2}{2m^*(s)} = \frac{4\pi}{v_r}\left[-\frac{1}{4\rho^3}d_1 d_2^* y_2^{12}(s) + (q_{12}q_{34}^* - g_{123}d_4^* - d_1 g_{234}^*) y_4^{1234}(s)\right] + \frac{\pi}{v_r \rho^3}d_1 d_2^* y_2^{12}(s)\sideset{}{'}\sum_{h\neq 0}F\left(\frac{|g_h|}{2\rho}\right)\times$$

$$\times\left\{1 - \frac{4}{3}\left(1 + \frac{g_h^2}{4\rho^2}\right) + \frac{4}{15}\left[2 + 2\frac{g_h^2}{4\rho^2} + \left(\frac{g_h^2}{4\rho^2}\right)^2\right]\right\} + d_1 d_2^* y_2^{12}(k)\frac{1}{15}\sideset{}{'}\sum_{h\neq 0}\frac{1}{|x_n|}H_2(\rho|x_n|). \tag{2.8}$$

[The functions F and H_2 are defined in Eqs. (2.2) and (2.4).]

We can see that in this approximation only the dipole−dipole interaction makes a contribution to the summed part of Eq. (2.8). This interaction decreases fairly slowly and, therefore, we can estimate Eq. (2.8) by replacing summation with integration in accordance with the formulas

$$\sideset{}{'}\sum_n \to \frac{4\pi}{v_g}\int_G^\infty g^2 dg, \qquad \sideset{}{'}\sum_n \to \frac{4\pi}{v_r}\int_R^\infty x^2 dx$$

(v_r and v_g are the unit-cell volumes in the direct and reciprocal lattices; R and G are the truncation radii; $v_r \approx 4\pi R^3$ and $v_g \approx 4\pi G^3/3$).

Such integration yields

$$\frac{\hbar^2}{2m^*(s)} = -\frac{4\pi}{v_r}d_1 d_2^* y_2^{12}(s)\left\{\frac{1}{\rho^2}\left[\frac{1}{10}\rho^2 R^2 + \left(\frac{1}{4} - \frac{1}{10}\rho^3 R^2\right)\Phi(\rho R) + \frac{7}{40}\Phi'(\rho R) + \frac{1}{120}\Phi'''(\rho R)\right]\right.$$

$$\left. + \frac{16\pi\rho}{v_g}\frac{\sqrt{\pi}}{2}\frac{1}{120}\Phi'''\left(\frac{G}{2\rho}\right)\right\} + \frac{4\pi}{v_r}(q_{12}q_{34}^* - g_{123}d_4^* - d_1 g_{234}^*)y_4^{1234}(s). \tag{2.9}$$

Here, d_1, q_{12}, and g_{123} are the matrix elements of the dipole, quadrupole, and the octupole moments [see Eqs. (1.9) and (1.12)]; y_2^{12} and y_4^{1234} are the Cartesian harmonic polynomials of Eq. (1.8); Φ is the error function of Eq. (2.2); ρ is a parameter which occurs in the Ewald transformation defined by Eq. (2.1). Equation (2.9) can be converted to the spherical moments by means of Eq. (A.6).

In the limit $\rho \to 0$, we obtain

$$\frac{\hbar^2}{2m^*(s)} = \frac{4\pi}{v_r}\left[-d_1 d_2^* y_2^{12}(s)\frac{1}{10}R^2 + (q_{12}q_{34}^* - g_{123}d_4^* - d_1 g_{234}^*)y_4^{1234}(s)\right]. \tag{2.10}$$

The formula suggested by Heller and Marcus [14] can be obtained from Eq. (2.10) ignoring the quadrupole−quadrupole and higher interactions.*

§ 3. Permittivity

The relationship between the induction **D** and the electric field **E** is of the form

$$D_1(\mathbf{x}, t) = \int d\mathbf{x}' dt'\, \varepsilon_{12}(\mathbf{x} - \mathbf{x}', t - t')\, E_2(\mathbf{x}', t'),$$

which transforms to the following expression if the Fourier representation is used:†

$$D(\mathbf{k}, \omega) = \varepsilon_{12}(\mathbf{k}, \omega)\, E_2(\mathbf{k}, \omega).$$

The theory (see, for example, [1]) yields the following expression for $\varepsilon_{12}(\mathbf{k}, \omega)$ of a crystal:

$$\varepsilon_{12}(\mathbf{k}, \omega) = \left(1 - \frac{4\pi}{V\omega^2} \sum_{n\alpha} \sum \frac{e^2}{m}\right)\delta_{12} - \frac{4\pi}{V\hbar}\sum_{\lambda}{}' \left[\frac{\langle j_1(\mathbf{k})\rangle_{0\lambda}\,\langle j_2(-\mathbf{k})\rangle_{\lambda 0}}{\omega^2(\omega - \omega_{\lambda 0})} - \frac{\langle j_2(-\mathbf{k})\rangle_{0\lambda}\,\langle j_1(\mathbf{k})\rangle_{\lambda 0}}{\omega^2(\omega + \omega_{\lambda 0})}\right]. \qquad (3.1)$$

Here, λ and 0 are the indices of the excited and ground states of the unperturbed system which will be discussed in the next section; $\omega_{\lambda 0} = \omega_\lambda - \omega_0$ is the difference between the frequencies of these states; V is the volume of the crystal in question; $j(\mathbf{k})$ is the Fourier component of the current density. In general, this component is

$$\mathbf{j}(\mathbf{k}) = \sum \frac{e}{2m}(\mathbf{p}e^{-i\mathbf{k}\mathbf{x}} + e^{-i\mathbf{k}\mathbf{x}}\mathbf{p})$$

(the summation extends to all particles in the crystal; **p** is the particle momentum operator). However, the coordinate of each particle belonging to a molecule $n\alpha$ can be written in the form

$$\mathbf{x} = \mathbf{x}_{n\alpha} + \mathbf{r},$$

where $\mathbf{x}_{n\alpha}$ is the radius vector of the equilibrium position of the center of gravity of the molecule; **r** is the internal coordinate. Therefore, the Fourier component of the current density can be written in the form

$$\mathbf{j}(\mathbf{k}) = \sum_{n\alpha}\mathbf{j}(\mathbf{k}, n\alpha)\, e^{-i\mathbf{k}\mathbf{x}_{n\alpha}},$$

where

$$\mathbf{j}(\mathbf{k}, n\alpha) = \sum \frac{e}{2m}(\mathbf{p}e^{-i\mathbf{k}\mathbf{r}} + e^{-i\mathbf{k}\mathbf{r}}\mathbf{p}). \qquad (3.2)$$

(Here and later the sum without an index applies to all the particles of a given molecule.) Therefore, we obtain

$$\varepsilon_{12}(\mathbf{k}, \omega) = \left(1 - \frac{4\pi}{V\omega^2}\sum_{n\alpha}\sum \frac{e^2}{m}\right)\delta_{12} -$$

$$- \frac{4\pi}{V\hbar}\sum_{n\alpha}\sum_{m\beta}\sum_{\lambda}{}' \left[\frac{\langle j_1(\mathbf{k}, n\alpha)\rangle_{0\lambda}\,\langle j_2(-\mathbf{k}, m\beta)\rangle_{\lambda 0}}{\omega^2(\omega - \omega_{\lambda 0})} - \frac{\langle j_2(-\mathbf{k}, m\beta)\rangle_{0\lambda}\,\langle j_1(\mathbf{k}, n\alpha)\rangle_{\lambda 0}}{\omega^2(\omega + \omega_{\lambda 0})}\right]\exp\left[i\mathbf{k}(\mathbf{x}_{m\beta} - \mathbf{x}_{n\alpha})\right]. \qquad (3.3)$$

---

*However, it should be noted that there are some misprints in the formulas [14].
†The Fourier component of a function $f(\mathbf{x}, t)$ is defined here as

$$f(\mathbf{k}, \omega) = \int d\mathbf{x}\, dt f(\mathbf{x}, t)\exp\left[-i(\mathbf{k}\mathbf{x} - \omega t)\right].$$

The matrix elements of the transitions in Eq. (3.3) can be expanded in terms of the matrix elements describing electric and magnetic transitions of a given multipole order. This expansion is made much easier by a preliminary transformation which separates the electric and magnetic interactions.

Let e_1 be one of three unit vectors directed along the coordinate axes. (We recall that the indices 1, 2, ... run over the values of x, y, and z.) When the operator

$$j_1(\mathbf{k}, n\alpha) = \sum \frac{e}{2m}(\mathbf{p}\mathbf{e}_1 e^{-i\mathbf{k}\mathbf{r}} + e^{-i\mathbf{k}\mathbf{r}}\mathbf{e}_1\mathbf{p}) \tag{3.4}$$

is modified by the identity

$$\mathbf{e}_1 e^{-i\mathbf{k}\mathbf{r}} = \nabla(\mathbf{e}_1\mathbf{r}\varphi^*(i\mathbf{k}\mathbf{r})) + ((\mathbf{e}_1 \times i\mathbf{k}) \times \mathbf{r})\,\varphi'^*(i\mathbf{k}\mathbf{r}), \tag{3.5}$$

where

$$\varphi(x) = \int_0^1 dt e^{xt} = \frac{1}{x}(e^x - 1) = 1 + \frac{1}{2}x + \frac{1}{6}x^2 + \dots,$$

$$\varphi'(x) = \int_0^1 dt\, t e^{xt} = \frac{1}{x}\left[1 + \left(1 - \frac{1}{x}\right)(e^x - 1)\right] = \frac{1}{2}\left(1 + \frac{2}{3}x + \frac{1}{4}x^2 + \dots\right), \tag{3.6}$$

we obtain

$$j_1(\mathbf{k}, n\alpha) = Q_1(\mathbf{k}, n\alpha) + M_1(\mathbf{k}, n\alpha). \tag{3.7}$$

Here,

$$\dot{Q}_1(\mathbf{k}, n\alpha) = \sum \frac{e}{2m}[\mathbf{p} \cdot \nabla(\mathbf{e}_1\mathbf{r}\varphi^*(i\mathbf{k}\mathbf{r})) + \nabla(\mathbf{e}_1\mathbf{r}\varphi^*(i\mathbf{k}\mathbf{r})) \cdot \mathbf{p}] =$$
$$= \frac{i}{\hbar}[H, Q_1(\mathbf{k}, n\alpha)], \quad Q_1(\mathbf{k}, n\alpha) = \sum e(\mathbf{e}_1\mathbf{r})\,\varphi^*(i\mathbf{k}\mathbf{r}), \tag{3.8}$$
$$M_1(\mathbf{k}, n\alpha) = \sum \frac{e}{2m}(-i\mathbf{k} \times \mathbf{e}_1)[(\mathbf{r} \times \mathbf{p})\varphi'^*(i\mathbf{k}\mathbf{r}) + \varphi'^*(i\mathbf{k}\mathbf{r})(\mathbf{r} \times \mathbf{p})].$$

(H is the total Hamiltonian of the crystal in question.) A similar representation can be obtained also for $j_2(-\mathbf{k}, m\beta)$.

It can be shown (details will not be given here) that the operator Q describes transitions induced by an electric field and M describes the corresponding effect of a magnetic field.*

For the sake of brevity we shall omit the arguments in expressions of the type given by Eq. (3.7). This should cause no misunderstanding because the index 1 in these expressions is always associated with the argument $(\mathbf{k}, n\alpha)$ and the index is always associated with the argument $(-\mathbf{k}, m\beta)$.

Substituting Eq. (3.7) into Eq. (3.3), using the equalities $\langle \dot{Q}_1 \rangle_{0\lambda} = -i\omega_{\lambda 0}\langle Q_1 \rangle_{0\lambda}$ and $\langle Q_2 \rangle_{\lambda 0} = i\omega_{\lambda 0}\langle Q_2 \rangle_{\lambda 0}$, the identities

$$\pm \frac{\omega_{\lambda 0}^2}{\omega^2(\omega \mp \omega_{\lambda 0})} = \mp \frac{1}{\omega} + i\frac{i\omega_{\lambda 0}}{\omega^2} \pm \frac{1}{\omega \mp \omega_{\lambda 0}},$$

$$\pm \frac{i\omega_{\lambda 0}}{\omega^2(\omega \mp \omega_{\lambda 0})} = -\frac{i}{\omega^2} + \frac{i}{\omega(\omega \mp \omega_{\lambda 0})},$$

---

*The transformation applied above is a special case of that derived in [15, 16] (see § 29 in [17] and the literature given there).

and applying the completeness theorem, we transform Eq. (3.3) to

$$\varepsilon_{12}(\mathbf{k}, \omega) = \varepsilon_{12}^{(a)}(\mathbf{k}, \omega) + \varepsilon_{12}^{(b)}(\mathbf{k}, \omega), \tag{3.9}$$

where

$$\varepsilon_{12}^{(a)}(\mathbf{k}, \omega) = \left(1 - \frac{4\pi}{V\omega^2} \sum_{n\alpha} \sum \frac{e^2}{m}\right)\delta_{12} + \frac{4\pi}{V\omega^2} \sum_{n\alpha m\beta} \sum \frac{i}{\hbar} \langle [\dot{Q}_1, Q_2] + [M_1, Q_2] + [M_2, Q_1] \rangle_{00} \exp[ik(\mathbf{x}_{m\beta} - \mathbf{x}_{n\alpha})], \tag{3.10}$$

$$\varepsilon_{12}^{(b)}(\mathbf{k}, \omega) = -\frac{4\pi}{Vh} \sum_{n\alpha m\beta} \sum_{\lambda} \sum' \left\{ \frac{\langle Q_1\rangle_{0\lambda} \langle Q_2\rangle_{\lambda 0}}{\omega - \omega_{\lambda 0}} - \frac{\langle Q_2\rangle_{0\lambda} \langle Q_1\rangle_{\lambda 0}}{\omega + \omega_{\lambda 0}} + \right.$$

$$+ \frac{\langle Q_1\rangle_{0\lambda} \langle M_2\rangle_{\lambda 0} - \langle M_1\rangle_{0\lambda} \langle Q_2\rangle_{\lambda 0}}{i\omega(\omega - \omega_{\lambda 0})} + \frac{\langle Q_2\rangle_{0\lambda} \langle M_1\rangle_{\lambda 0} - \langle M_2\rangle_{0\lambda} \langle Q_1\rangle_{\lambda 0}}{i\omega(\omega + \omega_{\lambda 0})} +$$

$$\left. + \frac{\langle M_1\rangle_{0\lambda} \langle M_2\rangle_{\lambda 0}}{\omega^2(\omega - \omega_{\lambda 0})} - \frac{\langle M_2\rangle_{0\lambda} \langle M_1\rangle_{\lambda 0}}{\omega^2(\omega + \omega_{\lambda 0})} \right\} \exp[ik(\mathbf{x}_{m\beta} - \mathbf{x}_{n\alpha})]. \tag{3.11}$$

When the commutators in Eq. (3.10) are calculated, we obtain the following expression:

$$\varepsilon_{12}^{(a)}(\mathbf{k}, \omega) = \delta_{12} + \frac{4\pi}{V\omega^2} \sum_{n\alpha} \sum \frac{e^2}{m} \langle \delta_{12} \{| \bar{\varphi}(i\mathbf{k}\mathbf{r})|^2 + (i\mathbf{k}\mathbf{r})[\varphi'(i\mathbf{k}\mathbf{r})\varphi^*(i\mathbf{k}\mathbf{r}) - $$

$$- \varphi'^*(i\mathbf{k}\mathbf{r})\varphi(i\mathbf{k}\mathbf{r})] - 1\} + [(k_1 r_2 + k_2 r_1)(\mathbf{k}\mathbf{r}) - r_1 r_2 k^2] | \varphi'^*(i\mathbf{k}\mathbf{r})|^2 \rangle_{00}. \tag{3.12}$$

Equations (3.11) and (3.12) are the final results of the transformation.

At low values of k Eq. (3.12) reduces to the following expression (which is taken to terms of the order of k^2):

$$\varepsilon_{12}^{(a)}(\mathbf{k}, \omega) \approx \delta_{12} + \frac{\pi}{V\omega^2} \sum_{n\alpha} \sum \frac{e^2}{m} \langle (k_1 r_2 + k_2 r_1)(\mathbf{k}\mathbf{r}) - (\mathbf{k}\mathbf{r})^2 - r_1 r_2 k^2 \rangle_{00}. \tag{3.13}$$

The transformation performed above removes those terms from the initial expression (3.3) in which the factor ω^{-1} is not associated with the factor k (or a higher power of k). When the initial expression (3.3) is employed, the mutual compensation of these terms is usually discovered only after application to some specific model (see, for example, [1]).

In the calculation of the matrix elements we can expand all the operators of Eq. (3.8) in terms of irreducible representations of a rotation group. The general aspects of this procedure are considered in [15, 16]. We shall simply show, by way of example, that in the long-wavelength approximation the general formulas of Eq. (3.8) yield well-known expressions for the operators of the three lowest multipole transitions. Expanding Eq. (3.8) in terms of small values of k and retaining terms up to k inclusive, we obtain

$$Q_1(\mathbf{k}, n\alpha) \approx \sum er_1 \left(1 - \frac{1}{2}(i\mathbf{k}\mathbf{r})\right) = \sum er_1 - \frac{ik_2}{2} \sum e\left(r_1 r_2 - \frac{1}{3} r^2 \delta_{12}\right) - \frac{ik_1}{6} \sum er^2,$$

$$M_1(\mathbf{k}, n\alpha) \approx (\mathbf{e}_1 \times i\mathbf{k}) \sum \frac{e}{2m}(\mathbf{r} \times \mathbf{p}). \tag{3.14}$$

The last term in the first expression describes transitions between states with the same moment and parity. These transitions are unimportant in most of the problems encountered in this field. The other operators describe the electric dipole and quadrupole and the magnetic dipole transitions.

We shall now go over to the exciton model in Eqs. (3.3) and (3.11). Substituting the wave functions of Eq. (1.2) and the exciton energy of Eq. (1.5) into Eq. (3.3), and utilizing the expression

$$\sum_n \exp\left[i\left(\mathbf{k} - \mathbf{q}\right)\mathbf{x}_n\right] = N\delta_{\mathbf{k},\mathbf{q}},$$

where N is the number of unit cells, we obtain

$$\varepsilon_{12}^{(b)}\left(\mathbf{k}, \omega\right) = -4\pi \frac{N}{V}\sum_\alpha \sum_\beta \left\{ \frac{\langle j_1(\mathbf{k}, \alpha)\rangle_{0f}\,\langle j_2(-\mathbf{k},\beta)\rangle_{f0}}{\omega^2[\hbar\omega - (\varepsilon_f - \varepsilon_0 + \mu'(\mathbf{k}))]}\,a_\alpha'(\mathbf{k})\,a_\beta'^*(\mathbf{k}) - \frac{\langle j_2(-\mathbf{k},\beta)\rangle_{0f}\,\langle j_1(\mathbf{k},\alpha)\rangle_{f0}}{\omega^2[\hbar\omega + \varepsilon_f - \varepsilon_0 + \mu'(-\mathbf{k})]}\,a_\alpha'^*(-\mathbf{k})\,a_\beta'(-\mathbf{k})\right\}.$$

$$(3.15)$$

The quantities designated by primes do not refer to Coulomb excitons but to mechanical excitons whose Hamiltonian $\Gamma_{\alpha\beta}'(\mathbf{k})$ is defined in the next section [Eqs. (4.5) and (4.6)].

The results of the substitution of the same model wave functions into Eq. (3.11) will not be given because they can be derived quite easily from Eq. (3.15).

§ 4. Macroscopic "Transition Field" and Hamiltonian of Mechanical Excitons

In the derivation of Eq. (3.1) for the permittivity the total macroscopic electromagnetic field was regarded as a small perturbation. This means that in the derivation of the equations pertaining to the unperturbed problem the internal macroscopic field should be subtracted from the corresponding microscopic field and included in the perturbation (see, for example, § 13 in [1]). In this exciton model the excited states of the unperturbed system are known as mechanical excitons [1].

We shall introduce the "reducible moments," defined as

$$\widetilde{q}_l^{12\ldots l} = \frac{1}{l!}\sum_i e_i \widetilde{y}_l^{12\ldots l}(\mathbf{r}_i),\tag{4.1}$$

where

$$\widetilde{y}_l^{12\ldots l}(\mathbf{r}) = x_1 x_2 \ldots x_l\tag{4.2}$$

[compare with Eqs. (1.7) and (1.8)]. In contrast to the multipole moments of Eq. (1.7), describing the interaction of a system of particles with a transverse field (div E = 0), the reducible moments of Eq. (4.1) describe the interaction with an arbitrary electric field. If div E = 0, the reducible moments can be replaced with the irreducible moments and vice versa without any change in the magnitude of the interaction.

Bearing this point in mind, we shall replace the irreducible moments in Eq. (1.10a) with the reducible moments and we shall rewrite the expression obtained as a product of two factors:

$$\Gamma_{\alpha\beta}(\mathbf{k}) = \sum_{l_1} e^{-i\mathbf{k}\mathbf{x}_\alpha}\langle \widetilde{\widetilde{q}}_{l_1}(\alpha)\rangle_{f0}\,\widetilde{y}_{l_1}(\nabla_\alpha)\cdot \sum_{n\beta}{}'\sum_{l_1} e^{i\mathbf{k}\mathbf{x}_{n\beta}}\langle \widetilde{q}_{l_1}(\beta)\rangle_{0f}\,\widetilde{y}(\nabla_{n\beta})\,\frac{1}{|\mathbf{x}_{n\beta} - \mathbf{x}_\alpha|}.\tag{4.3}$$

The second factor is the microscopic (intrinsic) "transition potential" which is generated by all molecules of type β, with the possible exception of the molecule located at $\mathbf{x}_\alpha$, where $\mathbf{x}_\alpha \neq \mathbf{x}_\beta$. The potential $\varphi^{\text{micr}}(\mathbf{x})$, generated by all the molecules of type β at an arbitrary point $\mathbf{x}$, can be found by omitting the prime in Eq. (4.3) and replacing $\mathbf{x}_\alpha$ with $\mathbf{x}$:

$$\varphi^{\text{micr}}(\mathbf{x}) = \sum_{n\beta}\sum_l e^{i\mathbf{k}\mathbf{x}_{n\beta}}\langle \widetilde{\widetilde{q}}_l(\beta)\rangle_{0f}\,\widetilde{y}_l(\nabla_{n\beta})\,\frac{1}{|\mathbf{x}_{n\beta} - \mathbf{x}_\alpha|}.\tag{4.4}$$

The macroscopic potential $\varphi(\mathbf{x})$, corresponding to Eq. (4.4), can be found by "smearing out" uniformly each point $\mathbf{x}_{n\beta}$ over a physically infinitesimal volume (in practice, over the volume of a unit cell v_r). This is equivalent to the transformation of the summation over n in Eq. (4.4) to the integration of $\mathbf{x}' = \mathbf{x}_{n\beta}$, where the point $\mathbf{x}' = \mathbf{x}$ is not excluded from the integration domain. Such integration produces (see [7])

$$\varphi(\mathbf{x}) = \frac{4\pi}{v_r k^2} e^{i\mathbf{k}\mathbf{x}} \sum_\beta \sum_l \langle \widetilde{\widetilde{q}}_l(\beta) \rangle_{0l} \, \widetilde{y}_l(-i\mathbf{k}). \tag{4.5}$$

The same result can be obtained by representing a crystal as a continuous medium the following densities of the moments:

$$Q_l(\mathbf{x}) = \frac{1}{v_r} e^{i\mathbf{k}\mathbf{x}} \langle \widetilde{\widetilde{q}}_l(\beta) \rangle_{0l}.$$

These moment densities correspond to the bound-charge density

$$\rho^{bc}(\mathbf{x}) = \sum_l \widetilde{y}_l(-\boldsymbol{\nabla}) \, Q_l(\mathbf{x}) = \frac{1}{v_r} e^{i\mathbf{k}\mathbf{x}} \sum_\beta \sum_l \langle \widetilde{\widetilde{q}}_l(\beta) \rangle_{0l} \, \widetilde{y}(-i\mathbf{k}).$$

The potential generated by the bound charges can be found from the equation

$$\nabla^2 \varphi(\mathbf{x}) = -4\pi\rho^{bc}(\mathbf{x})$$

and it is identical with Eq. (4.5). This method of deriving the macroscopic field is a well-established procedure in the case of dipole interactions (see, for example, § 30 in [12]).

The Hamiltonian of mechanical excitons is defined by

$$\Gamma'_{\alpha\beta}(\mathbf{k}) = \Gamma_{\alpha\beta}(\mathbf{k}) - \Gamma^{macr}_{\alpha\beta}(\mathbf{k}), \tag{4.6}$$

where $\Gamma^{macr}_{\alpha\beta}(\mathbf{k})$ is the energy of the interaction, at a point $\mathbf{x}$, between moments of the type $\langle \widetilde{q}_l(\alpha) \rangle_{l0} \exp(-i\mathbf{k}\mathbf{r})$ and the macroscopic field defined by Eq. (4.5).

We find that

$$\Gamma^{macr}_{\alpha\beta}(\mathbf{k}) = \sum_l e^{-i\mathbf{k}\mathbf{x}} \langle \widetilde{q}_l(\alpha) \rangle_{l0} \, \widetilde{y}_l(\boldsymbol{\nabla}) \, p(\mathbf{x}) = \frac{4\pi}{v_r k^2} \sum_{l_1} \sum_{l_2} i^{l_1-l_2} \langle \widetilde{\widetilde{q}}_{l_1}(\alpha) \rangle_{l0} \langle \widetilde{\widetilde{q}}_{l_2}(\beta) \rangle_{0l} \, \widetilde{y}_{l_1+l_2}(\mathbf{k}) =$$

$$= \frac{4\pi}{v_r k^2} [d_1 d_2' k_1 k_2 + i(\widetilde{7}_{12} d_3' - d_1 \widetilde{q}_{23}') k_1 k_2 k_3 + \ldots]. \tag{4.7}$$

[The matrix elements in the above equation are given in contracted notation analogous to Eq. (1.12) and the lower moments are denoted as in Eq. (1.9).]

A comparison of Eqs. (4.7) and (2.5) shows that, with the exception of the first (dipole) term, these two equations are different. The difference can be removed if only the reducible moments are used in $\Gamma_{\alpha\beta}(\mathbf{k})$ right from the beginning. However, this would imply introduction of expressions which do not actually occur in the theory employed because they compensate each other. There is no justification for such introduction. Therefore, the Ewald transformation and the separation of the macroscopic field should be regarded as independent operations.

The author is deeply grateful to V. M. Agranovich for his advice during the course of this work, and to A. G. Molchanov for his comments.

Appendix

The starting point of the theory is the two-center expansion of the Coulomb potential as a Taylor series in terms of the internal coordinates $\mathbf{r}_i$ and $\mathbf{r}_j$:

$$V(n\alpha, m\beta) = \sum_{i,j} \frac{e_i e_j}{|\mathbf{x}_{m\beta} + \mathbf{r}_j - \mathbf{x}_{n\alpha} - \mathbf{r}_i|} = \hat{L}(\boldsymbol{\nabla}_\alpha) \frac{1}{|\mathbf{x}_{m\beta} - \mathbf{x}_{n\alpha}|}, \tag{A.1}$$

$$\hat{L}(\nabla) = \sum_{l_1=1}^{\infty} \sum_{l_2=1}^{\infty} \tilde{\hat{q}}_{l_1}(n\alpha)\, \tilde{y}_{l_1}(\nabla) \cdot \tilde{\hat{q}}_{l_2}(m\beta)\,(-1)^{l_2}\, \tilde{y}_{l_2}(\nabla) = \sum_{l_1=1}^{\infty} \sum_{l_2=1}^{\infty} \tilde{\hat{q}}_{l_1}(n\alpha)\, \tilde{\hat{q}}_{l_2}(m\beta)\,(-1)^{l_2}\, \tilde{y}_{l_1+l_2}(\nabla). \qquad (A.2)$$

Here, $\tilde{\hat{q}}_l$ · are the reducible moments of Eqs. (4.1) and (4.2).

The Coulomb potential outside any poles satisfies the Laplace equation $\nabla^2\varphi = 0$. It follows that we can add to $\hat{L}(\nabla)$ any expression proportional to ∇^2. In particular, we can systematically replace in $\hat{L}(\nabla)$ the reducible quantities $\tilde{q}_l$, $\tilde{y}_l(\nabla)$ with the irreducible quantities $\hat{q}_l$, $y_l(\nabla)$ [see Eqs. (1.7) and (1.8)] and vice versa. We can thus write

$$\hat{L}(\nabla) = \sum_{l_1=1}^{\infty} \sum_{l_2=1}^{\infty} \hat{q}_{l_1}(n\alpha)\, y_{l_1}(\nabla) \cdot \hat{q}_{l_2}(m\beta)\,(-1)^{l_2} y_{l_2}(\nabla) = \sum_{l_1=1}^{\infty} \sum_{l_2=1}^{\infty} \hat{q}_{l_1}(n\alpha)\, \hat{q}_{l_2}(m\beta)\,(-1)^{l_2}\, y_{l_1+l_2}(\nabla). \qquad (A.3)$$

We shall use the formula

$$\sum_{\substack{\text{upper} \\ \text{index}}} \hat{q}_l^{12\cdots}\, y_l^{12\cdots}(\nabla) = \frac{4\pi}{(2l+1)!!} \sum_{m=-l}^{l} Q_{lm} \mathcal{Y}_{lm}^{*}(\nabla), \qquad (A.4)$$

which can be derived by comparing the known single-center Taylor-series expansions of the Coulomb potential with the spherical harmonic polynomials.

Substituting Eq. (A.4) into Eq. (A.3), we can make the substitution

$$\frac{(4\pi)^2}{(2l_1+1)!!\,(2l_2+1)!!}\, \mathcal{Y}_{l_1m_1}^{*}(\nabla)\, \mathcal{Y}_{l_2m_2}^{*}(\nabla) \to (4\pi)^{1/2} B_{l_1l_2}^{m_1m_2}\, \frac{1}{(2L-1)!!}\, \mathcal{Y}_{LM}^{*}(\nabla),$$

where

$$B_{l_1l_2}^{m_1m_2} = (4\pi)^{1/2}\, \frac{(2L-1)!!}{(2l_1+1)!!(2l_2+1)!!}\, \langle Y_{l_1m_1} Y_{l_2m_2},\, Y_{LM}\rangle^{*} =$$

$$= \left[\frac{(L+M)!\,(L-M)!}{(2l_1+1)\,(2l_2+1)\,(2L+1)\cdot(l_1+m_1)!\,(l_1-m_1)!\,(l_2+m_2)!\,(l_2-m_2)!} \right]^{1/2},$$
$$L = l_1 + l_2, \qquad M = m_1 + m_2. \qquad (A.5)$$

This is equivalent to the substitution

$$\sum_{\substack{\text{upper} \\ \text{index}}} \hat{q}_{l_1}(n\alpha)\, \hat{q}_{l_2}(m\beta)\, y_{l_1+l_2}(\nabla) \to (4\pi)^{1/2} \sum_{m_1,m_2} \hat{Q}_{l_1m_1}(n\alpha)\, \hat{Q}_{l_2m_2}(m\beta) B_{l_1l_2}^{m_1m_2}\, \frac{1}{(2L-1)!!}\, \mathcal{Y}_{LM}^{*}(\nabla). \qquad (A.6)$$

The result of application of the operator $\mathcal{Y}_{LM}^{*}(\nabla)$ to the Coulomb potential is given by the formula

$$\left[\frac{1}{(2L-1)!!}\, \mathcal{Y}_{LM}^{*}(\nabla_\eta)\, \frac{1}{|\mathbf{x}_{m\beta} - \mathbf{x}_{n\alpha} - \eta|} \right]_{\eta=0} = \mathcal{Y}_{LM}^{*}(\mathbf{x}_{m\beta} - \mathbf{x}_{n\alpha})\, \frac{1}{|\mathbf{x}_{m\beta} - \mathbf{x}_{n\alpha}|^{2L+1}}, \qquad (A.7)$$

which is obtained by expanding $|\mathbf{x}_{m\beta} - \mathbf{x}_{n\alpha} - \eta|^{-1}$ in terms of the spherical harmonic polynomials $\mathcal{Y}_{lm}(\eta)$ and by subsequent application of the equality

$$[\mathcal{Y}_{lm}^{*}(\nabla_\eta)\, \mathcal{Y}_{l'm'}(\eta)]_{\eta=0} = \frac{(2l+1)!!}{4\pi}\, \delta_{ll'} \delta_{mm'}.$$

The substitution of Eq. (A.6) into Eq. (A.3) and then into Eq. (A.1) gives, in combination with Eq. (A.7), the required two-center expansion in terms of the spherical multipole moments [8].

Literature Cited

1.   V. M. Agranovich and V. L. Ginzburg, Spatial Dispersion in Crystal Optics and the Theory of Excitons, Wiley, New York (1967).
2.   A. S. Davydov, Usp. Fiz. Nauk, 82:393 (1964).
3.   D. P. Craig and J. R. Walsh, J. Chem. Soc., p. 1613 (1958).
4.   D. P. Craig, L. E. Lyons, S. H. Walmsley, and J. R. Walsh, J. Chem. Soc., p. 389 (1959); D. P. Craig and S. H. Walmsley, Mol. Phys., 4:113 (1961).
5.   R. Silbey, J. Jortner, and S. A. Rice, J. Chem. Phys., 42:1515 (1965).
6.   H. Kornfeld, Z. Phys., 22:27 (1924).
7.   V. S. Mashkevich, Fiz. Tverd. Tela, 2:908 (1960).
8.   B. C. Carlson and G. S. Rushbrooke, Proc. Cambridge Phil. Soc., 46:626 (1950).
9.   J. O. Hirschfelder, C. F. Curtiss, and R. B. Bird, Molecular Theory of Gases and Liquids, Wiley, New York (1954).
10. A. R. Edmonds, Angular Momentum in Quantum Mechanics, Princeton University Press, Princeton, N. J., (1957).
11. P. P. Ewald, Ann. Phys. (Leipzig), 64:253 (1921).
12. M. Born and K. Huang, Dynamical Theory of Crystal Lattices, Oxford University Press (1954).
13. E. Jahnke and F. Emde, Tables of Functions with Formulas and Curves, Dover, New York (1943).
14. W. R. Heller and A. Marcus, Phys. Rev., 84:809 (1951).
15. Yu. K. Khokhlov, Zh. Eksp. Teor. Fiz., 26:576 (1954).
16. Yu. K. Khokhlov, Tr. Fiz. Inst. Akad. Nauk SSSR, 53:157 (1971). [Studies in Nuclear Physics, Consultants Bureau, New York (1973), pp. 151-160.]
17. M. E. Rose, Multipole Fields, Wiley, New York (1955).

QUENCHING OF THE PHOTOLUMINESCENCE BY ELECTRON IRRADIATION AND ENERGY TRANSFER

Sh. D. Khan-Magometova

An investigation was made of the role of free and localized excitons and of reabsorption in energy transfer and in luminescence. Much of the investigation was carried out by the method of preliminary irradiation of crystals with ionizing radiations and subsequent measurement of the photoluminescence parameters. The investigation yielded the relationship governing the photoluminescence quenching, leading to deformation of the luminescence spectra with increasing concentration of the quenching impurity. It was established that, in general, the process of transfer of the excitation energy should be of multistep nature.

INTRODUCTION

Much theoretical and experimental work has been done on molecular crystals. These crystals are of interest as objects of important investigations of various aspects of solid-state physics and because of their numerous practical applications.

A distinguishing feature of molecular crystals is the van der Waals intermolecular interaction. The van der Waals forces are much weaker than the interaction between atoms in a molecule and, therefore, the molecules in the lattice of a molecular crystal retain many of their free-state properties.

Molecular crystals include solidified inert gases (Ne, Ar, Kr, Xe), crystals consisting of molecules with saturated bonds (H_2, O_2, CH_4, etc.), and finally, numerous aromatic hydrocarbon crystals such as benzene, naphthalene, anthracene, etc.

Luminescent molecular crystals have been used for long time in detection of ionizing radiations by scintillation spectrometers and dosimeters.

Organic molecular crystals are frequently used as models of various biological objects (chloroplast membranes, nerve fibers, various parts of cells, etc.) which have quasicrystalline structures. The extreme complexity of structures and functions of biological objects makes it difficult to investigate them by purely physical quantitative methods and therefore modeling of biophysical processes in living structures by much simpler systems is invaluable.

Important properties of molecular crystals are manifested in the mechanisms of absorption, emission, and transfer of the excitation energy and, therefore, the majority of investigations have been concerned with these processes. Frenkel and Davydov [1-4] suggested a theory of the absorption and luminescence of molecular crystals. This theory was subsequently developed by many Soviet and other workers [5-10] and was found to describe and explain satisfactorily the principal experimental relationships. However, there are many subjects, such as the participation of excitons in luminescence, migration of excitation energy, and localization of

233

this energy which require further theoretical analysis and comprehensive experimental investigations.

The present paper describes mainly an investigation of energy transfer in luminescent molecular crystals (anthracene, stilbene, terphenyl, and others) during their optical excitation. The principal method used in this investigation was irradiation with electrons or β particles. This irradiation quenched the photoluminescence. The measured dependences of the luminescence efficiency on the absorbed dose, which were obtained at various wavelengths in the luminescence spectra, were used to draw conclusions about the exciton and photon (reabsorption) mechanisms of energy transfer in these crystals. It should be stressed that luminescence quenching by ionizing radiations is of intrinsic interest because of the wide use of molecular crystals in the detection and spectrometry of such radiations.

Chapter I is concerned with the models used in the description of energy migration and with the methods employed in investigations of energy transfer in molecular crystal.

Chapters II-V give descriptions of the experiments carried out by the present author, list the results obtained, and give their interpretation.

CHAPTER I

INVESTIGATION OF TRANSFER OF ELECTRON EXCITATION ENERGY IN MOLECULAR CRYSTALS

This investigation is concerned with the transfer of the electron excitation energy by singlet excitons and by reabsorption of luminescence. Therefore, we shall restrict our theoretical analysis to these two energy migration mechanisms.

In thin molecular crystals the excitation energy is transferred mainly in nonradiative manner via excitons. The efficiency of the exciton energy transfer can be judged on the basis of, for example, the data given in [11, 12]. A considerable impurity (exciton-induced) luminescence was observed at low temperatures at impurity concentrations of the order of 10^{-5} g/g. At room temperature the luminescence efficiency increased rapidly with increasing concentration up to 10^{-3} g/g but the rise was slight at higher concentrations. These data indicated that during its lifetime in a crystal (10^{-8} sec) an exciton can travel a distance of $\sim 10^3$-10^4 lattice constants.

In the case of relatively thick samples whose absorption and luminescence spectra overlap strongly we find that, in addition to the nonradiative migration via excitons, the photon energy transfer involving reabsorption of luminescence (emission of luminescence at one point and its subsequent absorption at another point [13-18]) is of considerable importance. Experiments show that reabsorption depends on the thickness of a crystal, the absorption coefficient, the angle of incidence of the exciting radiation, the temperature, and the nature of reflection of luminescence from the surfaces. The importance of reabsorption is usually estimated with the aid of an approximate solution of an integral equation. One of these approximations involves the use of an absorption coefficient corresponding to the luminescence wavelength reabsorbed most strongly. The subject of reabsorption of light will be considered in Chap. II.

The migration of excitons, which are scattered by phonons and lattice defects, can be described mathematically in the same way as the motion of multiply scattered particles, i.e., we can use the transport equation or the diffusion equation. As is known, the latter equation follows from the former under specified conditions.

The possibility of utilization of the transport equation is governed by the ratio of the dimensions of an exciton wave packet ($\sim 1/|\mathbf{k}|$) to the mean free path of excitons under scattering conditions (λ_s). The Boltzmann transport equation is applicable to the motion of free excitons if $|k|\lambda_s \gg 1$. Estimates derived by Agranovich and Konobeev [19] show that this condition is satisfied at room temperature but not at temperatures approaching 10^3-$10^{6\circ}$K because of the reduction in λ_s.

In some cases the migration of excitons can be described using the diffusion equation, which is simpler than the transport equation. The diffusion approximation is applicable if $l/\lambda_s \gg <$, where l is the diffusion length, or (which is practically equivalent) if during its lifetime an exciton (or some other quasiparticle) undergoes a sufficient number of collisions which alter its direction of motion. Estimates of the value of λ_s given in [19, 20] amount to ~(10-100)a, where a is the lattice constant. Thus, the condition $l/\lambda_s \gg 1$ is usually satisfied. Strictly speaking the diffusion approximation can be used only if a second condition is also satisfied: the value of λ_s should be much less than the characteristic geometrical dimensions of the region for which the diffusion equation is being solved. For example, in applying the cellular method the second condition is not obeyed and therefore we have to solve the transport equation (this is discussed later in the present chapter and in Chap. III).

In the case of localized excitons we have $\lambda_s \sim a$ and, since $l \sim 10^{-5}$-10^{-6} cm (this will be demonstrated later), the diffusion approximation can be applied.

A phenomenological theory of the diffusion of excitons was suggested and developed by Faidysh [21-23]. It has subsequently been used by many workers (see, for example, [24, 25]) in the solution of specific problems in energy migration in molecular crystals. In this method a crystal is regarded as a homogeneous and isotropic medium and the exciton density satisfies the standard diffusion equation. The surfaces of a crystalline sample are described by boundary conditions, i.e., an allowance can be made for the arrival of excitons on the face of a crystal, followed by reflection, de-excitation, or nonradiative annihilation. The probability of capture of an exciton by an impurity is assumed to be proportional to the impurity concentration and all the parameters occurring in the diffusion equation (the diffusion coefficient D, the diffusion length l, and the probability of the capture by impurities present in a unit concentration) are regarded as empirical constants which can be estimated only by comparing the results of the calculations with the experimental results. In this way the value of l in anthracene was estimated to be 0.46×10^{-5} cm [24] and $(0.7$-$1.5) \times 10^{-5}$ cm [23, 26-29].

A different description of the migration of excitons was used by Galanin and Chizhikova [12]. They employed a theory which has been developed for the Brownian motion or diffusion in liquids and has been applied to the quenching of fluorescence of solutions [30-36]. In this theory one regards impurity molecules as "black" and "gray" spherical particles: the black particles are assumed to quench completely the diffusing excitation and the gray particles quench this excitation only partially. The dependence of the probability of excitation transfer to an impurity molecule on the distance from it can be estimated theoretically in the case of resonant transfer. For example, in the case of the dipole—dipole interaction [37, 38], the probability is inversely proportional to the sixth power of the distance. Moreover, it is possible to deduce the dependence of this probability on some properties of the host and impurity molecules, such as their luminescence spectra, the excited-state lifetime, or the refractive index. Thus, although the probability is assumed to be proportional to the impurity concentration, it is possible to relate it to some physical parameters of the host and impurity molecules. It follows from [12, 31-36] that the relative quantum efficiency of luminescence in the presence of impurity molecules capable of absorbing the excitation energy is given by

$$A = \frac{I}{I_0} = \frac{1}{1 + 4\pi D p p_0 \tau_0 c} = \frac{1}{1 + Bc}, \tag{L.1}$$

where D is the diffusion coefficient of excitons; τ_0 is the excited-state lifetime; p_0 is the relative probability of energy transfer from the excited molecules of the host substance to the nearest impurity molecule; c is the impurity concentration. Faidysh et al. [26, 28] compared these two approaches to the transfer of energy by excitons and they showed that the two theories are in satisfactory agreement and that Eq. (I.1) describes only the experimental data obtained for low impurity concentrations.

Migration of excitons in impure crystals can also be described as outlined in the present paper and in [19]. We can represent a crystal with impurities by an assembly of "cells" and having solved the transport or the diffusion equation for each of these cells, we can estimate the dependence of the probability of "capture" of the excitation energy on the impurity concentration. In general, this probability is a nonlinear function of the impurity concentration (Chap. III and [39, 40]).

Many authors are of the opinion that the transfer of energy from the host substance to impurities in molecular crystals involves the participation of free and localized excitons.

The possibility of energy transfer by free excitons was considered theoretically by Agranovich, Konobeev, et al. [41-48]. These workers derived formulas and obtained estimates of the principal diffusion parameters of free excitons (mean free path, diffusion length, and diffusion coefficient). The temperature dependences of the diffusion coefficient and of the intensity of the fluorescence emitted by impure crystals were derived theoretically in [42-44]. Experimental investigations of these parameters were reported in [49-51] for naphthalene crystals. An analysis of the experimental data and a comparison with the theory led to the conclusion [42, 44] that the transfer of excitation energy in anthracene-doped naphthalene crystals takes place via free excitons.

The theory of diffusion of localized excitons was developed by Trlifaj [10, 52, 53] who solved the time-dependent Schrödinger equation on the assumption that the state of a crystal is nonstationary because the excitation which produces localized excitons is accompanied by a change in the equilibrium positions of the molecules and the motion of an excited state involves motion of a local deformation. In the zeroth approximation the solution of the Schrödinger equation obtained by Trlifaj is in agreement with the result of Davydov [2], obtained for a model of a crystal with molecules rigidly bound to their equilibrium sites.

Trlifaj obtained expressions for the diffusion parameters of localized excitons (l and D) and for the probability of transfer of the excitation energy. Trlifaj expressed these quantities in terms of the distribution of energy in the luminescence spectrum and in terms of the absorption cross sections of the host and the impurity molecules. The diffusion length, estimated by Trlifaj [10, 53], was 0.46×10^{-5} cm. The estimates obtained by Faidysh et al. on the basis of Trlifaj's formulas and their own experimental results [28] gave $l = 1 \times 10^{-5}$ cm. The values of D and l were determined experimentally for anthracene by Galanin and Chizhikova [12], who studied the dependences of the quantum efficiency and the excited-state lifetime in anthracene on the impurity (naphthacene) concentration: they found that $l = 0.35 \times 10^{-5}$ cm and D = 4.5 $\times$ 10^{-4} cm^2/sec. This value of l was close to that obtained by Trlifaj but this cannot be regarded as a proof that the energy is tranferred in anthracene solely by localized excitons. The diffusion of localized excited states has also been investigated by other workers [54, 55].

The relative importance of free and localized excitons in the transfer of energy in molecular crystals has not yet been finally determined. Although the theoretically calculated parameters of the energy transfer by free and localized excitons are quite different, both sets of parameters are quite close to the values found experimentally. Therefore, it is difficult to determine the role of the various excited states in energy transfer. Some information on this subject could be obtained by investigating the temperature dependences of the diffusion coefficients of free and localized excitons. For example, it is reported in [42, 44, 53] that the diffusion coefficients of free and localized excitations depend on temperature as $T^{-1/2}$ and as $\exp(-u/kT)$, respectively (here, u is the activation energy of diffusing excitons, related to the local deformation of the crystal lattice).

It is assumed in [19, 28, 56] that the available experimental data on the migration of energy in crystals support the hypothesis of the existence of excitons with properties intermediate between those of free and localized states. In view of this it is suggested in [19] that a critical review should be made of the theory of migration of various types of excitation in crystals and this review should be extended to the temperature dependences of the diffusion coefficients. The theory of energy transfer under conditions of intermediate coupling between excitons and phonons should also be developed.

A calculation of the mean energy-transfer time for neighboring molecules is given in [56]. In the case of free and localized excitons these transfer times are 10^{-14}-10^{-15} and 10^{-12} sec, respectively. The migration of energy via excitons with intermediate properties is deduced from the observation that the experimental value of the transfer time is 10^{-13} sec.

The results reported in the present investigation support the idea that successive transfer of the excitation energy from free to localized excitons plays an important role in photoluminescence (this is known as the multistep process, which is considered in Chap. IV and in [40, 57]).

It is worth stressing that every description of the migration of excitons in molecular crystals should make an allowance for the anisotropy of their properties because these crystals have lattices which are far from cubic. The efficiency of energy transfer may differ from one axis to another because the distances between molecules and their relative orientations vary with the direction. A unit cell of anthracene is a monoclinic prism with edges $a = 8.5$ Å, $b = 6$ Å, and $c = 11.14$ Å. The anthracene molecules are oriented so that along the a axis they are "face to face" and not "edge to edge" [24]. Bearing this point in mind and the different distances between molecules along different axes, we may conclude that in anthracene crystals the excitation energy should travel more freely between molecules located in neighboring planes parallel to the bc plane than along the planes themselves. It is also worth noting that the diffusion length l in the case of energy migration in the ab plane is longer than in the case of migration along the c axis. An allowance for the anisotropy in the description of energy transfer in molecular crystals meets with considerable mathematical difficulties and has not yet been made in theoretical investigations.

Let us now consider the experimental methods for investigating the migration of electron excitation energy, which are employed in studies of molecular crystals. It is usual to measure the efficiency, the average decay time, and the polarization of the luminescence at various concentrations of the luminescence centers in the form of impurities introduced into the lattice of the host crystal [12, 21-23, 26-29, 54, 58]. Such investigations have been carried out at various temperatures [27, 50, 51, 59-62]. The basic idea behind this method is that by varying the impurity concentration we can alter the average distance across which the excitation energy has to be transferred from the host to the impurity molecules. This alters the efficiency of energy transfer and the parameters which characterize the transfer process (luminescence efficiency, mean excited-state lifetime, and other properties of the host and impurity molecules).

The average distance traveled in a crystal by a diffusing excitation can be altered also by varying the thickness of a crystal one of whose plane faces carries an "excitation detector" which is a film of an impure crystal [24, 63]. Excitons generated by the absorption of light diffuse to this "detector" film. The luminescence efficiency of the "detector" can be used, in combination with the known thickness of the sample, to determine the parameters characterizing exciton diffusion.

The diffusion of excitons can also be studied in the following way [64]. Exciting radiation falls on a plane surface of a crystal whose thickness is sufficient for total absorption of the incident light. If the wavelength (absorption coefficient) is varied, the depth of penetration of the exciting radiation changes and, therefore, the "primary" excitons appear at different depths below the surface. In the case of nonradiative annihilation of excitons on the surface the luminescence efficiency should be a function of the exciton diffusion. This explanation is widely adopted in the interpretation of the "anticorrelation" of the luminescence efficiency and the absorption coefficient, which was observed in [14, 65]. We shall show (Chap. V) that this is not a satisfactory explanation of the "anticorrelation" observed for anthracene.

Zhevandrov [66] suggested that the dependence of the polarization of the luminescence on the polarization of the exciting radiation could be used to detect and study the migration of excitation energy in molecular crystals. A calculation which ignores the energy migration shows that the polarization of the luminescence should depend strongly on the orientation of the electric vector of the exciting radiation. Experimental studies of a large number of pure and mixed crystals show a complete absence of this dependence, which is interpreted as proving that energy is transferred between the molecules of the host lattice and the impurity molecules (in the latter case the migration occurs over distances of the order of 30 lattice constants).

When a molecular crystal is bombarded with ionizing radiation, the disturbed molecules can act as a quenching impurity which "captures" excitons. Therefore, energy transfer processes can be studied by employing the irradiation method of introducing a quenching impurity into a crystal [67-71]. The same method was used in [70] to find a rough value of the diffusion length in anthracene: the value obtained in this way was $\sim 0.9 \times 10^{-5}$ cm.

The main advantage of the irradiation method of introducing impurities, compared with the standard method of employing doped crystals, is the opportunity of varying the impurity concentration in the same sample because this ensures a high reproducibility and reduces the experimental error. This method was employed in the present study to introduce a quenching impurity into crystals and it was found that the shape of the photoluminescence spectra changed gradually with increasing impurity concentration (deformation of the spectra was observed). The relevant experiments and their interpretation are described in Chaps. III and IV.

The principal experimental methods for investigating energy transfer in molecular crystals utilize the variation of the impurity concentration or of the distance between the region where excitons are created and the region where they are captured. If a crystal is irradiated with low-energy electrons resulting from the radioactive decay of tritium, one can vary simultaneously the concentration of the quenching impurity and the thickness of the layer in which exciton diffusion takes place. An analysis of the experiments based on this arrangement made it possible to estimate the diffusion length in anthracene crystals and the importance of reabsorption of luminescence during migration of energy across these crystals (Chap. II and [25]).

CHAPTER II

USE OF β RADIATION OF TRITIUM IN INVESTIGATIONS OF ENERGY TRANSFER PROCESSES IN ANTHRACENE CRYSTALS

When crystals are irradiated with ionizing particles or quanta, the disturbed molecules formed in this way reduce the photoluminescence efficiency. This is known as the "degradation of luminescence."

The degradation of luminescence must be allowed for in the use of scintillation counters in dosimetry, radiometry, and spectrometry of ionizing radiations. The degradation of organic scintillators is utilized in dosimeters intended for large-dose measurements, for example, in high-power industrial and laboratory γ-ray sources [72, 73].

According to the published data [67, 70, 74, 75] the reduction in the luminescence efficiency as a result of irradiation is given by the expression

$$\frac{I}{I_0} = A(R) = \frac{1}{1+BR},\tag{II.1}$$

where A (R) is the ratio of the luminescence efficiency after irradiation to the same efficiency before irradiation (this is known as the relative photoluminescence efficiency); R is the absorbed radiation dose; B is a constant. This is the standard formula for the quenching of luminescence by foreign impurities. The quenching which occurs in molecular crystals under the influence of irradiation is closely related to the migration of energy in such crystals [67-70, 74, 75]. Impurities in the form of disturbed molecules quench the luminescence by capturing the migrating excitation energy. Therefore, the degradation of luminescence can, in principal, be used in investigations of energy transfer, exciton diffusion, and transport parameters of this diffusion.

In particular, it would be interesting to determine the diffusion length by this method and to estimate the importance of exciton energy transfer in the degradation of luminescence, which has not been studied very extensively.

The investigations described in the present chapter involved the irradiation of crystals by low-energy tritium electrons. This made it possible to vary simultaneously the concentration of the luminescence-quenching impurity and the distance between the region where excitons were generated and the region where they were captured, i.e., the thickness of the layer in which exciton diffusion occurred.

§1. Experimental Method

The ionizing radiation was generated in a zirconium−tritium target with a nomical activity of 7 curies, which was placed in contact with a crystal. The photoluminescence was measured before and after irradiation. The luminescence was excited and measured on the side opposite to that subjected to tritium-electron irradiation (Fig. 1). The dose received by the crystal was varied (by varying the duration of irradiation) between 10^6 and 10^8 rad. Crystals of different thickness (0.5-6 μ) was used and measurements were made of the relative luminescence efficiency A, i.e., of the ratio of the luminescence emitted by a crystal after irradiation to that emitted before irradiation.

The basis of the experiments is explained schematically in Fig. 2. If there were no energy transfer in a crystal, the reduction in the luminescence intensity after irradiation would be solely due to the local capture of excitons, i.e., it would be due to the capture by the molecules disturbed by β radiation and located at the points where excitons were created. This is shown as a shaded region in Fig. 2. Curve a represents the distribution with depth of the molecules disturbed by the β radiation; curve b is the reduction in the intensity of the exciting light (incident from the left) with increasing depth and the distribution of the probability of creation of excitons, which is proportional to this intensity. The horizontal arrows represent schematically the direction of migration of the excitation energy. The transfer of energy from the left to the right results in an additional reduction in the luminescence intensity because of the formation

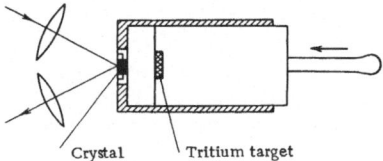

Fig. 1. Holder for a crystal whose luminescence was measured and which was irradiated with β rays from a tritium target.

Crystal   Tritium target

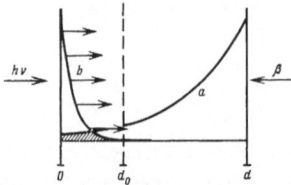

Fig. 2. Schematic representation of a
crystal of thickness d irradiated from
the right with H^3 electrons and illuminated from the left with exciting light.
The horizontal arrows show schematically the direction of migration of excitation energy. Curve a represents the
distribution of the β-ray-disturbed molecules and curve b represents the reduction in intensity of the exciting light with
depth (the probability of generation of excitons is proportional to this intensity).

of excitons in the region where the concentration of disturbed molecules is higher. This additional reduction is a function of the effective distance over which the excitation energy is transferred.

The measurements were carried out on anthracene crystals. The luminescence of each
crystal was measured on the excitation side (on the left in Fig. 2) before and after irradiation
with tritium electrons. The excitation was provided by a SVDSh-250 lamp whose light passed
through a monochromator and the filter. The wavelength of the exciting light was 365 nm. The
photoluminescence of anthracene crystals of different thicknesses was determined as a function
of the absorbed β-ray dose by measuring the luminescence intensity at the maxima of the luminescence spectrum of anthracene. The relatively small differences between the degradation of
the various luminescence maxima were ignored because, for the purpose of the present chapter,
it was sufficient to assume that the reduction in the photoluminescence intensity as result of irradiation did not involve any change in the spectrum. The problem of deformation of the photoluminescence spectra after irradiation is considered in Chap. III.

Thin crystals (0.5–6 μ thick) were grown in air from a powder prepared by crushing large
fragments of single crystals purified by zone melting. In the intervals between measurements
and irradiation these crystals were stored in an inert (nitrogen) atmosphere. The measurements
and the irradiation were carried out in air. It was known [56, 76, 77] that under these conditions
anthracene could become oxidized, particularly in the presence of strong ultraviolet radiation.
The oxidation products (photo-oxides) were known to have quenching properties. However, our
estimates (see, [78] and Chap. V) showed that the distribution of photo-oxides with depth in thin
(up to 4–5 μ) crystals was practically homogeneous. Moreover, the results of repeated measurements of the luminescence efficiency performed on the same sample were practically identical,
i.e., the concentration of photo-oxides remained practically constant. Thus, the influence of
this additional (apart from that produced by β-ray irradiation) quenching could be ignored.

The anthracene single-crystal films prepared in this way were bonded to thin aluminum
foil diaphragms. Both surfaces of the film were used: one for the excitation and observation
of the luminescence and the other for contact with the radioactive surface of the tritium target.

The thickness of the anthracene films was measured to within 2–3% by the method involving absorption of α radiation.

   This method (rather than optical interference) was used because the experimental conditions were such that it was not convenient to bring a film into optical contact with glass (both
surfaces of the film were used). The apparatus used in the thickness measurements [79] was
originally developed for measurements of the stopping power. The apparatus is shown schematically in Fig. 3a. A flat ionization chamber (3) had a front wall which was transparent to α radiation. This wall was made of a copper grid with apertures of $40 \times 40 \ \mu$ size and a transparency
of 0.2. A dc amplifier (1) and an ÉPP-09M electronic potentiometer (2) were used. A P^{210}
source of α rays (6, 7 in Fig. 3a) of about 10 microcurie activity was placed in front of the ionization chamber. A collimator (5) with an aperture of 0.5 mm diameter was placed between the
source and the sample (4). The sample was attached rigidly to the collimator. The electrodes
used in the ionization chamber were of 26 mm diameter; they were separated by a gap of 15 mm
across which a voltage of 2000 V was applied.

   A sychronous motor with a step-down gear unit, a micrometer screw, and a system of
guides (not shown in Fig. 3a) drove the ionization chamber and the amplifier at a constant velocity in the direction indicated by the arrow, i.e., opposite to the α-ray beam emerging from
the collimator (the beam diameter did not exceed 10 mm). The dependence of the ionization current on the chamber—collimator distance was recorded on the chart of the ÉPP-09M potentiometer. Since the chart was driven by the synchronous motor responsible for the motion of the
ionization chamber, the chart and the chamber were linked kinematically relative to the source.
The speed of the chamber was 35.4 times slower than the speed of the chart of the ÉPP-09M
potentiometer recorder.

   Figure 3b shows the time dependences of the potentiometer readings, proportional to the
ionization current, i.e., it shows the dependences of these readings on the displacement of the
chamber. Curve I was recorded without a sample and curve II with an anthracene sample in
position. The vertical lines on the right-hand side are the reference points relating to the
chamber—collimator distance. To the left of the reference points each curve is nearly horizontal (n = 0) until the α-ray beam reaches the boundary of the ionization region in the chamber
(grid electrode): at this point n begins to rise. When curves I and II are made to coincide, the
difference δ between the reference points becomes proportional to the air equivalent of the
sample under investigation, which is the thickness of a layer of air equivalent to the sample in
respect of the α-ray stopping power. The air equivalent of the sample was found quite easily

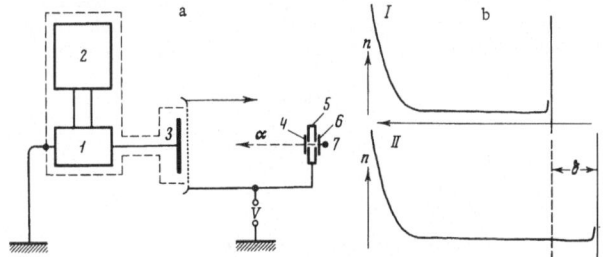

Fig. 3. a) Block diagram of the apparatus used in thickness
measurements: 1) dc amplifier; 2) ÉPP-09M potentiometer;
3) measuring electrode in ionization chamber; 4) sample; 5)
collimator; 6), 7) β-ray source. b) Dependences of the ion
current on the chamber-collimator distance, recorded with the
ÉPP-09M potentiometer: I) without sample; II) with sample
(δ is the shift of the curves on the potentiometer recorder
chart).

from the ratio of the speeds of the recorder chart and the ionization chamber (35.4) and the density of air. The thickness of the sample (in g/cm^2) was equal to the quotient of the air equivalent (in g/cm^2) and the relative α-ray stopping power of anthracene S_{air}^{ant} (relative to air). The average energy of α rays after passage through anthracene was 3-4 MeV and the value of S_{air}^{ant} found for this energy in [79] was 1.2 ± 0.02.

There was no degradation in the photoluminescence of anthracene under the action of α radiation during these thickness measurements.

The most important task was a very precise measurement of the distribution (with depth) of the anthracene molecules disturbed by β radiation. The concentration of such molecules was proportional to the dose absorbed during irradiation and, therefore, the problem could be reduced to finding the distribution of the dose in a film of a material of low atomic number Z (for example, anthracene) in contact with the active surface of a radioactive target.

The method and results of the measurements of the dose distributions and some conclusions were reported in [80] (the investigation reported in that paper was concerned specially with the dosimetric problems encountered in investigations of energy transfer in anthracene). Here, we shall simply give a brief description of the basic idea of the method.

The distribution of the dose in a solid film of a low-Z material lying on a target usually extends to the depth of the order of 1 μ and, therefore, it is quite difficult to determine this distribution. Therefore, the lower part (nearest to the target) of the solid film is replaced with air and the ionization current is measured. The ionization chamber employed in such measurements is shown schematically in Fig. 4.

A target A is placed in a cylindrical recess in the lower electrode (diameter 15 cm) and it is covered by an aluminum foil D of thickness 5 μ with an aperture Q (at the center of the target) of 5 mm diameter. The electrons from the target reach the chamber by passing through this aperture. The working volume of the chamber is the space between electrodes which are separated by h = 5 mm. The chamber and the target are placed in an evacuated enclosure in which the pressure p is varied from 10 to 743 torr. The dependence of the ion current I on the pressure p is determined. The working gas in the chamber, i.e., the layer of air above the target, has two functions: 1) it acts as a detector of the absorbed β-ray energy, and 2) it serves as a variable-density filter between the radiation source and the detector.

The task is to determine the dose rate at a depth x below the surface of a material irradiated with an infinitely wide β-ray beam generated by a source in contact with the material. When this geometry is employed the density of the material x (expressed in g/cm^2) does not af-

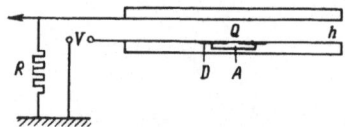

Fig. 4. Schematic representation of the chamber used in measurements of the dose distribution above a tritium target: A) active surface of the target; D) diaphragm covering the target; Q) aperture in the diaphragm; h) distance between the electrodes; R) input resistance of the amplifier; V) voltage applied to the chamber.

fect the dose rate and can have any value. Moreover, the material above the source can be in-homogeneous along the direction x: it may consist of several planar films parallel to the source and these films can have different densities provided the interaction with the radiation is the same in all the films. It is shown in [80] that under these conditions the radiation source and the detector are "dimensionally equivalent": the readings of a "point" detector of unit mass re-cording the radiation emitted by an infinitely wide source located at a distance x (g/cm^2) from the detector are equal to the readings of an infinitely wide plane-parallel detector of unit thick-ness (1 g/cm^2) located at the same distance from the source but recording the radiation emitted by a unit area of the source. The chamber electrodes (Fig. 4) should be made of low-Z mate-rial and its dimensions should be such that no electrons escape laterally even when p = 0. It was reported in [80] that this "layered geometry" can indeed be attained experimentally. The dose rate is calculated from the "dimensional equivalence" of the source and the detector. The relationship obtained for the absorbed dose rate at a depth x (g/cm^2) in an air-equivalent mate-rial is of the form

$$P_{ae} = \frac{\omega}{eh\beta Q}\frac{d\mathcal{J}}{dp}.$$ (II.2)

Here, ω is the work of formation of a pair of ions in air (34 eV) by β rays emitted from H^3; e is the electronic charge; Q is the area of the aperture in the diaphragm (cm^2); h is gap between the electrodes in the ionization chamber (cm); $\beta = \beta$ (T°) is a coefficient relating the pressure p (torr) and the density of air ρ: $\rho = \beta p$; $\mathcal{J}$ is the ion current in the chamber.

It was reported in [80] that the current flowing through a resistance R (Fig. 4) depended on the polarity of the electrodes and on the applied voltage. This was due to the fact that the electric field in the chamber deformed the electron tracks in the gas and thus altered the length of those parts of the tracks which passed through the gas. Moreover, the electric field in-creased or reduced the energy of the electrons moving in the gas, which altered the number of ion pairs generated in the chamber. Finally, the relatively slow rate of recombination of ions in the chamber also contributed to the dependence of the current on the voltage.

An analysis of the components of the current made it possible to determine the pressure dependence of the true ion current (the true current was considered to be the total charge of ions of one sign generated in the chamber per unit time in the absence of a field between the electrodes [80]). Moreover, a correction was made for the "transport" current of β electrons emitted from the open surface Q of the tritium target. The function $\mathcal{J}(p)$, where $\mathcal{J}$ is the true ion current, was used in combination with Eq. (II.2) to find the distribution of the dose in the air-equivalent material. This distribution was converted to the distribution in anthracene P(x) with the aid of the average ratio of the mass stopping powers of anthracene and air (this ratio was 1.14). The distributions of the dose in air and anthracene were nearly exponential for some of the radioactive targets. The distribution of the absorbed dose rate in anthracene in contact with a 7-curie target was found to be of the form

$$P(x) = P_0 e^{-\gamma x},$$ (II.3)

where P_0 = 912 rad/sec and γ = 1.77 ± 0.07 μ^{-1}. This distribution was obtained for the central part of the target, i.e., in the region where anthracene crystals were located during irradiation.

§2. Experimental Results and Estimates of the Diffusion

Length of Excitons in Anthracene Crystals

The results of measurements of the relative luminescence efficiency A as a function of the crystal thickness d are plotted in Fig. 5. The continuous curves join the points obtained by

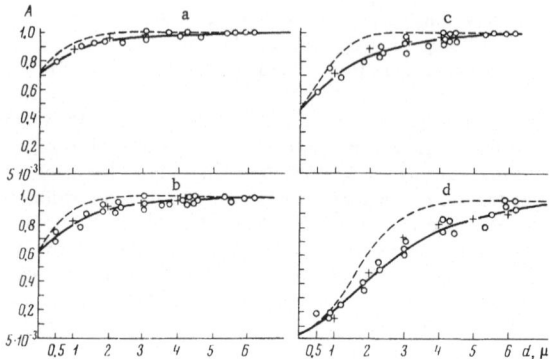

Fig. 5. Dependences of the relative quantum efficiency A
on the thickness of crystals d, recorded for different
maximum doses R_{max} (rad): a) 1.65×10^6; b) 2.66×10^6;
c) 5.34×10^6; d) 7.3×10^7. The circles represent the
average values of A obtained for three maxima of crystals
of similar thickness. The continuous curves are plotted
to join these circles.

averaging the degradation of the luminescence at three maxima in the spectra of different crystals of the same thickness. These points are represented by circles. The dependences plotted in Figs. 5a-5d correspond to different maximum doses (R_{max}) absorbed through the surfaces of the crystals placed in contact with a tritium target.

The values of R_{max} for Figs. 5a-5d are, respectively, 1.65×10^6, 2.66×10^6, 5.34×10^6, and 7.3×10^7 rad. The diffusion length l can be estimated from the one-dimensional equation describing the steady-state diffusion of excitons in an isotropic medium:

$$D\frac{d^2n}{dx^2} - \left(\frac{1}{\tau} + \alpha R(x)\right)n + kIe^{-kx} = 0. \tag{II.4}$$

Here, n (x) is the exciton concentration; D is the diffusion coefficient of excitons; τ is the average lifetime of excitons in an unirradiated crystal; R (x) is a function which describes the distribution of the dose with depth in anthracene (R is the dose in rad); α is the probability of capture of an exciton (per unit time) by anthracene molecules disturbed by a dose of 1 rad; kI is the rate of generation of excitons per unit volume; I is the intensity of light reaching the surface of a crystal (x = 0); ks is the effective absorption coefficient of the exciting light. To start with we shall ignore the reabsorption of luminescence and consider only sufficiently thin crystals. The function R (x) can be written in the form $R(x) = Pte^{-\gamma d} e^{\gamma x} = R_0 e^{\gamma x}$, where P = 912 rad/sec, t is the duration of irradiation in sec, $\gamma = 1.77$ (1/μ), d is the thickness of a crystal in μ, and R_0 is the dose in rad at x = 0 (on the surface of a crystal opposite to that which is irradiated with β rays, as shown in Fig. 2).

We shall consider the conditions at the boundaries x = 0 and x = d. We shall use the boundary conditions of the type $\frac{dn}{dx}\big|_{x=0} = \frac{dn}{dx}\big|_{x=d} = 0$, representing exciton-reflecting faces of a crystal. The question of boundary conditions is still a controversial subject. For example, the diffusion equation was solved and the value of l for excitons was found in [24] with the aid of the conditions $\frac{dn}{dx}\big|_{x=d=0} = 0$ implying total reflection of excitons from the faces of a crystal; similar boundary conditions were employed by Toyozawa [20]. Other workers [21-23, 81] have used $n|_{x=d=0} = 0$, which implies annihilation of excitons at the surfaces of a crystal. In general, we can have boundary conditions of the type $b\frac{dn}{dx} + n(x) = 0$, where b = const. Gallus and Wolf [63] found l for phenantrene crystals with anthracene impurities by assuming that the conditions dn/dx = 0 and n = 0 were equally likely. They found, depending on the selection of the boundary conditions, two values of the diffusion length l and these values differed by a factor of 2.

The boundary conditions selected in the present investigation correspond to the hyperbolic formula (II.1) for the degradation of the luminescence of crystals of any thickness. We shall consider a thin crystal which is irradiated uniformly across its thickness with ionizing particles. In this case the diffusion equation for excitons is

$$D\frac{d^2n}{dx^2} - \left(\frac{1}{\tau} + \alpha R\right)n + kIe^{-kx} = 0, \text{ where } R = \text{const}.$$

We shall use this equation to find the relative luminescence efficiency

$$A = \frac{\int_0^d n dx}{\int_0^d n|_{R=0} dx} = \frac{1}{1 + \alpha\tau R} \frac{I(1 - e^{-kd}) + Dn'(d) - Dn'(0)}{(1 - e^{-kd}) + Dn'|_{R=0}(d) - Dn'|_{R=0}^{(n)}}. \tag{II.5}$$

It is difficult to see how the second factor in the above expression can be close to unity for doses ranging from 10^4 to 10^8 rad, unless it is assumed that the boundary conditions are $n'|_0 = n'|_d = 0$: in this case it is found that $A = 1/(1 + \alpha\tau R)$. The published values of the absorption coefficient of anthracene at λ = 365 nm range from 5 μ^{-1} [82] to 12 μ^{-1} [83]. We shall use 8.2 μ^{-1} employed in [24]. Bearing in mind that the exciting light falls at an angle of 60° to the surface of the crystal and that the refractive index of anthracene at λ = 365 nm is 1.5 [83], we find that the effective value of the absorption coefficient is k = 10 μ^{-1}.

If we use the diffusion equation (II.4), we obtain the following expression for the relative luminescence efficiency

$$A = k\int_0^d \frac{e^{-kx}dx}{1 + \alpha\tau R(x)} - \frac{D}{I}\int_0^d \frac{dn}{dx} \frac{d}{dx}[(1 + \alpha\tau R(x))^{-1}]dx. \tag{II.6}$$

We obtain Eq. (II.6) from Eq. (II.4) if we integrate by parts, use the boundary conditions specified above, and assume that $1 - e^{-kd} \approx 1$ for all thicknesses d $\gg \sim 0.2 \cdot \mu$.

The first integral on the right-hand side of Eq. (II.6) gives that relative luminescence efficiency which would be obtained in the absence of any energy transfer process. The integrand in the second integral on the right-hand side of Eq. (II.6) is always ≥ 0 and, this integral (which will be denoted by Δ) represents the reduction in the relative luminescence efficiency because of the diffusion of excitons. We note that the value of the first integral in Eq. (II.6) is quite close to $1/(1 + \alpha\tau R_0)$ because the function R (x) in the denominator of the integrand varies much more slowly than the numerator due to the smallness of the argument of the exponential function in R (x). The first integral approaches its upper limit $1/(1 + \alpha\tau R_0)$ when the dose R_0 or the thickness d are reduced. Let us assume that d → 0. In this case the second integral in Eq. (II.6) vanishes because n' tends to zero in the integrand and the first integral assumes its maximum value $1/(1 + \alpha\tau R_0)$ [in approaching this limit the right-hand side of Eq. (II.6) must be divided by $(1 - e^{-kd}) \neq 1$ before the limit d → 0 is reached]. Thus,

$$\lim A|_{d\to 0} = \frac{1}{1 + \alpha\tau R_0} = \frac{1}{1 + \alpha\tau R_{max}}.$$

Using the above expression we can find the value of $\alpha\tau$ for anthracene irradiated with tritium β rays. The continuous curves of Figs. 5a-5c extrapolated to d → 0 give the values of $\lim A_{d\to 0}$, which are 0.71 (for R_{max} = 1.65 × 10⁶ rad), 0.60 (for R_{max} = 2.66 × 10⁶ rad), and 0.44 (R_{max} = 5.34 × 10⁶ rad). The corresponding values of $\alpha\tau = \dfrac{1}{R_{max}}\left(\dfrac{1}{\lim A}-1\right)$ are 2.48 × 10⁻⁷, 2.52 × 10⁻⁷, and 2.48 × 10⁻⁷ rad⁻¹. Hence, the most probable value of $\alpha\tau$ is (2.5 ± 0.5) × 10⁻⁷ rad⁻¹ and, therefore, the curve in Fig. 5d (R_{max} = 7.3 × 10⁷ rad) can be extended to d = 0.

No reliable data have yet been published on the values of $\alpha\tau$ for various types of ionizing radiation and different crystals (see, for example, [84]). Measurements of $\alpha\tau$ for anthracene bombarded with $Sr^{90}-Y^{90}$ β rays were carried out by the present author and it was found that $\alpha\tau$ = (5 ± 1) × 10⁻⁷ rad⁻¹, i.e., the hard radiation emitted by $Sr^{90}-Y^{90}$ produced a stronger degradation of the luminescence than did the soft β rays emitted by H^3. This was probably due to the relatively low radiochemical efficiency of the radiations which give rise to a high ionization density.

Having found $\alpha\tau$, we can calculate the first integral on the right-hand side of Eq. (II.6). The dashed curves in Fig. 5 represent the function

$$k\int_0^d \frac{e^{-kx}dx}{1+\alpha\tau R_0 e^{\gamma x}}$$

plotted against the crystal thickness d. We shall now estimate the diffusion length l_0. For each thickness d we can find from Fig. 5 the quantity

$$\Delta(d) = \frac{D}{I}\int_0^d \frac{dn}{dx}\frac{d}{dx}[(1+\alpha\tau Re^{\gamma x})^{-1}]\,dx$$

from the difference between the ordinates of the dashed and the continuous (experimental) curves. In the expression for Δ (d) the integrand includes an unknown function dn/dx. Since our aim is to obtain approximately l_0 without solving Eq. (II.4), we can replace the unknown function dn/dx by a simpler expression which does not differ too greatly from the actual form of dn/dx. This simpler expression is the derivative of the solution of the equation

$$D\frac{d^2m}{dx^2} - \left(\frac{1}{\tau}+\alpha\bar{R}\right)m + kIe^{-kx} = 0,$$

which differs from Eq. (II.4) by the constancy of $\bar{R}$, the average value of the dose. The derivative dm/dx of the solution m (x) of the above equation subject to the boundary conditions $m'|_0 = 0$ and $m(\infty) = 0$ is given by

$$\frac{dm}{dx} = \frac{k^2I\tau}{1+\alpha\tau\bar{R}-k^2l^2}\left\{\exp\left(-\frac{\sqrt{1+\alpha\tau\bar{R}}x}{l}\right) - \exp(-kx)\right\},$$

where $l = \sqrt{D\tau}$. Substituting dm/dx into the formula for Δ (d), we obtain

$$\Delta(d) = \frac{k^2l^2\alpha\tau R_0\gamma}{1+\alpha\tau\bar{R}-l^2k^2}\left\{\int_0^d \frac{e^{-kx}e^{\gamma x}dx}{(1+\alpha\tau R_0 e^{\gamma x})^2} - \int_0^d \frac{\exp\left(-\frac{\sqrt{1+\alpha\tau\bar{R}}x}{l}\right)e^{\gamma x}dx}{(1+\alpha\tau R_0 e^{\gamma x})^2}\right\} \approx$$

$$\approx \frac{k^2l^2\alpha\tau R_0\gamma}{1+\alpha\tau\bar{R}-l^2k^2}\left\{\frac{1}{(1+\alpha\tau R_0)^2 k} - \frac{l}{(1+\alpha\tau R_0)^2\sqrt{1+\alpha\tau\bar{R}}}\right\}. \tag{II.7}$$

In deriving this approximate expression for Δ (d) we have made use of the fact that in the first integral the function e^{-kx} depends much more strongly on x than does $e^{\gamma x}/(1+\alpha\tau R_0 e^{\gamma x})^2$,

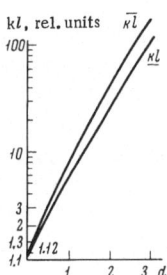

Fig. 6. Dependence of the value of kl on the thickness d (R_{max} = 5.34 × 10^6 rad).

and the fact that $(1 - e^{-k\prime l}) \approx 1$ for all thicknesses d > ~0.2. In the second integral we have assumed that the function $\exp(-\sqrt{1 + \alpha\tau\bar{R}}\, x/l)$ varies faster than does $e^{\gamma x}/(1 + \alpha\tau R_0 e^{\gamma x})^2$, and that $[1 - \exp(-\sqrt{1 + \alpha\tau\bar{R}}\, d/l)] \approx 1$. The validity of these assumptions depends on the value of l, which is not yet known. We note that the simplifications made in Δ (d) are permissible if l is of the order of 0.1 μ. Transforming the last formula for Δ (d), we obtain

$$kl \approx \frac{2\sqrt{1 + \alpha\tau\bar{R}}}{\sqrt{1 + \dfrac{4\alpha\tau R_0 \gamma}{\Delta(1 + \alpha\tau R_0)^2 k}} - 1}.$$ (II.7a)

The values of kl calculated from Eq. (II.7) for the maximum dose of 5.34 × 10^6 rad are plotted as a function of the thickness d in Fig. 6. The curve designated by kl is computed on the assumption that $\bar{R} = R_0$, where R_0 is the dose reaching the rear (opposite to the irradiated) face of the crystal whereas the curve designated by $\overline{kl}$ is calculated on the assumption that $\bar{R} = R_{max}$, where R_{max} is the maximum dose reaching the front (irradiated) face. When d → 0 these two curves given the same extrapolated value of kl_0, equal to 1.12. Similar slightly curved dependences with approximately the same slopes are obtained also for the other three series of measurements corresponding to maximum doses of 1.65 × 10^6, 2.66 × 10^6, and 7.3 × 10^7 rad. In all these cases the kl and $\overline{kl}$ curves intersect at d = 0. The values of kl extrapolated to d → 0 for these three cases are, respectively, 1.38, 1.30, and 1.35. The average value of kl_0 for the four series of measurements is kl_0 = 1.3, which corresponds to l_0 = 0.13 μ. We note that the simplifications made in the derivation of Eq. (II.7) are invalid for d < 0.2 μ and, consequently, the values of kl calculated from Eq. (II.7) cannot, strictly speaking, be extrapolated to d → 0. However, if we calculate the integrals in Eq. (II.7) on the assumption that d < 0.2 μ and apply $kl_0 = \lim|_{d\to0} kl$, we find that the values of kl_0 obtained in this way are very close to those computed with the aid of Eq. (II.7a).

Since l found in this way depends strongly on the thickness d (Fig. 6), we have assumed that the diffusion length is the value of l extrapolated to d → 0. The dependence l (d) may appear for two reasons. The first reason is the inaccuracy of the calculation of l as a result of the approximations made in the derivation of Eq. (II.7a), including the replacement of the function dn/dx with the function dm/dx corresponding to homogeneous irradiation. The error associated with the use of dm/dx instead of dn/dx decreases monotonically with decreasing d. Therefore, the extrapolation of l to d → 0 is necessary for the reduction of the error in l_0. Equation (II.7a) is derived on the assumption that l is small. We can see from Fig. 6 that the assumption can also lead to an error when Eq. (II.7a) is used to calculate l and this error increases with d. Since l → 0.13 μ for d → 0, extrapolation once again reduces error in l_0 to a reasonable value. The second reason for the increase of l with the thickness is the possibility of energy transfer not only by exciton diffusion but also by reabsorption of the luminescence (photon transfer), which increases in importance with thickness. Since the reabsorption is ignored in the original equation (II.4), the value of l should increase with the thickness even when it is calculated rigorously from Eq. (II.4) and from experimental data without any simplifying assumptions. In this case the "diffusion length" l is effectively due to exciton diffusion and to reabsorption; once again this value should be equal to the true diffusion length in the limit $d \to 0$. Thus, there is every reason to assume that the true diffusion length is l_0, which is the value of l extrapolated to $d \to 0$.

The diffusion length found by extrapolation to zero thickness ($l = 1.3 \times 10^{-5}$ cm) is in good agreement with the values given by other workers who have used different methods for estimating l (Chap. I). Some of the published values of l for anthracene are listed below:

$l \cdot 10^{-5}$, cm . . .	0.46	1.5	0.7	0.35	0.46*	1.1*	0.9	1.3	2.0	0.43*
Reference . . .	[24]	[26—29]	[23]	[12]	[10]	[27]	[70]	[25]	[64]	[55]

*These values of l are calculated using the theory of migration of localized excitations.

The surfaces of thin anthracene films which were used in estimating the values of l coincided with the ab crystallographic plane and the geometrical conditions were such that the migration of energy was strongest at right-angles to the plane of the sample. Since the films were anisotropic along this direction, the obtained values of l were somewhat lower than the "average" (measured over all directions) diffusion length.

§ 3.　Role of Reabsorption of Luminescence in Energy Transfer in Anthracene Crystals

Reabsorption of luminescence in a crystal gives rise to secondary emission. If the absorption and luminescence spectra overlap, an allowance must be made for the participation of reabsorption in excitation energy transfer, i.e., for the emission of luminescence at one point and absorption followed by emission at another. Reabsorption plays an important role even in the case of relatively thin anthracene crystals.

We shall now consider the experimental curves in Figs. 5a-5d making an allowance for energy transfer by reabsorption of luminescence.

If we consider crystals 2-6 μ thick, we can interpret the curves in Fig. 5 ignoring exciton diffusion completely. The exciting light is almost completely absorbed in a thin surface layer of thickness $d_0 \approx 0.3$-0.5 μ (Fig. 2). All the "primary" excitons are generated in this layer. In crystals 2-6 μ thick this surface layer lies on the side opposite to that bombarded with rays and it is practically undamaged by these rays. In this layer the quantity $1/[1 + \alpha \tau R(x)]$ is close to unity and the concentration of the primary excitons is practically the same as in an unirradiated crystal. In the case of crystals which are only a few microns thick, a significant degradation of the luminescence is observed and this indicates an efficient energy transfer (to the right in Fig. 2) over distances of the order of 1 μ in the direction of the strongly irradiated layers. Thus, the degradation of the luminescence of "thick" (2-6 μ) crystals should be due to transfer of energy over distances much larger than l_0 and it is not necessary to invoke exciton diffusion to explain this energy transfer.

We shall now consider successive transformations of the exciton $\to$ photon $\to$ exciton $\to \dots$ type in a planar unirradiated crystal 2-6 μ thick on the assumption that: a) each exciton is converted into a luminescence photon with a probability of unity; b) the probability that a luminescence photon emerges from the crystal through either of its faces is δ and the probability of its absorption in the crystal is $1 - \delta$; c) when a luminescence photon is absorbed in the crystal, the probability of formation of an exciton is q. Generally speaking, the value of δ should depend on the position at which the luminescence photon is generated, on the photon energy, and on the thickness d of the crystal, whereas q should depend on the photon energy. We shall ignore these dependences and use average values of δ and q. We can easily show that the fraction Q of the luminescence emerging from such a crystal is given by the expression

$$Q = \delta + \delta(1-\delta)q + \delta(1-\delta)^2 q^2 + \ldots = \delta + \frac{\delta q(1-\delta)}{1-q(1-\delta)},$$ (II.8)

where the first term δ represents the first-generation photons and the term $\delta q(1-\delta)/[1-q (1-\delta)]$ represents the second- and later-generation photons. The value of q, which is the quantum efficiency averaged out over the luminescence spectrum of anthracene, can be found as described in [82]. Integration of the curves given in that paper yields $q \approx 0.7$. The value of Q for an anthracene crystal about $2\ \mu$ thick is $Q \approx 1$ [82] whereas for a crystal 2-6 μ thick it is at least 0.7-0.8. If we use Eq. (II.8), we find δ lies between 0.48 (Q = 0.75) and 1.00 (Q = 1).

Let us now consider the influence of a preliminary irradiation with β rays. If we use ξ to denote the ratio of the number of the second- or later-generation photons to the number of the first-generation photons, $\xi = q(1-\delta)/[1-q(1-\delta)]$, we find that the expression for the luminescence efficiency becomes

$$A = \frac{[(1+\alpha\tau R)^{-1}]_{d_0} + \xi[(1+\alpha\tau R)^{-1}]_d}{1+\xi}.$$ (II.9)

The expressions in the square brackets in the numerator are average values of the function $[1 + \alpha\tau R(x)]^{-1}$, which represents the probability of the de-excitation of excitons for a radiation dose R (x) and thicknesses in the range $0-d_0$ and $0-d$, respectively. These average values depend on the distribution across the thickness of the crystal of those excitons which give rise to luminescence photons. The concentration of excitons should always decrease away from the illuminated surface, i.e., from the left to the right in Fig. 2, and therefore the function $(1 + \alpha\tau R (x))^{-1}$ should decrease along the same direction. Thus, the formula for A includes the average values of $[(1 + \alpha\tau R)^{-1}]_{d_0}$ and $[(1 + \alpha\tau R)^{-1}]_d$ which are never smaller than the simple average values of the function $(1 + \alpha\tau R (x))^{-1}$, in the same range of depths. Denoting the average values by $\overline{(1 + \alpha\tau R (x))^{-1}}$, we find that $[(1 + \alpha\tau R)^{-1}]_{d_0} \geqslant \overline{(1 + \alpha\tau R)_{d_0}^{-1}}$ and $[(1 + \alpha\tau R)^{-1}]_d \geqslant \overline{(1 + \alpha\tau R)_d^{-1}}$. If we consider $R_{max} = 1.65 \times 10^6$ rad and $R_{max} = 2.66 \times 10^6$ rad for crystals of thickness $d > 2\ \mu$, $R_{max} = 5.34 \times 10^6$ rad for crystals of $d > 2.5\ \mu$, and $R_{max} = 7.3 \times 10^7$ rad for crystals with $d > 4\ \mu$, we find that for these doses and crystals the values of $\overline{(1 + \alpha\tau R)_{d_0}^{-1}}$ corresponding to $d_0 = 0.5\ \mu$ are in the range 0.97-0.99. For these doses and thicknesses we can simplify Eq. (II.9) to

$$A = \frac{1 + \xi[(1 + \alpha\tau R)^{-1}]_d}{1+\xi}.$$ (II.10)

A relatively large fall in A was observed for the doses and thicknesses given in the preceding paragraph. According to Eq. (II.10), this could only be due to a reduction (because of irradiation) of the yield of the "daughter" (second- and later-generation) photons. We shall now use Eq. (II.10) to find the lowest possible values of A. These values are obtained for $\xi = \xi_{max}$ and for the lowest possible values of the function $[(1 + \alpha\tau R)^{-1}]_d$, given by

$$\overline{(1 + \alpha\tau R (x))_d^{-1}} = \frac{1}{d}\int_0^d (1 + \alpha\tau R_0 e^{\gamma x})^{-1}\,dx = 1 - \frac{1}{\gamma d}\ln\frac{1 + \alpha\tau R_{max}}{1 + \alpha\tau R_0}.$$

Here, ξ_{max} corresponds to the smallest value of the probability δ equal to 0.48, for which $\delta_{max} = 0.57$. If we use Eq. (II.9) and substitute $[(1 + \alpha\tau R)^{-1}]_{d_0} = \overline{(1 + \alpha\tau R)_{d_0}^{-1}}$ and $[(1 + \alpha\tau R)^{-1}]_d = \overline{(1 + \alpha\tau R)_d^{-1}}$, we obtain the values represented by crosses in Fig. 5. We can see that these values are in satisfactory agreement with the experimental curves. Thus, the observed fall in the luminescence efficiency A of crystals 2-6 μ thick can be explained by assuming that the emission

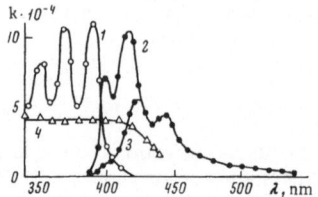

Fig. 7. Spectrum of the absorption
coefficient k of crystalline anthracene
(1), luminescence spectrum of an anthracene single crystal (2), luminescence spectrum of polycrystalline
anthracene (3), and dependence of the
relative quantum efficiency of anthracene on the wavelength of exciting
light (4) (taken from [111]).

of the "daughter" photons is practically homogeneous over the whole crystal. This is related
to the assumption that the effective absorption coefficient of the luminescence k_1 is quite small
(not exceeding 0.1-0.2 μ^{-1}). Moreover, luminescence may undergo multiple internal reflection
from the surfaces of the crystal. Then, the value of ξ approaches ξ_{max} . The assumption that
the effective absorption coefficient of luminescence is small is not in conflict with the available
data on the absorption spectrum of anthracene and on the dependence of the quantum efficiency
on the wavelength of the exciting light [12, 82]. According to these data (Fig. 7) there is a range
of wavelengths ($\sim$ 420-450 nm) in which $k_1 < 0.2 \mu^{-1}$ but the luminescence efficiency and intensity are quite high.

The average refractive index (obtained ignoring anisotropy) of anthracene is n = 1.85,
which is greater than $\sqrt{2}$ so that the critical angle α for the crystal—air interface is less than
45°. If we consider [15] a planar uniformly emitting crystal, we find that when n > $\sqrt{2}$ only that
fraction of luminescence leaves a crystal which lies within two cones with a total vertex angle
of 2α. The light emitted along other directions suffers total internal reflection and is gradually
absorbed in the crystal. The solid angle of each of these cones is $2\pi(1 - \cos\alpha)$ and the fraction
of light which emerges from the whole surface of the crystal is $(1 - \cos\alpha)$ whereas a fraction
equal to $\cos\alpha$ is absorbed in the crystal. In practice, part of the latter fraction is scattered
and escapes through lateral surfaces.

An allowance for the total internal reflection (but not for scattering) in an anthracene
crystal gives a probability of 0.16 for the emergence of light from the two planar surfaces (n =
1.85), which is only one-third of the value we have assumed for the probability of the emergence
of photons from such a crystal. In the calculation of values of kl by means of Eq. (II.7a) we
have obtained kl of the order of 10 for thicker crystals (Fig. 6). Since Eq. (II.4) ignores reabsorption, the value of l corresponding to large d should be $\sim 1/k_1$. This may be regarded as
additional evidence in support of the assumption that the effective absorption coefficient of luminescence is small.

The present section can be summarized by saying that the role of reabsorption (photon
transfer) can be ignored only for crystals with thicknesses not exceeding 1 μ. A satisfactory
quantitative explanation of the degradation of the luminescence of crystals thicker than 2 μ can
be provided only on the assumption that the effective absorption coefficient of luminescence
does not exceed 0.2 μ^{-1} and that multiple reflection of luminescence takes place at the surface
of a crystal.

Fig. 8. Dependence of τ on the thickness d: 1) experimental results; 2)-4) results of calculations based on Eq. (II.11) and $k_0 = 5 \times 10^3$, 2×10^3, and 1×10^3 cm^{-1}, respectively.

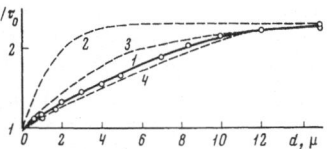

§ 4. Influence of Reabsorption on the Fluorescence

Decay Time of Anthracene

In the preceding section we found that the results obtained after tritium-electron irradiation of thin abthracene films could be explained only if we assumed that the luminescence was reflected from the plane faces of a thin crystal ("trapping" of the luminescence, which increases the uniformity of the emission across the thickness).

We also found it necessary to assume that the effective absorption coefficient of luminescence in anthracene is quite small, not exceeding $\sim (1\text{-}2) \times 10^3$ cm^{-1}.

These assumptions are not generally accepted. In many investigations dealing with the influence of reabsorption on the luminescence efficiency and decay time the multiple reflection is ignored completely* and the effective (corresponding to maximum reabsorption) absorption coefficient of luminescence is assumed to be $(4\text{-}5) \times 10^3$ cm^{-1} [17, 18].

In view of this it seemed desirable to measure the dependence of the luminescence decay time on the thickness of anthracene crystals (in the range 0.1-15 μ) and to interpret the results obtained. No experimental investigation of this kind has yet been carried out for anthracene.

Thin anthracene films were in optical contact with silica glass substrates and their thicknesses were measured with an MII-5 interference microscope to within 0.02 μ. The fluorescence decay time was measured with a phase fluorometer and special care was taken to ensure that samples were always subjected to the same conditions. This was done by constructing a special holder. The results of measurements are plotted in Fig. 8, where curve 1 joins the experimental points.

The theory which makes an allowance for the influence of the thickness of a crystal on its de-excitation time ignores the reflection of luminescence from the internal faces of a crystal. Nevertheless, in rough semiquantitative estimates we can use an expressing given in [85]† and derived as a result of development of a theory of reabsorption of luminescence [16, 86-88]. This expression gives the de-excitation time $\tau (z_0)$ in the case of a plane-parallel crystal (the notation is the same as in [85]):

$$\tau (z_0) = \frac{1 - \frac{q}{2}}{1 - q} - \frac{2q}{k^2} \left\{ \frac{(2 + k)^2 - (2 - k)^2 [z_0 (2 + k) + 1]}{(2 + k)^2 - (2 - k)^2 [z_0 (2 + k) + 1]\, e^{-2z_0 k}} \right\} e^{-z_0 k}. \tag{II.11}$$

Here, $z_0 = k_0 d$, where k_0 is the absorption coefficient corresponding to the maximum frequency reabsorbed in the crystal; d is the thickness of the crystal;

$$q = \int_0^\infty E(\nu)\, \eta(\nu)\, d\nu, \text{ where } E(\nu) = \rho(\nu) \Big/ \int_0^\infty \rho(\nu)\, d\nu,$$

---

*One of the exceptions is the monograph of Birks [15], who discusses the possibility of the "trapping" of luminescence in scintillation crystals.

†The author is grateful to Yu. V. Konobeev for an opportunity to read his paper before publication.

$\rho(\nu)$ is the density of the probability of emission of a luminescence quantum per unit time along any direction; $\eta(\nu)$ is the molecular quantum efficiency in the case of excitation with light of frequency ν; $K = 2\sqrt{1-q}$. It is clear from Eq. (II.11) that $\tau \rightarrow 1$ when $z_0 \rightarrow 0$ so that Eq. (II.11) gives the relative (compared with a very thin crystal) de-excitation time.

We calculated τ from Eq. (II.11) using the quantum efficiency q averaged out over the luminescence spectrum; the value of this efficiency was calculated on the basis of the data in [82] and it was 0.72. The results of calculations for $k_0 = 5 \times 10^3$ cm^{-1} are represented by the dashed curve 2 in Fig. 8. We can see that at low values of d this curve rises much faster than the experimental curve 1. The discrepancy between curves 1 and 2 can be due to two factors: 1) the internal reflection of luminescence, ignored in the derivation of Eq. (II.11); 2) an incorrect effective absorption coefficient of luminescence. Let us consider these factors separately.

The internal reflection of light should increase the thickness of a crystal traversed by luminescence to an effective value d_{eff}, which is larger than the geometrical thickness d: $d_{eff} > d$. Therefore, if we assume that the effective absorption coefficient of luminescence is really close to 5×10^3 cm^{-1}, we find that the dependence of τ/τ_0 on d calculated making an allowance for internal reflection would be even steeper at low values of d than curve 2. In other words, an allowance for the "trapping" of luminescence in a crystal combined with a value of k_0 equal to $\sim (4-5) \times 10^3$ cm^{-1} would give rise to a curve which would differ from the experimental dependence even more strongly than curve 2.

We shall now consider the second possibility and, ignoring the "trapping" of luminescence because of internal reflections, we shall assume that the effective absorption coefficient is $k_0 < (4-5) \times 10^3$ cm^{-1}. Curves 3 and 4 in Fig. 8 are plotted using Eq. (II.11) and the values of k_0 of 2×10^3 and 1×10^3 cm^{-1}, respectively. Good agreement with the experimental curve is obtained if it is assumed that $k_0 = (1.1-1.2) \times 10^3$ cm^{-1}. Thus if we ignore the internal reflection of luminescence, we find approximately the same value of k_0 as was obtained in the preceding section as the upper limit of the effective absorption coefficient of luminescence.*

In the preceding section it was shown that a satisfactory explanation of the radiation-induced degradation of the luminescence of thin anthracene single crystals could be provided only if it was assumed that the ratio ξ of the number of the "daughter" photons to the number of the first-generation photons was close to its maximum value (about 0.6). The large contribution of the "daughter" photons to the total yield is due to the relatively strong "reflection trapping" of luminescence in a crystal. In view of this, it is reasonable to assume that d_{eff}/d is at least 2. The results of calculations of the dependence τ (d) based on Eq. (II.11) which ignores reflection agree with the experimental results (curve 1 in Fig. 8) if it is assumed that $k_0 = (1.1-1.2) \times 10^3$ cm^{-1}. If we postulate that $d_{eff}/d = 2$, we find that the effective absorption coefficient of luminescence in thin anthracene single crystals is

$$\frac{(1.1 - 1.2) \cdot 10^3}{2} \text{ cm}^{-1} \approx 0.6 \cdot 10^3 \text{ cm}^{-1}.$$

The photons most strongly reabsorbed in the case of strong reflection trapping of luminescence should have the frequencies corresponding mainly to the maximum of the product $E(\nu)\eta(\nu)$ rather than the frequencies governed by the absorption and luminescence spectra. In the case of weaker reflection trapping of the luminescence the effective value of k should be af-

---

*When the manuscript of the present paper was completed, Yu. V. Konobeev kindly acquainted the author with an improved, compared with Eq. (II.11), relationship for the dependence of τ/τ_0 on d and with calculations showing that good agreement with out experimental data (Fig. 8) would be obtained if the internal reflection of luminescence were ignored and the absorption coefficient assumed to be $k_0 = 2 \times 10^3$ cm^{-1}.

fected more strongly by the absorption spectrum. The wavelength corresponding to the maximum of $E(\nu)\,\eta(\nu)$ is about 410 nm [82]. The results of measurements of the absorption of light in anthracene crystals in the $400 \leq \lambda \leq 430$ nm range are given in [89]. According to these results the absorption coefficient at $\lambda = 410$ nm at 25°C are 0.12×10^3 and 1.3×10^3 cm^{-1} for the a and b components of polarized light. The half-sum of these values (about 0.7×10^3 cm^{-1}) is close to the value 0.6×10^3 cm^{-1}, which we have just obtained. This can be regarded as confirmation of the occurrence of luminescence trapping because of the internal reflection and because of the low value of the effective absorption coefficient of luminescence in anthracene crystals.

CHAPTER III

DEFORMATION OF PHOTOLUMINESCENCE SPECTRA OF ANTHRACENE CRYSTALS DUE TO INCREASE IN IMPURITY CONCENTRATION AS A RESULT OF ELECTRON BOMBARDMENT

The luminescence of molecular crystals may be quenched by molecules which are disturbed by ionizing radiation and whose concentration is proportional to the absorbed dose. Introduction of quenching impurities into a crystal by irradiation is convenient because it makes it possible to vary this concentration within wide limits in the same sample and this can be done in a gradual but carefully controlled manner. Therefore, in the case of investigations of the dependences of the luminescence efficiency, shape of the luminescence spectra, and other properties on the impurity concentration in a crystal the radiation method has self-evident advantages over the methods involving measurements on many samples characterized by different impurity concentrations. The methodological superiority of the radiation method is demonstrated by the fact that such a fine concentration-dependent effect as the deformation of the photoluminescence spectra (discussed in the present chapter) can be observed after irradiation of molecular crystals.

Another advantage of the radiation method is the opportunity to produce either a homogeneous distribution throughout a sample or a strongly inhomogeneous but known distribution (Chap. II). We produced a homogeneous distribution of quenching impurities in anthracene by bombarding this compound with electrons generated in an accelerator.

§ 1. Experimental Method

In investigations of changes in the photoluminescence spectra of irradiated molecular crystals a special attention was paid to the reproducibility of the irradiation and measurement conditions. A sample was placed in a holder which was located in a quartz Dewar flask in a fixed position on a heat sink. The position of the Dewar flask with the sample was always the same relative to the apparatus used in measurement of the luminescence spectra. The reproducibility of the measurement conditions was checked by placing a scintillator on the heat sink on the opposite side to the sample (Fig. 9a). This scintillator acted as a standard and it had a luminescence maximum in the same region as the crystals under the investigation. When the Dewar flask was rotated about its axis, the investigated sample and the scintillator standard passed alternately in front of the slit. The measurements were carried out at room and liquid nitrogen temperatures (the latter temperature was measured with a thermocouple). The excitation conditions were kept constant by stabilizing the light flux from a DRSh-250 lamp with the aid of the circuit described in [90]. The apparatus used in measurements of the luminescence spectra is shown schematically in Fig. 9b. We investigated anthracene, stilbene, naphthalene, p-terphenyl, and p-quaterphenyl single crystals. These crystals were ground down to a thickness of 0.5 mm and the face which was illuminated (and which also emitted luminescence) was polished. The opposite rough face was blackened with soot. In the case of anthracene some measurements

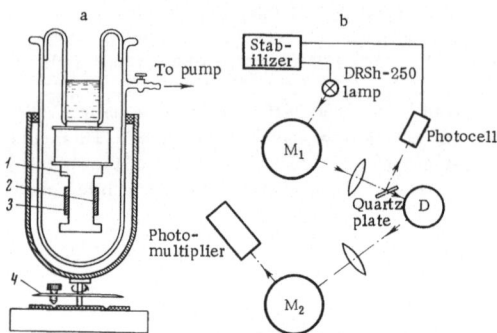

Fig. 9. Schematic representation of the apparatus used in measurements of the photoluminecence spectra. a) Quartz Dewar flask: 1) heat sink; 2) sample; 3) standard; 4) device for locking the flask in the "sample" or "standard" positions. b) Apparatus used to stabilize light flux (D is the Dewar flask; M_1 and M_2 are monochromators).

were made also on thin single-crystal films. All crystals were bonded to a foil frame and clamped in a holder. The luminescence spectrum of each sample was measured before and after irradiation with electrons accelerated in a Van de Graaff generator. After the first irradiation the sample was placed again in the Dewar flask and its luminescence spectrum was recorded; it was irradiated for the second time (this increased the absorbed dose) and the procedure was repeated. During the irradiation the sample with the holder were placed in a special chamber (Faraday cylinder) which was employed in order to measure continuously the electron current during irradiation. The value of this current and the irradiation period were used in a calculation of the total absorbed dose received by each sample. The Faraday cylinder is shown schematically in Fig. 10. This cylinder consisted of a graphite block with a cylindrical recess which contained a crystal 4 attached to a thin-foil frame. The Faraday cylinder was placed near the exit window of the acceleration tube of the Van de Graaff generator and the electron beam which passed through the cylinder diaphragm reached the sample. The electron current reaching the Faraday cylinder and the electron energy (1 MeV) were kept constant during irradiation. The relationship between the density of the current reaching the cylinder and the absorbed dose rate, taken from [91], was derived as follows.

Let $\mathcal{J}$ be the current measured with a galvanometer connected to the Faraday cylinder and $j = \mathcal{J}/Q$ be the current density (Q is the area of the aperture in the diaphragm screening the cylinder). The absorbed dose rate at the outer surface of the sample P (0) can be written in the form $P(0) = (1 + \Delta) \Pi dE/dx$. Here, Π is the electron current density, dE/dx is the stopping power of the sample, and the factor $(1 + \Delta)$ is introduced to allow for the increase in the dose rate in the surface layer because of the electrons back-scattered from the sample and from the graphite. The electron current density is $\Pi = j/e = \mathcal{J}/Qe$ (where e is the electronic charge), the stopping power dE/dx of anthracene for 1 MeV electrons is 1.76 MeV $\cdot$ cm$^2 \cdot$ g^{-1} [92], and the factor $(1 + \Delta)$ is about 1.10 [91]. If the diameter of the aperture in the diaphragm is 5 mm, the area of this aperture is Q = 0.196 cm^2 and then $P(0) = 6.15 \times 10^{19} \times \mathcal{J}$ MeV $\cdot$ g$^{-1} \cdot$ sec^{-1} or $P(0) = 9.85 \times 10^{11} \times \mathcal{J}$ rad/sec, where the current density $\mathcal{J}$ is expressed in amperes. In the case of irradiation of low-atomic-weight materials by electrons of 1 MeV energy the dose rate (for normally incident beams) increases almost linearly between the irradiated surface and a depth

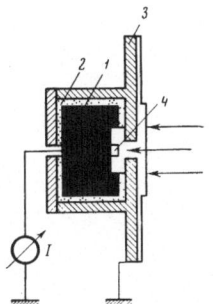

Fig. 10. Faraday cylinder used in measurements of the electron current during irradiation in a Van de Graaff accelerator: 1) graphite; 2) Plexiglas; 3) aluminum; 4) crystal.

of 1.5-2 mm where a maximum of the dose rate is observed (this maximum is about 1.5 times the surface dose rate). In our samples the average dose rate over the thickness (0.5 mm) was 1.15 of the surface rate $\bar{P} = 1.15P(0) = 11.3 \times 10^{11} \times \mathcal{J}$ rad/sec.

The last expression was used in the irradiation experiments. The heating of the samples was prevented by keeping the electron current below 0.1 μA. The dose $\bar{R}$ absorbed during irradiation was deduced from the duration of irradiation t: $\bar{R} = \bar{P}t$.

The dose was determined to within at least $+10\%$. The maximum absorbed dose was about 10^8 rad.

§ 2. Experimental Results

Figures 11a-11d show the photoluminescence spectra of anthracene, stilbene, terphenyl, and quaterphenyl crystals obtained before and after irradiation. It is interesting to note that the ratios of the amplitudes of the various maxima are affected by irradiation, particularly in the case of large doses. In other words, the rate of degradation varies along the spectrum and irradiation deforms the photoluminescence spectra. The luminescence efficiency at the edges of the spectra (at the short- and long-wavelength ends) is reduced by irradiation to a smaller extent than in the central parts of the spectra. This can also be seen in Fig. 12 where the experimental points represent the relative luminescence efficiency A plotted as a function of the dose R at different wavelengths. The value of A (R) at λ = 391 and 445 nm decreases with increasing R more slowly than at λ = 415 nm (this wavelength corresponds to midpoints of the spectra). Another important result, which does not follow directly from Figs. 11 and 12, is the observation that all the crystals (with the possible exception of naphthalene) obey the relationship

$$A(R) = \frac{1}{1 + BR}, \qquad B = \alpha\tau \tag{III.1}$$

only in the short-wavelength part of the spectrum and the deviations from Eq. (III.1) gradually increase with increasing λ.

The dependence of the luminescence efficiency on the dose or on the concentration of the quenching impurity is influenced strongly by the energy transfer mechanism. The relationship (III.1) is obeyed if the energy is transferred in a single step and if the probability of de-excitation is a linear function of the impurity concentration. In view of this it would be interesting to consider the origin of the experimentally observed deviations from Eq. (III.1).

An analysis of the results obtained indicates that irradiation resulting in significant degradation of the luminescence does not alter greatly the absorption of light in the fundamental region and in the region of the luminescence spectrum [75, 93].

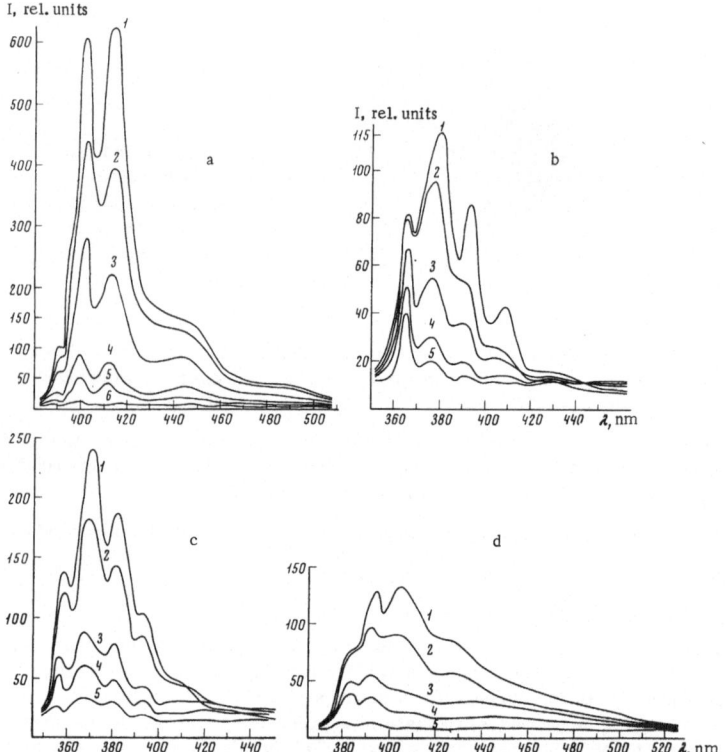

Fig. 11. Photoluminescence spectra of anthracene (a), stilbene (b), n-terphenyl (c), and n-quaterphenyl (d) crystals before and after electron irradiation. (a, b): 1) Before irradiation; 2)-6) after irradiation with doses of 5×10^5, 10^6, 5×10^6, 10^7, and 5×10^7 rad, respectively. (c, d): 1) Before irradiation; 2)-5) after irradiation with doses of 5×10^5, 10^6, 5×10^6, and 5×10^7 rad, respectively.

It is natural to assume that the deviations from the hyperbolic formula of Eq. (III.1) are due to a departure from the linear relationship between the probability of annihilation of an excited state and the impurity concentration. However, if we consider the wide range of impurity concentrations, such a dependence is unlikely to be correct. In the case of high impurity concentrations the relationship between the de-excitation probability and impurity concentration should be more complex. Agranovich and Konobeev [86] obtained a nonlinear relationship for this dependence.

If we assume that the probability of de-excitation is not affected greatly by irradiation, it follows from Eq. (III.1) that in an irradiated crystal this probability is equal to the sum of the probability for an unirradiated crystal $1/\tau$ and the probability $B/\tau R$ ($B/\tau = \alpha$) associated with irradiation, which should be proportional to the dose (τ is the average excited-state lifetime in an unirradiated crystal).

Irradiation results in irreversible damage to a crystal and the concentration of disturbed molecules is proportional to the absorbed dose. On the other hand, excited states in a crystal

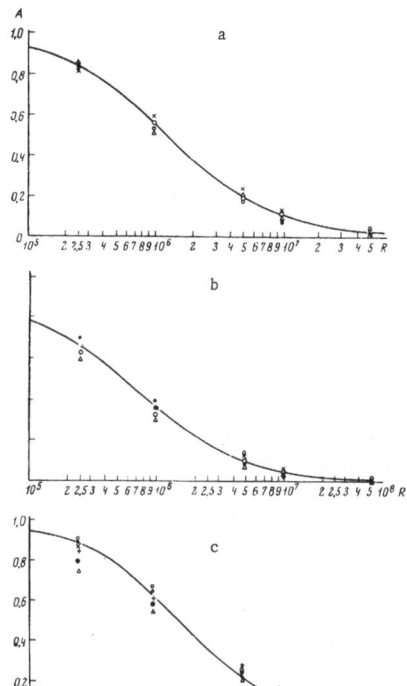

Fig. 12. Dependences of the relative efficiency A of the luminescence of anthracene crystals on the absorbed dose R. The points represent experimental results obtained for several batches of samples. The continuous curves were calculated using Eq. (III.1). a)-c) $\lambda = 391$, 415, and 445 nm, respectively.

can migrate and their migration can be regarded as diffusion. It is natural to assume that the disturbed molecules capture and quench diffusing excitons, i.e., they act as exciton traps. However, in this case the concentration of the disturbed molecules is proportional to the probability of exciton capture only if the "spheres of action" of the disturbed molecules do not overlap. We shall denote the concentration of the disturbed molecules by c. We shall assume that every molecule of this kind is located at the center of a spherical cell and that these cells fill completely the whole crystal without overlap. This is of course impossible but, for the sake of simplicity, we shall ignore the overlap of the cells and the statistical nature of the distribution of the disturbed molecules. We shall assume that all the cells are identical and that the volume of each cell is approximately equal to $1/C = 4/3\pi r^3$ (r is the cell radius). The volume of a "sphere of action" of a disturbed molecule can be estimated from the product σl, where σ is the exciton-capture cross section of a disturbed molecule and l is the diffusion length of excitons. The "spheres of action" of the disturbed molecules do not overlap if $\sigma l \gg 1/C$ or, in other words, $\rho^2 l/r^3 \ll 1$ or

$$\frac{\rho_0^2}{r_0^3} \ll 1, \tag{III.2}$$

where ρ is the radius of the capture cross section, $\rho_0 = \rho/l$, and $r_0 = r/l$.

If the "spheres of action" of the disturbed molecules overlap, the departure from proportionality between their concentration and the capture probability should give rise to deviations

TABLE 1

	Anthracene					Stilbene				
λ, nm	385	400	415	445	485	357	377	392	407	425
η	1	0.89	0.81	0.75	0.50	1	0.99	0.74	0.7	0.3

	Terphenyl					Quaterphenyl			
λ, nm	358	365	370	393	430	383	393	430	500
η	1	0.93	0.88	0.83	0.66	1	0.9	0.88	0.65

from Eq. (III.1). Thus, generally speaking, Eq. (III.1) is approximate. Deviations from Eq. (III.1) were observed experimentally in [93]. In this investigation polycrystalline samples of anthracene, naphthalene, fluorene, naphthacene, diphenyl—terphenyl, and quaterphenyl were ir-radiated with Co^{60} γ rays (doses up to R ~ 10^8 rad) and the photoluminescence spectra were re-corded before and after irradiation. The principal conclusion reached in [93] was that Eq. III.1) should be replaced by a "more general" expression $(1 + BR^\eta)^{-1}$, where the exponent η is usually slightly less than unity. It was found that the values of η and, therefore, of B (not given in [93]) were different for different maxima in the luminescence spectra. In the short-wave-length parts of the luminescence spectra the values of η were equal to unity but with increasing λ the parameter η decreased significantly (to 0.7-0.8 at the long-wavelength edge). Hence, dif-ferent maxima in the luminescence spectra should be degraded by irradiation at different rates or, in other words, the original spectrum should become deformed after irradiation. The lumi-nescence spectra of irradiated samples were not given in [93] and therefore it was difficult to establish the nature of the deformation from the data obtained in that investigation. If, follow-ing [93], we replace our dependence A (R) with the "generalized" expression $(1 + BR^\eta)^{-1}$, we find that a satisfactory agreement with our data for the long-wavelength parts of the lumine-nescence spectra can be achieved only by assuming that $\eta < 1$, whereas at the short-wavelength edges of the spectra it is usually found that $\eta \sim 1$. This result is in qualitative agreement with the data reported in [93]. Table 1 gives the values of η obtained for different maxima and parts of our luminescence spectra. The dependence $\eta(\lambda)$ for naphthalene is weak.

The description of the luminescence degradation by means of the expression suggested in [93] has no physical meaning for $\eta \neq 1$, i.e., for these values it is purely formal.

§3. Cellular Method Calculation of Average Probability of Capture of Excitation Energy by Impurities

We shall now consider one possible explanation of the experimental results which is based on the refinement and reduction to more specific levels of the ideas on the luminescence degra-dation put forward in the preceding sections. We shall use the cellular method, which has been employed in the theory of nuclear reactors [94]. We shall regard an irradiated crystal as an assembly of identical spherical cells of radius r (cm). A disturbed molecule ("black" for mi-grating excitation) is located at the center of each cell. This molecule is a sphere of radius ρ (cm). The absorbed dose R is measured in rads (1 rad = 100 ergs/g). The relationship be-tween R and r is therefore

$$\frac{1}{C} = \frac{4}{3} \pi r^3 = \frac{1.6 \cdot 10^{-8}}{GR} \quad \text{or} \quad \frac{4}{3} \pi r_0^3 = \frac{1.6 \cdot 10^{-8}}{GR\rho^3} \,. \tag{III.3}$$

Here, G is the "yield" of disturbed molecules per 1 MeV of the absorbed energy and the density of the crystal is assumed to be unity.

Migrating excitons in crystals are frequently described by the diffusion equation [21-24]. This is permissible because according to the published estimates, $\lambda_s \ll l$, where λ_s is the mean free path of excitons under scattering conditions. However, the diffusion description of energy migration within a single cell may be incorrect because the condition $\lambda_s \ll \rho$ may not be satisfied. Therefore, it is preferable to replace the diffusion equation by the more general transport equation [95]

$$v\mu \frac{\partial \psi}{\partial x} + v\frac{1-\mu^2}{x}\frac{\partial \psi}{\partial \mu} + \frac{v}{\lambda_s}\psi - \frac{1}{2}\frac{v}{\lambda_s}\psi_0 + a\psi = \frac{f}{2}.$$ (III.4)

Equation (III.4) applies to steady-state spherically symmetrical conditions. Here, x is the distance to the center of the cell. The quantities $|\vec{v}|$ (exciton velocity) and λ_s are assumed to be constant; $\mu = \cos\theta$, where θ is the angle between the velocity $\vec{v}$ and the radius vector x. The scattering is isotropic and $a = 1/\tau$. The distribution function $\psi(x, \mu)$ is defined so that $4\pi x^2 \psi(x, \mu)dxd\mu$ is equal to the number of excitons located at distances from x to x + dx from the center of the cell, and the directions of the velocities of these excitons are within the range from μ to $\mu + d\mu$. The symbol f denotes the density of exciton sources, i.e., the number of excitons generated per unit time in unit volume by the exciting light and $\psi_0(x) = \int_{-1}^{1} \psi(x, \mu)d\mu$ is the volume density of excitons at a distance x from the center of a cell. The first two terms in Eq. (III.4) give the rate of reduction in the number of excitons as a result of collisionless motion, whereas the third and fourth terms represent the increase and decrease in the number of excitons as a result of scattering, per unit intervals of x and μ. The meaning of the other terms in Eq. (III.4) is self-evident.

We shall solve approximately Eq. (III.4) by expanding the function $\psi(x, \mu)$ as a series in the Legendre polynomials $\mathcal{P}_k(\mu)$, $\psi(r, \mu) = \sum_k \mathcal{P}_k(\mu) \, \psi_k(x)$, restricting each expansion to the first three terms (k = 0, 1, 2):

$$\psi(x, \mu) = \frac{1}{2}\psi_0(x) + \frac{3}{2}\psi_1(x)\mu + \frac{5}{2}\psi_2(x)\frac{1}{2}(3\mu^2 - 1).$$ (III.5)

The boundary conditions are

$$\int_0^1 \psi(\rho, \mu)\mu d\mu = 0 \qquad \text{and} \qquad \int_{-1}^1 \psi(r, \mu)\mu d\mu = 0,$$ (III.6)

where the first condition represents the absence of an exciton flux out of a "black" sphere of radius ρ and the second condition represents the fact that the net exciton flux at the boundary of a cell vanishes because of symmetry. If we substitute Eq. (III.5) into Eq. (III.4) and subject the equation obtained in this way to the operations $\int_{-1}^{1} \ldots \mathcal{P}_k(\mu)d\mu$, we obtain three equations from which all three functions $\psi_0(x)$, $\psi_1(x)$ and $\zeta_2(x)$ can be found.

The equilibrium intensity of the luminescence emitted by a crystal depends on the density of excitons given by $\psi_0(x)$ and, therefore, it is sufficient to find this function. The solution of this problem obtained on the assumption that $D = v\lambda_s/3$, $l^2 = D/a$, and $\lambda_s/l \ll 1$, where D is the diffusion coefficient, yields

$$\psi_0(x) = \frac{f}{a}\left[1 - \frac{C_k\rho}{x}(e^{x/l} - \beta e^{-x/l})\right],$$ (III.7)

where

$$\beta = \frac{1 - r_0}{1 + r_0}\, e^{2r_0},$$

$$C_k = \left(1 + \frac{2\lambda_s}{3\rho} + \frac{\lambda_s^2}{2\rho^2}\right)^{-1} C_d, \qquad C_d = (e^{\rho_0} - \beta e^{-\rho_0})^{-1}. \tag{III.8}$$

The number of excitons within each cell is

$$4\pi \int_\rho^r \psi_0(x)\, x^2 dx = \frac{f}{a}\, \{q - 4\pi\rho l^2 C_k\, [\beta e^{-r_0}(1 + r_0) - \beta e^{-\rho_0}(\rho_0 + 1) + e^{\rho_0}(1 - \rho_0) - e^{r_0}(1 - r_0)]\}, \tag{III.9}$$

where $q = 4/3\pi\,(r^3 - \rho^3)$ is the volume of each cell minus the volume of a "black" sphere, and the average volume density of excitons is

$$\frac{f}{a}\left\{\frac{q}{Q} - \frac{3\rho_0}{r_0^3} C_k[\quad]\right\},$$

where $Q = 4/3\pi r^3$ is the total volume of one cell.

Since $q/Q \approx 1$, the relative luminescence efficiency can be written in the form

$$A(R) = 1 - \frac{3\rho_0}{r_0^3} C_k[\quad], \tag{III.10}$$

where the square brackets denote the expression enclosed in the same brackets in Eq. (III.9). If we use Eq. (III.3), we can rewrite A (R) in the form

$$A(R) = 1 - 2.5 \cdot 10^8 \pi G l^3 \rho_0 R C_k[\quad]. \tag{III.11}$$

The quantity β in Eqs. (III.10) and (III.11) depends on r and consequently, on the dose R. Thus, the right-hand side of Eq. (III.11) is a function R and this equation should describe, within the limitations imposed by the simplifying assumptions, the reduction in the luminescence intensity as a result of irradiation. It follows from the above expressions that the probability of exciton capture per unit time, which is $a = 1/\tau$ for an unirradiated crystal, can be expressed by the following formula which makes an allowance for the presence of quenching impurities:

$$\tau^{-1}\left\{1 - \frac{3\rho_0}{r_0^3} C_k[\quad]\right\}^{-1}, \tag{III.12}$$

i.e., the presence of these impurities increases this probability by a factor $1/\{\quad\}$.

Thus, an increase in the capture probability with increasing concentration of quenching impurities is described by an expression which is generally nonlinear with respect to the impurity concentration.

The diffusion approximation can be used if $\lambda_s \ll \rho$; in this case $1 + \frac{2\lambda_s}{3\rho} + \frac{\lambda_s^2}{2\rho^2} \approx 1$ and the constant C_k in Eq. (III.8) and in subsequent formulas should be replaced with C_d. The formula (III.11) for the relative luminescence efficiency, obtained by solving the transport equation (III.4), is also retained in the diffusion approximation. Once again we must replace C_k with C_d. Clearly, Eq. (III.11) can be derived for $\lambda_s \ll \rho$ much more simply by solving the standard diffusion equation $D\left(\frac{\partial^2 \psi_0}{\partial x^2} + \frac{2}{x}\frac{\partial \psi_0}{\partial x}\right) - a\psi_0 + f = 0$ for a cell, subject to the boundary conditions $\psi_0(\rho) = 0$ and $\psi_0(r) = 0$.

The "kinetic" and the "diffusion" constants C_k and C_d differ by the factor $1 + \frac{2}{3}\frac{\lambda_s}{\rho} + \frac{1}{2}\left(\frac{\lambda_s}{\rho}\right)^2$. The number of terms in this factor depends on the number of terms used in the expansion of Eq. (III.5). If only one term is retained in this expansion, we find that $C_d/D_k = 1$; when two terms are kept we obtain $\frac{C_d}{C_k} = 1 + \frac{2}{3}\frac{\lambda_s}{\rho}$, where three terms remain, we have $\frac{C_d}{C_k} = 1 + \frac{2}{3}\frac{\lambda_s}{\rho} + \frac{1}{2}\left(\frac{\lambda_s}{\rho}\right)^2$; and so on.

Figure 13 shows S-shaped dependences of C [ ] on r_0 plotted for $\rho_0 = 10^{-1}$, 10^{-2}, 10^{-3}, and 10^{-4}. Here, $\lim_{r_0 \to 0}$ [ ] $C_d = 0$, and $\lim_{r_0 \to \infty}$ [ ] $C_d = 1 + \rho_0$. The curve for $\rho_0 = 10^{-1}$ differs in shape from the curves plotted for $\rho_0 = 10^{-2}$, 10^{-3}, and 10^{-4}. Near the plateau the curve for $\rho_0 = 10^{-2}$ is similar in shape to the corresponding parts of the curves for $\rho_0 = 10^{-3}$ and $\rho_0 = 10^{-4}$: these parts of the curves are identical apart from a horizontal shift. The curves for $\rho_0 = 10^{-3}$ and $\rho_0 = 10^{-4}$ are of identical shape over wider ranges of the argument r_0 and the dependence for $\rho_0 = 10^{-4}$ in Fig. 13 represents simply the dependence for $\rho_0 = 10^{-3}$ shifted to the left. As the parameter ρ_0 decreases, the shape of the dependences of C_d [ ] on r_0 approaches ever more closely, i.e., over a wider range of the argument, some universal curve. This curve can be easily found by writing the quantity C_d [ ] of Eq. (III.8) in the form of a fraction:

$$C_d [\quad] = \frac{e^{r_0 - \rho_0}(r_0 - 1)(\rho_0 + 1) + e^{-(r_0 - \rho_0)}(r_0 + 1)(1 - \rho_0)}{e^{r_0 - \rho_0}(r_0 - 1) + e^{-(r_0 - \rho_0)}(r_0 + 1)}. \tag{III.13}$$

Let us expand the exponential functions in the numerator and the denominator of this fraction retaining only terms of the third power in $(r_0 - \rho_0)$. Then, the numerator reduces to $2/3 r_0^3 + 1/3 r_0^4 \rho_0 - r_0^3 \rho_0^2 - 2/3 \rho_0^3 + r_0^2 \rho_0^3 - 1/3 r_0 \rho_0^4$. Let us now consider the graph for $\rho_0 = 10^{-3}$ in Fig. 13. If we ignore very small values of C_d [ ] and the values close to the plateau, where C_d [ ] ≈ 1, we find that the order of magnitude of the argument r_0 is 10^{-1}. The first term in the numerator of Eq. (III.13) is of the order of 10^{-3} and the other terms are of the order of 10^{-7} or less. Expansion of the exponential functions in the denominator of Eq. (III.13) yields $2/3 r_0^3 + 2\rho_0 - r_0^2 \rho_0 + 1/3 \rho_0^3$, where the first and second terms are of the order of 10^{-3} and the third and fourth terms are of the order of 10^{-5} and 10^{-9}, respectively.

When the condition (III.2) is satisfied, the following relationship is obeyed quite accurately:

$$C_d [\quad] \approx \frac{2/3\, r_0^3}{2/3\, r_0^3 + 2\rho_0} = \frac{1}{1 + 3\frac{\rho_0}{r_0^3}}. \tag{III.14}$$

Similar results are obtained when larger numbers of terms are retained in the expansion of the exponential functions in the numerator and denominator. Moreover, similar conclusions follow if the analysis is extended to curves for $\rho_0 = 10^{-2}$ and $\rho_0 = 10^{-4}$ in Fig. 13. In general, the condition (III.2), which specifies the absence of overlaps between the "spheres of action" of the disturbed molecules, is sufficient to ensure that Eq. (III.14) is satisfied. If Eq. (III.14) can be used, the expression (III.10) for the relative luminescence efficiency can be simplified to

$$A(R) = \frac{1 + 3\frac{\rho_0}{r_0^3}\left[1 - \left(1 + \frac{2\lambda_s}{3\rho} + \frac{\lambda_s^2}{2\rho^2} + \cdots\right)^{-1}\right]}{1 + 3\frac{\rho_0}{r_0^3}}. \tag{III.15}$$

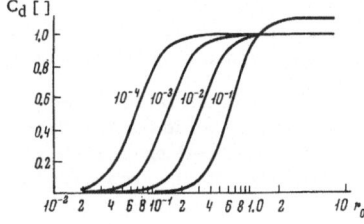

Fig. 13. Dependence of C_d [ ] on r_0 for different values of ρ_0, indicated alongside the curves.

If, moreover, we also have $\lambda_s/\rho \ll 1$, we find that Eq. (III.15) reduces to Eq. (III.1) where $3\rho_0/r_0^3 = 2.5 \cdot 10^8 \pi Gl^3\rho_0 = 4\pi D\rho c\tau$, which is in agreement wi th the results of investigations of quenching in solutions [12, 31–36]. Thus, the relationship derived for A (R) is a generalization of the standard hyperbolic formula for luminescence quenching.

§ 4.   Comparison with Experimental Data

Since the expression obtained for A (R),

$$A\ (R) = 1 - 2.5.10^8 \pi Gl^3\rho_0 RC_k\ [\ \ ],$$

contains many unknown parameters, it is not easy to compare this expression with the experimental results. We usually do not know the values of G, l, and ρ or the value of the ratio $C_d/C_k = 1 + \frac{2}{3}\frac{\lambda_s}{\rho} + \frac{1}{2}\left(\frac{\lambda_s}{\rho}\right)^2 + \dots$ In view of this we limited our comparison to anthracene because the value of l was known for this crystal. The results of measurements at several wavelengths λ in the luminescence spectrum of anthracene were used to plot, on a double logarithmic scale, the dependences of $[1-A\ (R)]/R$ on R (experimental points). Next, the same scale was used to plot the theoretical dependences of C_d [ ] on r_0^3 for $\rho_0 = 10^{-1}$, 10^{-2}, 10^{-3}, and 10^{-4}. The experimental points were shifted so as to coincide with one of the theoretical curves.* It was found that the experimental points corresponding to $\lambda = 391$ nm fitted well the theoretical dependence for $\rho_0 = 10^{-3}$ (and, consequently, the theoretical curves for $\rho_0 = 10^{-4}$, 10^{-5}, etc.) but could not be fitted at all to the curve corresponding to $\rho_0 = 10^{-1}$. The experimental points did not fit too well the curve corresponding to $\rho_0 = 10^{-2}$. Thus, in order to describe the degradation of the short-wavelength luminescence of anthracene it was necessary to assume that $\rho_0 \lessgtr 10^{-3}$. Similarly, it was established that the dose dependence of the degradation of the $\lambda = 445$ nm luminescence agreed with the thoretical curve for $\rho_0 = 10^{-1}$ whereas the middle part of the spectrum of anthracene (412 nm) was best fitted by the curve corresponding to $\rho_0 = 10^{-2}$.

In the course of the fitting of the experimental points to the theoretical curves we compared also the values plotted along the axes of the experimental and theoretical graphs. When the abscissas were made to coincide, the dose R was compared with the corresponding value of r_0^3 in accordance with Eq. (III.3). This made it possible to estimate the product Gl^3. When the ordinates were made to coincide, we could estimate

$$2.5 \cdot 10^8 \pi Gl^3\rho_0 \left(1 + \frac{2}{3}\frac{\lambda_s}{\rho} + \frac{1}{2}\frac{\lambda_s^2}{\rho^2}\right)^{-1}$$

of Eq. (III.11). Knowing ρ_0 and Gl^3, we were able to determine the order of magnitude of the ratio C_d/C_k. At all wavelengths the ratio C_d/C_k was of the order of unity. This indicated that the relationship (III.1) was valid provided the condition (III.2) was satisfied, i.e., provided there was no overlap between the "spheres of action." Table 2 lists the value of ρ_0 and of Gl^3 estimated for anthracene in the way described above.

The diffusion length l in anthracene is about 10^{-5} cm and since it varies with the wavelength λ (Table 2), we should regard this value as the average over the luminescence spectrum. Knowing $(\sqrt[3]{Gl})_{\lambda_1}$ and $(\sqrt[3]{Gl})_{\lambda_2}$ (Table 2) and the average value of l over the spectrum, we can find G. The value obtained in this way is $G \approx 200$ MeV^{-1}, which is in satisfactory agreement

---

*The experimental and the theoretical curves were fitted with the aid of curves plotted on transparent paper. The curves being compared were shifted parallel to the coordinate axes and the abscissas of the two figures were directed in opposite directions because $R \propto 1/r_0^3$. The graphs for $\rho_0 = 10^{-5}$, 10^{-6}, etc. were practically coincident with the graphs for $\rho_0 = 10^{-4}$ and therefore they were excluded from the analysis.

TABLE 2

Luminescence wavelength of anthracene, nm	ρ_0	Gl^3, cm^3/MeV	l, cm	ρ, cm
391	10^{-3}	10^{-12}	$1.7 \cdot 10^{-5}$	$1.7 \cdot 10^{-8}$
412	10^{-2}	$2 \cdot 10^{-13}$	10^{-5}	10^{-7}
445	10^{-1}	$0.6 \cdot 10^{-14}$	$3 \cdot 10^{-6}$	$3 \cdot 10^{-7}$

with the radiochemical yield of p terphenyl $G_{chem} = 600$ MeV^{-1} [93]. According to Eq. (III.3) the highest concentration of the disturbed molecules is 6×10^{17} cm^{-3}, corresponding to R = 5 × 10^7 rad. Having found G, we can easily calculate the values of l and ρ listed in Table 2 which also depend on the luminescence wavelength.

At the short-wavelength edge of the spectrum the reduction in the luminescence intensity obeys well Eq. (III.1). An estimate of the constant B gives $B = 2.5 \times 10^8 \; \pi G l^3 \rho = 7.7 \cdot 10^{-7}$ rad^{-1}. For the sake of comparison we ought to mention that the value of B obtained for an Sr^{90}–Y^{90} source was $(5 \pm 1) \times 10^{-7}$ rad^{-1} [25].

The results obtained can be checked by calculating the value of A for anthracene by means of Eq. (III.11) using the parameters listed in Table 2. The curves calculated in this way are plotted in Fig. 12. We can see that Eq. (III.11) and the listed values of G, l, and ρ provide a satisfactory description of the radiation-induced reduction in the luminescence intensity, including the experimentally established difference between the rates of degradation in different parts of the spectrum. The dependence of the rate of degradation on the wavelength can be explained only if we assume different values of ρ_0 for different wavelengths and this leads to a dependence of l and ρ on λ. It would seem that the experimental points and theoretical curves can be made to agree if we assume that $\rho_0 \approx 10^{-2}$ for all the wavelengths λ, because the precision of the determination of ρ_0 is relatively low. It is then found that l and ρ are independent of λ and the results of calculations based on Eq. (III.11) show no difference between the rates of degradation along the spectrum, which does not agree with the experimental data.

The main question that arises in the interpretation of the results obtained is whether the dependences l (λ) and ρ (λ) reflect at all and if so to what extent the physical aspects of the luminescence mechanisms or whether these dependences are of purely formal nature. The present author is of the opinion that over a wide range of the luminescence wavelengths (except, possibly, the longest ones) the dependences l (λ) and ρ (λ) are due to the presence of several types of exciton in crystals. The transfer of electron excitation energy in molecular crystals can involve free and localized excitons (Chap. I). It is natural to assume that the diffusion lengths l and the capture cross sections of the disturbed molecules $\sigma = \pi \rho^2$ will be different for free and for localized excitons because of the difference between their interactions with the lattice and the value of l should be larger for the free excitons whereas the opposite should be true of the value of σ.

The mechanism of de-excitation of the free and localized exciton states has been investigated much less thoroughly than the migration of these excitons. However, there are some experimental data supporting the view that the short-wavelength edges of the luminescence spectra of molecular crystals such as anthracene are due to the de-excitation of free excitons [11, 96–104]. The main contribution to the middle parts of the luminescence spectra is made by the impurity and defect luminescence, i.e., by the de-excitation of localized excited states [56, 66, 75].

Evidently, the values of l and σ estimated in the present chapter are the average effective values since in any crystal excitons may be transformed from one type to the other (this happens in the multistep transfer of the excitation energy). It follows that the obtained values of l and σ

depend on whether the excited state which finally gives rise to a photon of wavelength λ can be regarded (during its lifetime in a crystal) as a localized or a free exciton. The effective values of l and ρ (Table 2) vary monotonically with λ over the whole luminescence spectrum and therefore at different wavelengths λ the relative contributions of the free and localized excited states are different. The interpretation of the dependences l (λ) and ρ (λ) in the region between the middle parts of the spectra and the long wavelengths is somewhat more difficult because irradiation can produce D luminescence centers, investigated by Faidysh et al. [26]. The luminescence of these centers occurs mainly at long wavelengths. One may expect that the results are affected by the influence of the reabsorption of luminescence because, in spite of the special measures (covering of the rear surface with a layer of soot), such reabsorption could not be excluded completely in our measurements. However, the influence of reabsorption cannot be the dominant effect. The results of measurements on anthracene crystals of different thickness do not vary greatly with the thickness. Before irradiation the luminescence spectra of all the crystals are sharp, without any significant reduction in the intensity of the short-wavelength maxima because of reabsorption.

Moreover, the dependence of the effective value of l on λ as a result of reabsorption should be basically different from that observed. The experimentally determined short-wavelength luminescence is generated just below the surface of a crystal and if multiple reabsorption occurs this luminescence must be due to the de-excitation of the first and possibly second or third generations of excitons. The long-wavelength luminescence which is generated at greater depths, should include a considerable contribution from the later ("daughter") generations of excitons. Therefore, the reabsorption-governed effective length of l should increase with increasing λ.

§ 5. Comparison of the Effect of Quenching Impurities Generated by Irradiation with the Effect of Luminescent Impurities

Changes in the luminescence spectra of some molecular crystals with increasing impurity concentration were observed when quenching impurities were generated by irradiation with ionizing particles. Similar deformation of the luminescence spectra of polycrystalline powders was described in [93], but no interpretation was given of the results obtained.

The deformation of luminescence spectra was first observed when quenching impurities were introduced into crystals by irradiation simply because the irradiation method was most convenient. We mentioned earlier that the irradiation method can be used to vary the impurity concentration in the same sample and this makes it possible to reproduce exactly the conditions in the measurements of the luminescence spectra. The deviations of the dependence of the luminescence efficiency of the host substance on the impurity concentration from the hyperbolic form and the corresponding changes in the shapes of the spectra should be also observed for impurities which are not generated by irradiation and, in particular, for luminescent impurities.

The published concentration dependences of the relative luminescence efficiency are usually integrated over the spectra or are given for specific wavelengths (maxima). However, in one specific case [54] the luminescence spectra of anthracene are plotted as a function of the concentration of phenazine and acridine impurities. Figure 14, taken from [54], shows eight luminescence spectra of anthracene doped with phenazine. These spectra were obtained at 95°K for crystals $2\,\mu$ thick, which were excited with light of $\lambda_{exc} = 360$ nm. The concentrations of phenazine corresponding to the eight spectra were as follows:

Spectrum No.	1	2	3	4	5	6	7	8
Conc. mole/mole ...	0	$1.5 \cdot 10^{-6}$	$5 \cdot 10^{-6}$	$1.5 \cdot 10^{-5}$	$5 \cdot 10^{-5}$	$1.9 \cdot 10^{-4}$	$5 \cdot 10^{-4}$	$1.5-10^{-3}$

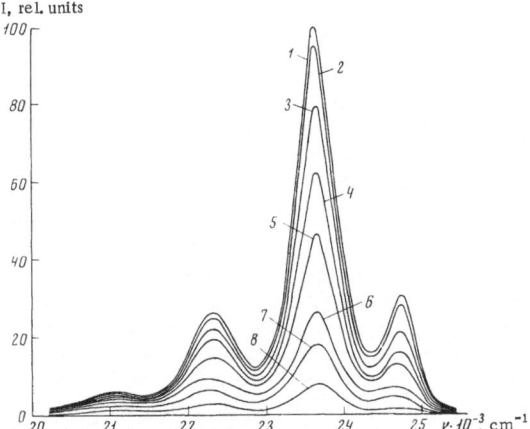

Fig. 14. Influence of the concentration of phenazine
(the values of the concentration are tabulated in
text) on the photoluminescence spectra of phenazine-
doped anthracene [54]; λ_{exc} = 360 nm.

It is worth noting that as we go over from spectrum 1 to spectrum 8, the energy distribution over the spectrum changes primarily because of the rapid fall of the luminescence intensity
in the middle part of the spectrum. Deviations from Eq. (III.1) also occur in this part of the
spectrum. If, as in § 2, we describe the fall in the luminescence efficiency at various wavelengths by the simple formula $(1 + BC^\eta)^{-1}$, we find that values $\eta \approx 1$ are obtained even for short
wavelengths (λ < 405 nm). The values of η estimated for the spectra in Fig. 14 are as follows:

λ, nm	405	425	450	475
η	0.9	0.7	0.6	0.6

We can see that these values are quite close to those obtained for irradiation-generated impurities (Table 1). The experimental data given in [54] on the reduction in the luminescence intensity of anthracene with increasing concentration of luminescent impurities are insufficient
for estimating the values of ρ and l corresponding to different wavelengths, as has been done in
§ 4. However, if we compare the reduction in the relative luminescence efficiency deduced from
Fig. 14 with the theoretical curves of Fig. 12, we find that a dose of the order of 10^6–10^7 rad
gives rise to the same reduction in the luminescence efficiency as phenazine present in concentrations of the order of 10^{-4} mole/mole. If we use Eq. (III.3) to convert the dose R into the concentration of the disturbed anthracene molecules, we find that the quenching effect of such disturbed molecules is stronger (by up to one order of magnitude) than the effect of the phenazine
molecules. The curve in Fig. 12a (λ = 391 nm) can be compared with the reduction in the luminescence efficiency at λ = 400 nm in Fig. 14, the curve in Fig. 12b (λ = 415 nm) with the reduction in the efficiency at λ = 417 nm in Fig. 14, and the curve in Fig. 12c (λ = 445 nm) with the
reduction in the luminescence intensity at λ = 445 in Fig. 14. When the experimental points
taken from [54] are plotted against our theoretical curves, the scatter is quite large and the
comparison itself is not quite correct because the theoretical curves are not universal but they
depend not only on l but also on ρ, i.e., on the radii of the capture cross sections. Nevertheless,
we can draw the conclusion that the nonlinearity in the reduction of the efficiency and the deformation of the spectra are the same for impurities generated by irradiation and those added to
crystals.

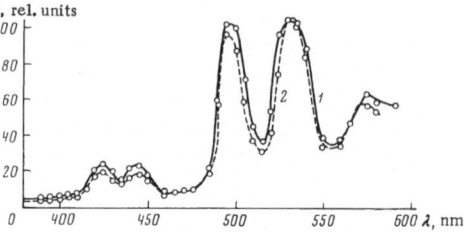

Fig. 15. Fluorescence spectrum of a naphthacene-doped anthracene single crystal ($c = 10^{-4}$
g/g) obtained before (1) and after (2) irradiation
with a 2×10^6 rad dose of Sr^{90}-Y^{90} β rays.

The present author investigated, together with Zhevandrov and Gribkov [75], the degradation of the luminescence (under the action of β rays emitted by an Sr^{90}-Y^{90} source) of pure anthracene and of anthracene doped with naphthacene. It was found that the degradation of the
luminescence of the naphthacene-doped crystals was much weaker than the degradation of the
luminescence of the pure crystals (for the same doses). Figure 15 shows the luminescence
spectra of an anthracene crystal containing 10^{-4} g/g of naphthacene and recorded before and after irradiation with a dose of $\sim 2 \times 10^6$ rad. The luminescence efficiency of the irradiated doped
anthracene was A (R) $\approx$ 0.7-0.8. Irradiation of pure anthracene with the same dose gave A(R) $\approx$
0.15-0.2 for the anthracene luminescence maximum (see, for example, Fig. 12b). If it was assumed that the reduction in the efficiency as a result of irradiation and introduction of naphthacene into anthracene could be described by a linear relationship governing the excitation capture probability, the relatively weak degradation of the luminescence of the doped crystals could
not be explained completely [75]. Estimates based on the additivity of the capture probabilities
of the radiation-disturbed molecules and the naphthacene molecules gave A (R) $\approx$ 0.5, which was
considerably smaller than the experimentally obtained value of 0.7-0.8. This discrepancy was
attributed in [75] to "luminescence shielding," similar to that observed for plastics [105] and
complex biological molecules [106].

The weaker degradation of the luminescence of the doped crystals, compared with the
degradation of pure anthracene, could also be explained in a different way, utilizing the generally nonlinear relationship between the excitation capture probability and the impurity concentration considered in the present chapter. If it is assumed that the quenching effect of the naphthacene molecules on the luminescence of the anthracene host is similar to the effect of the molecules disturbed by β rays, it is found that the 10^{-4} g/g concentration of naphthacene corresponds
to an "initial" dose of the order of 10^7 rad. The actual irradiation effectively adds another $\sim 10^6$
rad to this "dose." Therefore, the luminescence efficiency decreases by not more than 20-30%
(Fig. 12b, $\lambda = 415$ nm). This is in agreement with the experimental results reported in [75].

CHAPTER IV

DEPENDENCE OF AVERAGE EXCITED-STATE LIFETIME AND OF EFFICIENCY OF ANTHRACENE PHOTOLUMINESCENCE ON RADIATION-INDUCED IMPURITY CONCENTRATION

In the preceding chapter we interpreted the experimental results on the luminescence efficiency and spectra of crystals bombarded with ionizing radiation on the assumption that several types of excitation (in particular, free and localized excitons) may exist in these crystals.
The deformation of luminescence spectra as a result of irradiation and the consequent dependences of the effective values of the diffusion length l and the capture cross section σ on the
wavelength λ in the luminescence spectrum were explained on the assumption that transition could

take place from one type of excited state to another i.e., we assumed a generally multistep mechanism of excitation energy transfer.

The mechanism of excitation transfer in a crystal affects not only the luminescence efficiency and the shape of the spectrum but also the law governing the decay of luminescence and the dependences of the average decay time on the impurity concentration C, the temperature $T°$, the luminescence wavelength λ, etc. It is also known that the very simple derivation of the hyperbolic formula

$$A\ (C) = (1 + (B_1 C)^{-1} \tag{IV.1}$$

for the concentration dependence of the luminescence efficiency yields also a similar relationship for the concentration dependence of the average decay time

$$\frac{\tau(C)}{\tau(0)} = (1 + B_1 C)^{-1}. \tag{IV.2}$$

It is natural to assume that the mechanisms responsible for the departure from the hyperbolic concentration dependence of the luminescence efficiency should be manifested also in the concentration dependence of the average decay time τ, i.e., in other words we may expect the above formula for τ (C) to be approximate.

We shall use these considerations to study the average lifetime τ of the excited state in anthracene single crystals irradiated with electrons generated in a Van de Graaff accelerator.

It will be useful to interpret the experimental data not only on the concentration dependence of τ but also on the concentration dependence of the integral luminescence efficiency, which will be considered from a somewhat different point of view than in the preceding chapters. This will be done making use of the additional measurements of the radiation-induced reduction in the integral relative luminescence efficiency of anthracene.

§ 1. Experimental Method and Results Obtained

We used a batch of anthracene single crystals of thicknesses ranging from 0.1 to 500 μ. The electron-beam irradiation method and the dosimetric measurements were described in the preceding chapter. The average luminescence decay time τ was measured with a phase fluorometer* and the values of τ were obtained in the wavelength range 400-460 nm. A modulated excitation was provided at the wavelength 313 nm. The measurements of τ at different luminescence wavelengths λ would have provided valuable information on the excitation transfer mechanism. Unfortunately, such measurements could not have been made sufficiently accurately with the available apparatus. The dependence of τ of unirradiated crystals on their thickness was due to the reabsorption of luminescence and was considered in Chap. II (§ 4). The holder used in the fluorometer measurements ensured reproducibility of the position of a crystal in the fluorometer after irradiation.

The upper part of Fig. 16 shows one of the obtained dependences of the relative luminescence decay time, i.e., of the ratio $\tau(R)/\tau_0$, on the absorbed electron radiation dose R (in rad). The values of τ plotted in this figure were obtained for the same crystal. The dependences of $\tau(R)/\tau_0$ on the dose R obtained for crystals of different thickness agreed within the experimental error. Apart from the average luminescence decay time, we determined also the integral relative luminescence efficiency A (R)/A_0 (this was the average efficiency over the whole spectrum). The curves obtained for crystals of different thickness were similar to the curves

---

*The author is grateful to L. A. Tumerman and O. F. Borisova for providing an opportunity to carry out the fluorometer measurements.

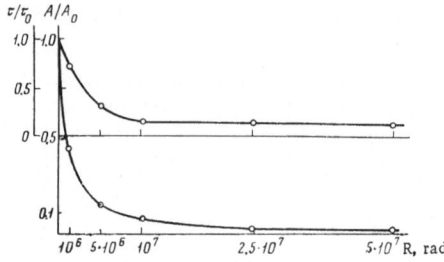

Fig. 16. Dependences of the relative luminescence decay time τ/τ_0 and of the relative
efficiency A/A_0 on the absorbed dose R.

plotted in the lower part of Fig. 16, which represents one of the dependences of A (R)/A_0 obtained in this way. The integral luminescence efficiency A (R)/A_0 was measured inside a sphere
coated internally with magnesium oxide (the reflection coefficient of MgO was ~97% in the wavelength range from 400 to 500 nm). The sample was excited with a KDSSh-1000 lamp whose light
was passed through a monochromator. The wavelength of the exciting light generated in this
way was λ = 313 nm. The luminescence was passed through a SZS-9 filter whose transmission
spectrum was located at the luminescence wavelengths of anthracene and detected with a photomultiplier. We can see from Fig. 16 that the nature of the dose dependences of τ (R)/τ_0 and
A (R)/A_0 differed at doses exceeding 10^7 rad: in this range τ/τ_0 was practically constant
whereas the luminescence efficiency continued to decrease. A similar dependence of τ on the
impurity concentration was reported in [12]. Obviously, the concentration dependence of the
average luminescence decay time does not normally obey a hyperbolic formula.

§2.   Discussion of Experimental Dependences of τ

on R and of Deviations from Hyperbolic Law

of Luminescence Degradation

We shall interpret the results of measurements of τ and the results reported in the preceding chapter (deviations from the hyperbolic quenching law and deformation of the spectra)
on the assumption that the mechanism of energy transfer in molecular crystals is generally of
multistep nature.

The recent investigations (see [11, 107-109] and Chap. I) have provided evidence in support of the following mechanism of de-excitation in molecular crystals.

The absorbed light generates excited exciton states in the lower exciton bands (these are
known as free excitons). The free excitons diffusing across the lattice can be de-excited (intrinsic exciton luminescence) or they can become localized near lattice defects or impurity
molecules. Some of the energy of the free excitons is lost during localization. The localized
excited states can also become de-excited (defect luminescence) or they can be annihilated by
nonradiative capture. The localized excitons may be captured by defects and nonluminescent
impurities present in "pure" samples as well as by defects and impurities introduced by a deliberate treatment (these may be the molecules disturbed by irradiation). The impurity and defect luminescence contribute mainly to the middle part of the luminescence spectrum of anthracene whereas the free-exciton luminescence predominates at short wavelengths and makes contributions to other parts of the spectrum (these contributions may be due to transitions from the
exciton band to the vibrational sublevels of the ground state [11]).

Our measurements were carried out at room temperature. The strongest intrinsic exciton luminescence band of anthracene was observed in the spectral region ~2.5 × 10^4 cm^{-1} and
the nearest defect luminescence band was located at ~2.4 × 10^4 cm^{-1} (see [11, 56]). The gap between the bottom of the exciton band and the defect levels was about 3kT at room temperature

and therefore no exchange of energy took place between these levels and the excited states of the host lattice. Thus, the defect and the impurity luminescence played an important role at room temperature.

In view of the difference between the energies of the free and localized excitons and because of the possible difference between the reactions of these excitons with impurities, the mechanism described above can explain the deformation of luminescence spectra resulting from an increase in the concentration of the quenching impurity. Other consequences of the multi-step energy transfer process include a nonexponential decay law and deviations of the concentration dependence of τ from Eq. (IV.2). Both consequences are observed experimentally.

It is quite difficult to describe the migration of excitions in crystals by the cellular method making allowance for the multistep nature of the energy transfer process. Therefore, we shall make this allowance within the framework of a simpler description of energy transfer which is provided by the balance equations.

We shall start by considering transient (decay) conditions and we shall assume that there are N free excitons in a band at a time t = 0. We shall use the following notation: α_r is the probability of radiative de-excitation of free excitons per unit time; $\alpha_d C_d'$ is the probability of localization of free excitons near defects or accidental impurities (C_d' is the concentration of such defects and impurities in the volume being considered); $\alpha_i C_i$ is the probability of localization near delibarately introduced impurities (C_i is the concentration of such impurities). The concentration of such impurities). The concentration of free excitons n decreases exponentially with time in accordance with the law:

$$n\,(t) = N \exp\,[-(\alpha_r + \alpha_d C_d' + \alpha_i C_i)\,t]. \tag{IV.3}$$

We shall now derive the equation of balance for the localized excitons. We shall assume that the localization of the free excitons can occur at defects or at deliberately introduced impurities and, therefore, there will be two groups of localized excitons. The number of excitons localized at defects (and accidental impurities) will be denoted by m_d. These excitons will be described by the following equation:

$$\frac{dm_d}{dt} = n\alpha_d C_d' - m_d\,(\beta_r + \beta_d^d C_d'' + \beta_d^i C_i). \tag{IV.4}$$

Here, β_r is the probability of radiative de-excitation of the localized excitons;[*] $\beta_d^d C_d''$ is the probability that an exciton localized at a defect suffers nonradiative capture by a defect or an accidental impurity (C_d'' is the concentration of defects or accidental impurities at which this process can take place); $\beta_d^i C_i$ is the probability of capture of a localized exciton by a deliberately introduced impurity.

The corresponding equation for the excitons localized near deliberately introduced impurities (m_i) is of the form

$$\frac{dm_i}{dt} = n\alpha_i C_i - m_i\,(\beta_r + \beta_i^d C_d'' + \beta_i^i). \tag{IV.5}$$

Here, $\beta_i^d C_d''$ is the probability that an exciton localized near a deliberately introduced impurity is captured by a defect or an accidental impurity; β_i^i is the probability of capture by a deliberately introduced impurity of an exciton localized near this impurity.

---

[*]Strictly speaking, we should consider the radiative de-excitation of defects to which the localized excitons transfer their energy.

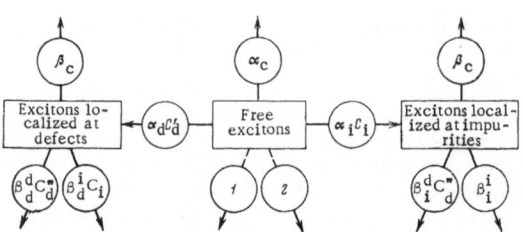

Fig. 17. Schematic representation of the transformation of electron excitation energy in molecular crystals.

In contrast to the probability of localization of excitons near defects used in Eq. (IV.4), the probability β_i^i should be practically independent of C_i [107, 109]. The assumed energy conversion scheme is shown schematically in Fig. 17. Apart from the localization and the radiative de-excitation (known also as radiative decay), the free excitons can also undergo nonradiative capture or dissociation. For some of these processes the capture probability is independent of C_i and, therefore, these processes cannot be separated experimentally from the self-localization of the free excitons (process $C_d C_d^l$) followed by nonradiative capture. Other processes have probabilities proportional to C_i and, therefore, these processes are inseparable from the localization ($\alpha_i C_i$) followed by capture ($\beta_i^d C_d^{\prime\prime}$ or β_i^i). Consequently, we must ignore these processes. If we solve Eqs. (IV.4) and (IV.5) using Eq. (IV.3) and the initial conditions $m_d(0) = m_i(0) = 0$, we obtain

$$m_d = \frac{N\alpha_d C_d'}{\beta_d - \alpha}(e^{-\alpha t} - e^{-\beta_d t}),$$

$$m_i = \frac{N\alpha_i C_i}{\beta_i - \alpha}(e^{-\alpha t} - e^{-\beta_i t}). \tag{IV.6}$$

Here,

$$\alpha = \alpha_r + \alpha_d C_d' + \alpha_i C_i,$$
$$\beta_d = \beta_r + \beta_d^d C_d'' + \beta_d^i C_i,$$
$$\beta_i = \beta_r + \beta_i^d C_d'' + \beta_i^i$$

The luminescence decay law is governed by the function $\alpha_r n(t) + \beta_r [m_d(t) + m_i(t)]$, which is the sum of three exponential functions.

At low values of C_i the decay time τ decreases with increasing impurity concentration whereas at high values of C_i the decay is effectively governed only by the function $m_i(t)$, which is nearly exponential:

$$m_i(t) \approx N e^{-\beta_i t} \text{ when } C_i \to \infty.$$

The decay time τ at high values of C_i is close to $1/\beta_i$ and is independent of the impurity concentration.

We shall now consider the steady-state luminescence and solve the equation of balance for the free and localized excitons.

For the free excitons we obtain

$$n\alpha_r = \frac{N\alpha_r}{\alpha} = \frac{N\alpha_r}{\alpha_r + \alpha_d C_d' + \alpha_i C_i}, \tag{IV.7}$$

whereas the corresponding equation for the localized excitons is

$$(m_d + m_i)\beta_r = \frac{N\beta_r}{\alpha}\left(\frac{\alpha_d C_d'}{\beta_d} + \frac{\alpha_i C_i}{\beta_i}\right).$$

(IV.8)

In these expressions N represents the steady-state number of the free excitons generated in the volume being considered per unit time whereas n and m represent the corresponding equilibrium numbers. Using Eqs. (IV.7) and (IV.8), we obtain expressions for the luminescence efficiency entirely due to the free or the localized excitons.

For the free excitons, the luminescence efficiency is

$$A_f = 1 \left/ \left(1 + \frac{\alpha_i}{\alpha_r + \alpha_d C_d'}C_i\right)\right.,$$

(IV.9)

which corresponds to Eq. (IV.1).

For the localized excitons, we obtain

$$A_1 = A_f\left(\frac{1}{1 + \frac{\beta_d^i}{\beta_r + \beta_d C_d''}C_i} + \frac{\alpha_i C_i /\alpha_d C_d^1}{\frac{\beta_r + \beta_1^d C_d'' + \beta_1^i}{\beta_r + \beta_d C_d''}}\right)$$

(IV.10)

The form of Eqs. (IV.9) and (IV.10) suggests a qualitative explanation of the observed deformation of luminescence spectra. If we attribute the short-wavelength luminescence to the free excitons and the middle part of the spectrum to the localized excitons, we find the experimental value of the ratio A_1/A_f is less than unity for all doses below 10^7 rad. Moreover, at low doses this ratio decreases rapidly with increasing dose (starting from $A_1/A_f = 1$ for zero dose) and then it begins to rise slowly with increasing dose approaching unity. This dependence follows also from the empirical formulas $(1 + B_1C^\eta)^{-1}$ and $(1 + B_2C^\eta)^{-1}$ with $\eta < 1$ and $B_2 > B_1$ given in §2 (Chap. III). We select the coefficient of C_i in Eqs. (IV.9) and (IV.10) so as to ensure agreement between the experimental results and the calculated functions $A_f(C_i)$ and $A_1(C_i)/A_f(C_i)$.

The intensity of the long-wavelength luminescence decreases with increasing C_i more slowly than in the middle part of the spectrum. We have mentioned in Chap. III that this can be due to the formation of radiation defects (D centers [56, 76]) whose luminescence corresponds to the long-wavelength part of the spectrum of anthracene.

The decay time τ reaches saturation at high quenching impurity concentrations because the probability of nonradiative capture of excitons localized near impurity molecules is practically independent of the concentration C_i. At high values of C_i the free excitons are localized very rapidly at impurities and then the decay time becomes independent of C_i and equal to $\sim 1/\beta_i$.

The rapid localization of the free excitons at impurity molecules should also give rise to saturation of the luminescence efficiency. The relative luminescence efficiency is given by

$$L \propto \frac{N}{\alpha}\left(\alpha_r + \frac{\beta_r\alpha_d C_d'}{\beta_d} + \frac{\beta_r\alpha_i C_i}{\beta_i}\right),$$

(IV.11)

which can be integrated as follows:

$$L \propto \int_0^\infty (\alpha_r n + \beta_r m_d + \beta_r m_i)\,dt.$$

As $C_i \to \infty$, $L \to \sim N\alpha_r/\beta_i = \text{const}$. However, the luminescence efficiency L and the average decay time τ should reach saturation at different impurity concentrations. The average value of τ for the whole spectrum is given by the expression

$$\tau = \frac{\int_0^\infty t\,(\alpha_r n + \beta_r m_d + \beta_r m_i)\,dt}{L}. \tag{IV.12}$$

Integration of the above equation gives

$$\tau = \frac{\left[\alpha_r + \frac{\alpha_d C_d' \beta_r}{\beta_d}\left(1 + \frac{\alpha}{\beta_d}\right) + \frac{\alpha_i C_i \beta_r}{\beta_i}\left(1 + \frac{\alpha}{\beta_i}\right)\right]}{\alpha\left[\alpha_r + \frac{\alpha_d C_d' \beta_r}{\beta_d} + \frac{\alpha_i C_i \beta_r}{\beta_i}\right]}. \tag{IV.13}$$

The values of τ for the various components of the luminescence spectrum can be defined as follows:

$$\tau_i = \frac{\int_0^\infty t\alpha_r n\,dt}{L_i},$$

$$\tau_{m_d} = \frac{\int_0^\infty t\beta_r m_d\,dt}{L_{m_d}},$$

$$\tau_{m_i} = \frac{\int_0^\infty t\beta_r m_i\,dt}{L_{m_i}},$$

where $L_i = \int_0^\infty \alpha_r n\,dt$, etc. Integration of the above expressions yields

$$\tau_i = \frac{1}{\alpha}, \quad \tau_{m_d} = \frac{1}{\alpha} + \frac{1}{\beta_d}, \quad \tau_{m_i} = \frac{1}{\alpha} + \frac{1}{\beta_i}. \tag{IV.14}$$

The expression in parentheses in Eq. (IV.11), which is proportional to αL, begins to rise linearly with increasing C_i beginning from some concentration C_i^*, which corresponds to the saturation of the efficiency L.

The quotient obtained by dividing the two expressions in square brackets in Eq. (IV.13) also begins to rise linearly with increasing C_i and this happens at some concentration C_i^{**} (as $C_i \to \infty$, the numerator increases quadratically with C_i whereas the denominator increases linearly). In the $C_i > C_i^{**}$ range we should have $\tau = \text{const}$. In general, $C_i^* \neq C_i^{**}$ but since we do not know many of the parameters in the above formula, we cannot say which of these concentrations should be higher, i.e., we cannot predict a priori whether τ or L will reach saturation earlier. In our measurements the saturation of L or of A could not be established reliably because of the low value of the luminescence efficiency and high impurity concentrations. If this saturation were observed at all, it must have occurred at doses R > $(3-4) \times 10^7$ rad, i.e., at impurity concentrations 3-4 times higher than those necessary for the saturation of τ ($C_i^*/C_i^{**} \geq 3$-4). Similar experimental results were reported in [12], when it was found that $C_i^{**} \approx 10^{-4}$ g/g for naphthacene-doped anthracene and the saturation of L (which was not established definitely in [12]) indicated that the ratio C_i^*/C_i^{**} should be ≥ 4-5.

The results of measurements on several single-crystal and polycrystalline samples of "pure" anthracene are reported in [110]. The luminescence spectra were determined at 20 and −196°C and the spectra of the fluorometric phase were obtained, i.e., the values of τ were measured at different luminescence wavelengths. It was found that the decay time τ varied strongly within the luminescence spectrum (oscillations of the fluorometric phase were observed). A comparison of the phase and the fluorescence spectra [110] indicated that the observed luminescence of anthracene was due to at least two different types of luminescence center with different fluorescence spectra and very different (a factor of 7) values of τ. The luminescence of the centers of the first type (n_1) had maxima at $\nu = 2.5 \times 10^{-4}$, 2.37×10^{-4}, and 2.23×10^{-4} cm^{-1}. The luminescence of the other centers (n_2) was observed at $\nu = 2.41 \times 10^{-4}$ and 2.27×10^{-4} cm^{-1}. The value of τ for the n_1 luminescence was about 1 nsec and that for the n_2 luminescence was about 7 nsec. Two possibilities were considered in [110]: 1) both centers are excited independently; 2) one type of center is excited predominantly but energy is transferred to the other centers during the excited-state lifetime. We can see that the results reported in [110] and the interpretation invoking energy transfer (the second possibility) are in agreement with our description of the energy transfer process if it is assumed that the n_1 luminescence corresponds to the radiative de-excitation of free excitons (purely electronic transitions) and to vibronic transitions in anthracene crystals, and the n_2 luminescence corresponds to the de-excitation of defects at which excitons are localized.

The present chapter can be summarized by saying that an allowance for the multistep nature of the transfer of excitation energy is sufficient to explain qualitatively all the experimental results including the deviations from the hyperbolic quenching formula, the deformation of the luminescence spectra of crystals containing quenching impurities, and the dependence of τ on the concentration of these impurities. The explanation of the deformation of the spectra based on the overlap of the "spheres of action" (Chap. III) and the explanation postulating multistep energy transfer are compared in Conclusions.

CHAPTER V

DEPENDENCE OF QUANTUM EFFICIENCY OF LUMINESCENCE OF ANTHRACENE ON EXCITING-LIGHT WAVELENGTH

Wright [14, 65] and other workers [78] found that the quantum efficiency of luminescence L of some molecular crystals (particularly, anthracene) depends on the exciting-light wavelength λ. Oscillations of the absorption coefficient k of the exciting light with the wavelength λ give rise to oscillations L and the functions k (λ) and L (λ) are anticorrelated (the higher the value of k the lower is L and the maxima of k correspond to minima of L and conversely). The Wright effect occurs in crystals whose thickness is sufficient for practically complete absorption of the exciting light and the luminescence is usually observed on the side opposite to the excited face.

The generally accepted explanation of the Wright effect is based on the assumption that excitons are annihilated nonradiatively on or near the surface. An increase in the depth of penetration of the exciting light (corresponding to lower values of k) should reduce the fraction of the quenched (not radiatively de-excited) excitons and, consequently, it should increase the relative quantum efficiency L. If we know the spectra k (λ) and L (λ) and solve the diffusion equation for excitons, we can estimate the thickness x_0 of the surface layer in which excitons are quenched completely. The published estimates of x_0 are $\sim 10^{-5}$ cm [21] and $< 10^{-4}$ cm [65], and it is assumed that the quenching layer is produced by photo-oxidation reactions.

It is pointed out in [21, 65] that the Wright effect can also be explained by the reabsorption of luminescence (no quantitative estimates are given in [21, 65]). An increase in the depth

of penetration of the exciting light and, consequently, of the depth of the region where luminescence is generated should increase the reabsorption and enhance the intensity of the luminescence observed on the side opposite to the excited face.

The present chapter describes experiments [78] which were carried out in order to determine the mechanism of the Wright effect. This mechanism is interesting not only from the
scientific and practical points of view but also in relation to some of the problems considered
in Chap. II. If it were possible to explain the Wright effect by nonradiative annihilation of excitons on the surface of a crystal, it would be necessary to reject the hypothesis of exciton-re-
flecting boundaries represented by the conditions $n'|_{x=0} = n'|_{x=d} = 0$, employed in § 2 (Chap. II).
The measurements of the luminescence of thin anthracene single crystals and the irradiation of
these crystals with tritium β rays (Chap. II) were carried out in air. If a thin quenching layer
with a high impurity-concentration gradient formed near the surface of a crystal as a result of
photo-oxidation reactions (as suggested in the proposed explanations of the Wright effect), the
presence of this layer would be reflected in Eq. (II.4) and in the subsequent formulas in § 2
(Chap. II) used to estimate the diffusion length of excitons in anthracene.

§ 1.   Experimental Method and Results Obtained

The special feature of our method was the measurement of the quantum efficiency $L(\lambda)$
for the same anthracene crystal under identical conditions, except for the state of its front (excited) surface: in some measurements this surface was oxidized, i.e., the crystal was stored
for a long time in an oxidizing atmosphere without any protection from light, whereas in other
measurements this surface was "clean," i.e., the layer containing photo-oxides was removed
by applying a solvent in an inert atmosphere. The thickness of the removed layer exceeded
10^{-4} cm (the cleaning process was repeated many times).

The measurements were carried out using an apparatus which included two monochromators (Fig. 18). The investigated anthracene single crystal was placed in a special hermetically
sealed optical chamber through which gaseous nitrogen was circulated continuously. The chamber contained a device for washing the surface layer in the inert nitrogen atmosphere. The crystal was excited with a DKSSh-1000 xenon lamp whose continuous spectrum was passed through
one of the monochromators (M_2). The second monochromator (M_1) selected the required luminescence maximum of anthracene. The experiments indicated that the effect under investigation
was practically independent of the luminescence wavelength. The luminescence intensity was
measured with a photomultiplier. The excitation spectrum was also recorded. This spectrum

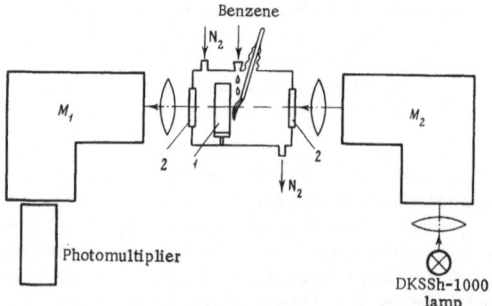

Fig. 18. Apparatus used in measurements of the
dependence of the quantum efficiency of luminescence on the wavelength of the exciting light: 1)
investigated crystals; 2) quartz windows.

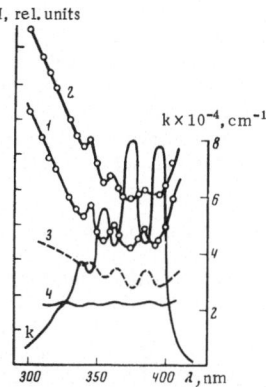

Fig. 19. Luminescence excitation
spectra (L) and absorption spectrum (k) of anthracene single crystals: 1) experimentally obtained
excitation spectrum of an "oxidized"
crystal; 2) corresponding excitation
spectrum of a "washed" crystal (the
ordinate scale is the same for curves
1 and 2); 3) excitation spectrum calculated on the assumption of total
quenching of excitons on the front
surface of a crystal; 4) excitation
spectrum obtained making allowance
for the reabsorption of primary luminescence in a crystal (curves 3 and 4
are not renormalized to the scale of
curves 1 and 2).

was normalized to the same number of incident quanta at all wavelengths by introducing corrections for the energy distribution in the spectrum emitted by the xenon lamp. In this way we
found the relative quantum efficiency of the luminescence as a function of the wavelength of the
exciting light, L (λ).

By way of example, Fig. 19 shows two L (λ) spectra obtained for one of the investigated
crystals. Here, curve 1 represents the spectrum obtained for an oxidized surface and curve 2
is the spectrum of a washed crystal. Curve k represents the absorption spectrum k (λ) taken
from [111]. The anticorrelation of the k (λ) and L (λ) spectra can be seen quite clearly in Fig.
19. However, the anticorrelation between L and k is not perfect: the same values of k at the
short- and the long-wavelength edges of the absorption spectrum correspond to different values
of L.

§2. Discussion of Results

A comparison of curves 1 and 2 in Fig. 19 shows that they are of practically the same
shape and that washing of the surface layer increases only slightly the luminescence intensity
without any change in the shape of the spectrum, i.e., there are no significant changes in the
dependence of the luminescence efficiency on the wavelength of the exciting light. This does not
agree with the assumption that the photo-oxide quenching layer at the surface is thin.

We may postulate that the structure of the L (λ) spectra is due to the annihilation of excitons not in the surface layer but on the surface itself. In this case any washing of a crystal would not change the boundary conditions for the diffusing excitons and would not alter the spectrum L (λ). If we adopt this hypothesis, we find it difficult to expalin the increase in the luminescence intensity which results from washing of the surface of anthracene. Moreover, we can show that the shape of the spectrum L (λ) obtained on this assumption differs considerably from that found experimentally. If the suggested hypothesis is adopted, the diffusion equation for excitons becomes

$$Dn'' - \frac{1}{\tau} n + I_0 k e^{-kx} = 0. \tag{V.1}$$

Having solved this equation subject to the boundary conditions $Dn' + \varkappa n |_{x=0} = 0$, and $n |_{x=\infty} = 0$, we obtain the following expression for the bulk value of the luminescence efficiency:

$$L \propto \int\limits_0^\infty n dx = \frac{I_0 \tau}{1 - l^2 k^2} \left[1 - kl \frac{k \frac{D}{\varkappa} + 1}{\frac{1}{l} \frac{D}{\varkappa} + 1} \right], \tag{V.2}$$

where $l = \sqrt{D\tau}$ is the diffusion length of excitons. The clearest structure in the L (λ) spectrum is observed for $D/\varkappa \to 0$, i.e., when all the excitons suffer nonradiative annihilation on the front surface (x = 0). In this case we obtain $L \propto 1/(1 + kl)$. This relationship is represented by the dashed curve (3) in Fig. 19. The depth of oscillations of $L_3 (\lambda)$ at wavelengths λ corresponding to the central part of the absorption spectrum is more or less equal to the depth of oscillations of the experimentally obtained spectra L(λ). However, at the short- and long-wavelength ends of the absorption spectrum k (λ) corresponding to $k (\lambda) - 3000 \text{ Å} \leqslant \lambda \leqslant 3500 \text{ Å}$ and $4000 \text{ Å} \leqslant \lambda \leqslant 4200 \text{ Å}$ the dependence $L_3(\lambda)$ is much weaker than that found experimentally. Thus, even the assumption of complete quenching of excitons on the front surface of a crystal is not in agreement with the experimental results.

Possible amplitudes of the oscillations of the luminescence intensity as a result of reabsorption of the luminescence can be estimated by considering the absorption of the first-generation luminescence photons (for these photons the oscillations should be strongest). An approximate estimate can be obtained by dividing the luminescence spectrum of anthracene into two regions, one of which is assumed to be reabosrbed and the other not reabsorbed. Bearing in mind the nature of the absorption spectrum of anthracene (Fig. 7), we shall postulate that the reabsorbed photons are of wavelengths $\lambda \leq 425$ nm whereas the photons that are not reabsorbed correspond to wavelengths $\lambda \geq 425$ nm. The absorption coefficient k_1 is small in the part of the spectrum which does not suffer reabsorption and therefore, the probability of trapping of the corresponding phonons in a crystal (because of multiple reflection) can be regarded as independent of the depth x of the creation of these photons below the front (excited) surface. Elementary calculations yield the following expression for the fraction of the nonreabsorbed photons which are trapped by reflection:

$$1 - \frac{1}{2} \left(1 - \sqrt{1 - \frac{1}{n^2}} \right), \tag{V.3}$$

where n is the refractive index (the value of this index for the nonreabsorbed photons is about 1.9 [112, 113]). It is assumed that the photons emitted in the backward direction, i.e., into a solid angle 2π (half of the total solid angle) facing the luminescence detector (on the left hand side in Fig. 18), are completely absorbed because of the considerable thickness of the crystal in this direction.

We shall now calculate the probability of reflection trapping of the photons corresponding to the reabsorbed part of the spectrum. We shall assume that these photons are generated at a point O located at a distance x from the front surface S (Fig. 20). The probability of emergence

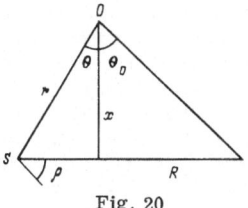

Fig. 20

of those photons which lie within the solid angle $d\Omega = 2\pi\rho d\rho \cos\theta/r^2$, supported by a ring-shaped area $2\pi\rho d\rho$ of the surface S is $e^{-k_1 r}$, where k_1 is the absorption coefficient. The average probability of emergence of any reabsorbed luminescence photon generated at the point O is $\frac{1}{4\pi}\int\limits_{(\Omega)} e^{-k_1 r}d\Omega$, where integration is carried out right up to the angle θ_0 at which total internal reflection takes place at the front face; since $\cos\theta - x/r$, the probability is given by the expression

$$\frac{k_1 x}{2}\int\limits_{k_1 x}^{k_1\sqrt{R^2+x^2}}\frac{e^{-z}dz}{z^2} = \frac{k_1 x}{2}\left\{\frac{e^{-k_1 x}}{k_1 x} - E_1(k_1 x) - \frac{\exp(-k_1\sqrt{R^2+x^2})}{k_1\sqrt{R^2+x^2}} + E_1(k_1\sqrt{R^2+x^2})\right\}, \qquad \text{(V.4)}$$

where $E_1(z) = \int\limits_z^\infty \frac{e^{-y}dy}{y}$ is the exponential integral. Substituting $k_1\sqrt{R^2+x^2} = \frac{k_1 x}{\cos\theta_0} = \frac{k_1 x}{\sqrt{1-\frac{1}{n^2}}}$, we obtain

$$W_{em}(x) = \frac{1}{2}\left\{e^{-k_1 x} - k_1 x E_1(k_1 x) - \sqrt{1-\frac{1}{n^2}}\exp\left(-\frac{k_1 x}{\sqrt{1-\frac{1}{n^2}}}\right) + k_1 x E_1\left(\frac{k_1 x}{\sqrt{1-\frac{1}{n^2}}}\right)\right\}.$$

The probability of emission of luminescence photons varies with depth in accordance with the expression

$$\frac{e^{-kx}}{\int\limits_0^\infty e^{-kx}dx} = ke^{-kx}.$$

Therefore, the average probability of emergence from a crystal of any (generated at a depth x) reabsorbed photon is

$$\int\limits_0^\infty ke^{-kx}W_{em}(x)\,dx.$$

Integration and application of the relationship

$$\int\limits_0^\infty E_1(x)e^{-\mu x}x dx = \frac{1}{\mu^2}\ln(1+\mu) - \frac{1}{\mu(1+\mu)}$$

yields the following expression for the fraction of the first-generation photons absorbed in the crystal:

$$1 - \frac{1}{2}\left\{1 - \sqrt{1-\frac{1}{n^2}} - \frac{k_1}{k}\ln\left(1+\frac{k}{k_1}\right) + \frac{k_1}{k}\ln\left(1+\frac{k}{k_1}\sqrt{1-\frac{1}{n^2}}\right)\right\}.$$

When $k_1 \to 0$, the above expression reduces (as expected) to Eq. (V.3) which represents the nonreabsorbed photons. In the case of anthracene the probabilities that the energy of luminescence photons lies in the reabsorbed or the nonreabsorbed region are approximately equal and the average value of the quantum efficiency for the nonreabsorbed region is 0.27 (this value is obtained by numerical integration of the efficiency and of the luminescence spectrum of Fig. 7 [83, 111]). If we use these results and assume that the average optical constants of the reabsorbed photons are $k_1 = 5 \times 10^4$ cm^{-1} and $n = 1.7$, we obtain the dependence of the luminescence intensity on the wavelength of the exciting radiation represented by curve 4 in Fig. 19. The relatively weak oscillations of this curve indicate that the reabsorption of the luminescence is of little importance in the Wright effect.

The magnitude of the oscillations of the experimentally obtained dependences L (λ) and the increase in the luminescence efficiency without a change in the form of the spectrum L (λ) after washing of the surface of the crystal can only be explained if we assume that quenching centers penetrate relatively deep into a crystal and have an approximately exponential distribution with depth. If these quenching centers are due to the presence of photo-oxides, such a distribution is in agreement with the diffusion of oxygen into anthracene.

We shall assume that the distribution of the quenching substance is proportional to $\propto e^{-\gamma x}$, and we shall use the experimentally obtained spectra L (λ) (Fig. 19) to estimate the order of magnitude of γ. If we ignore the influence of reabsorption on the efficiency spectrum L (λ) and assume that this spectrum is governed solely by exciton migration, we can estimate γ by supplementing Eq. (V.1) with the term $-pne^{-\gamma x}$ and then solving the resultant equation. Here, p is the probability of annihilation of an exciton per unit time at that concentration of the quenching substance which is found on the surface $x = 0$. However, an estimate can also be obtained by a less accurate but simpler method. The diffusion length of excitons l does exceed $\sim 10^{-5}$ cm [24-26]. If we restrict ourselves to the range of low values of the absorption coefficient of the exciting light k (λ), i.e., to the range of wavelengths λ where $l \ll 1/k$, and if we also assume that $l \ll 1/\gamma$, we can express the exciton concentration n in the form

$$n \approx \frac{I_0 k \tau e^{-kx}}{1 + p\tau e^{-\gamma x}}. \tag{V.5}$$

The luminescence efficiency is $L \propto \int\limits_0^\infty ndx = I_0 k\tau \int\limits_0^\infty \frac{e^{-kx}dx}{1 + p\tau e^{-\gamma x}}$. It is natural to assume that $\tau p \gg 1$, and, if moreover, $k > \gamma$, the efficiency L is of the form $L \propto I_0 k/p (k-\gamma)$. Thus, under these assumptions the quantity L $(k-\gamma)/k$ should be constant, i.e., independent of k, for each experimentally determined spectrum L (λ).

We estimated the value of γ for the short-wavelength parts of the L (λ) spectra (the parts corresponding to low values of k) and we found that $(0.3 \pm 1.0) \times 10^4$ cm^{-1}. The quantity L $(k-\gamma)/k$ was constant (to within 5-6%) for each L (λ) spectrum in the range k = $(0.8-2.5) \times 10^4$ cm^{-1}.

We can summarize the foregoing discussion by saying that neither the nonradiative annihilation of excitons on the surface nor the reabsorption of luminescence can explain the Wright effect which occurs in anthracene crystals. This effect is probably due to a relatively deep penetration of quenching centers into a crystal and an approximately exponential distribution of these centers. The absence of a definite relationship between k and L is due to a fall in the quantum efficiency with increasing λ. The effective depth of penetration of quenching compounds into anthracene crystals is $1/\gamma \sim 3 \times 10^{-4}$ cm. A similar value was obtained in an investigation of the photoconductivity of anthracene crystals [114].

An analysis of the conditions applicable to the experiments described in § 2 (Chap. II) suggests that photo-oxidation of the investigated crystals should have no significant influence on the validity of the conclusions drawn from the results of these experiments. Repeated measurements of the luminescence efficiency carried out on crystals which were not subjected to β irradiation gave reproducible results, which indicated that the distribution of photo-oxides below the surface did not vary with time. Bearing in mind that the photo-oxidation was practically identical on both sides of films, we may conclude that the distribution of the quenching compounds with depth should not differ greatly from the homogeneous distribution. The maximum possible difference between the concentrations of photo-oxides on the two surfaces and in the central plane was found by adding the ordinates of two exponential functions $\propto e^{-\gamma x}$, where $\gamma = 0.3 \times 10^4 \, \text{cm}^{-1}$: it was found that this difference was 27% for crystals 5 μ thick, 18% for crystals 4 μ thick, and 11 and 6% respectively, for crystals 3 and 2 μ thick. A satisfactory allowance for the quenching effect of these oxides was made in the diffusion equation (II.2) by the term $1/\tau n$ (the rate of annihilation of excitons in an unirradiated crystal), where $1/\tau$ is independent of the coordinate x.

CONCLUSIONS

The investigations described in the present paper and the results obtained relate mainly to:

1) the mechanism of exciton energy transfer in molecular crystals;
2) some features of energy transfer by the reabsorption of luminescence;
3) some methodological problems such as the dosimetry in the case of electron bombardment of crystalline samples, an allowance for the photo-oxidation of anthracene, measurement of the thickness of single-crystal films, etc.

Let us briefly consider the results obtained.

A study of the influence of increasing the impurity concentration C on the luminescence spectra of the host substances in mixed crystals indicated that the standard hyperbolic formula for the quenching by impurities is approximate. The physical picture of energy transfer leading to the relationship $A (C) = (1 + BC)^{-1}$ is extremely simple: excited states migrating in a crystal are captured by impurities with a probability which is strictly proportional to the concentration C whereas the probabilities of all the other excitation annihilation processes (including radiative de-excitation) are constant and independent of the concentration C. In general, this relationship is not obeyed and it follows that this picture of energy migration is only approximate. However, the results of measurements of the dependences of the luminescence efficiency on the impurity concentration, including those obtained at different wavelengths λ, are insufficient for identification of the reasons why the simple picture of energy transfer is inadequate. In general, this inadequacy may be due to two reasons. Obviously, irrespective of the details of the migration mechanism, the direct proportionality between the capture probability and the concentration of the quenching centers cannot be obeyed at high concentrations and the linear dependence at all values of C does not represent the simplest possible situation. In general, the dependence of the capture probability on the concentration of "traps" should be nonlinear. It can reduce to the linear dependence only at fairly low impurity concentrations. The purely kinetic relationship obtained for the capture probability in Chap. III by the cellular method is quite general and applicable to any type of diffusion accompanied by capture (it applies to the diffusion of electrons, neutrons, radicals, etc.). The results of calculations show that the capture probability is proportional to the impurity concentration at values of C for which the "spheres of action" of individual traps do not overlap, i.e., the values satisfying the inequality $(1/C)/\sigma l > 10^{-2}$-10^{-1}. Significant deviations from this proportionality begin at $(1/C)/\sigma l \approx 10^{-1}$ (at this value the "spheres of action" begin to overlap). In the case of irradiated anthracene, deviations from the proportionality begin at $C \approx 10^{-4}$-10^{-3} mole/mole.

We shall now compare the results of calculations obtained by the cellular method with the experimental data on the degradation of luminescence by assuming initially (the physical interpretation will be given later) that only the nonlinearity of the capture probability is responsible for deviations from the standard hyperbolic quenching relationship.

Only one of the possible explanations of deviations from the hyperbolic quenching formula is considered in Chap. III and the results given in that chapter reduce to the calculation of the values of l and σ which can give agreement withe the experimental results if the only reason for deviations from this formula is the overlap of the "spheres of action" of individual quenching centers. The dependences $l(\lambda)$ and $\sigma(\lambda)$ obtained for irradiated samples are undoubtedly due to the dependence of the degradation on λ, i.e., due to the deformation of the spectra of the irradiated crystals.

A further circumstance, which is not reflected in that energy scheme which leads to the simple dependence of the luminescence efficiency of the host substance on the impurity concentration, is the existence of several types of excitation state. If we assume that these states can transform from one into another and that the probability of capture of these states by a given quenching impurity are different (even when each of these probabilities are proportional to the impurity concentration C), and that they give rise to luminescence of different wavelengths, we naturally find that even simple equations can lead to deviations from the standard quenching formula and to deformation of the luminescence spectra with increasing quenching impurity concentration. Thus, apart from the overlap of the "spheres of action," a multistep energy transfer mechanism and differences in the behavior of various excited states can, in principle, explain the experimental data on the degradation of luminescence.

There is as yet no definite evidence that the observed deviations from the standard quenching formula are due to the multistep transfer mechanism and the existence of different excited states and, therefore, it is not possible to estimate the relative importance of each of these factors. We can only say that an allowance just for the multistep transfer mechanism is sufficient to explain all the experimental results on the deformation of the luminescence spectra and on the impurity-concentration dependence of the average de-excitation time τ (Chap. IV). If we assume that under the conditions employed in our experiments only the multistep mechanism is relevant to the observed nonlinearities, we find that this is not in conflict with the results reported in Chap. III. However, the only acceptable interpretation of the dependences $l(\lambda)$ and $\sigma(\lambda)$ reported in Chap. III is that based on the existence of free and localized excitons and on conversion of the former into the latter. If this explanation is adopted, the overlap of the "spheres of action" in the central and long-wavelength parts of the luminescence spectra becomes only apparent because it is due to the fact that the probability of capture of the excitons localized near impurity molecules should be independent of the concentration of these molecules (Chap. IV). The conclusion that the probability of capture of some localized excitons is practically independent of the impurity concentration C follows from the constancy of τ at high values of C. Thus, we may assume that the estimates given in Chap. III and IV lead to conclusions which are in basic agreement and which support the occurrence of the multistep energy transfer in crystals, although the estimates have been obtained by different methods.

The results obtained on reabsorption can be stated simply by saying that the effective absorption coefficient k_0 of the luminescence generated in anthracene does not exceed $(1-2) \times 10^3$ cm^{-1} (this coefficient is several times smaller than the values quoted in the literature) and that reabsorption of the luminescence in planar anthracene crystals a few microns thick leads (because of the smallness of k_0 and because of internal reflection) to a practically homogeneous volume distribution of the secondary luminescence "sources." These two conclusions are interrelated and they follow from the results of investigation of the reduction in the luminescence efficiency of anthracene under the action of weakly penetrating β rays emitted by tritium targets (Chap. II). In other words, the values of the relative efficiency A (R) obtained in Chap. II can be explained only if we assume that k_0 is small and, moreover, that "daughter" luminescence photons are trapped in a crystal.

The estimates made in §2 of Chap. II confirm these conclusions on the reabsorption of luminescence in anthracene. For example, the dependence τ (d), where d is the thickness of the crystal, obtained without allowance for internal reflection of luminescence (this is known as the q approximation) agrees with the experimental results only for $k_0 \approx (1-2) \times 10^3$ cm^{-1} whereas an allowance for the trapping as a result of internal reflection reduces this value by a factor of 2-3, i.e., down to $\sim 0.6 \times 10^3$ cm^{-1}. On the other hand, if the luminescence and the quantum efficiency spectra are used to identify the wavelengths of the anthracene luminescence which are reabsorbed most strongly and this is done making allowance for reflection trapping, it is found that the value of k_0 is close to 0.6×10^3 cm^{-1}.

The principal method used in the present investigation to study energy transfer in molecular crystals was the irradiation with fast electrons or with β rays. Introduction of quenching impurities into crystals by irradiation was found to be a very convenient method which ensured a good reproducibility of the conditions during measurements carried out on the same sample. The required concentration of the quenching (radiation-disturbed) molecules could be obtained quite simply by varying the absorbed energy (dose). The absorbed energy could be controlled most simply and conveniently by the use of a beam of constant-energy electrons generated in an accelerator. In this case the electron-beam current could be measured with the aid of a Faraday cylinder in which the investigated sample was inserted. This was the method employed in the present study (Chap. III).

Irradiation of crystals with particles penetrating only to a depth of ~ 1 μ enabled us to create steep gradients of the quenching impurity concentration. This made it possible to study energy transfer processes relatively close to the surface (at depths of $\sim 0.1\mu$) as well as at greater depths. The source of soft β rays was a tritium target. We estimated the effective diffusion length of excitons in anthracene (1.3×10^{-5} cm) and were able to draw some conclusions on the reabsorption of luminescence (Chap. II). We were also able to interpret the Wright effect, which is the dependence of the luminescence efficiency on the wavelength of the exciting light. The investigation (described in Chap. V) demonstrated that this dependence cannot be explained by nonradiative annihilation of excitons on the surface of a crystal or by reabsorption of luminescence. The effect is likely to be due to strong penetration of quenching centers (generated by photo-oxidation) into a crystal so that an approximately exponential distribution of the concentration with depth is established. The effective depth of penetration of such quenching centers is $\sim 3 \times 10^{-4}$ cm.

A study of the degradation of luminescence of anthracene under the influence of electron irradiation yielded the efficiency of electrons of various energies represented by the coefficient $B = \alpha\tau$ in the formula A (R) = $(1 + BR)^{-1}$. The values of these coefficients for the H^3 and Sr^{90}–Y^{90} sources were 2.5×10^{-7} and 5×10^{-7} rad^{-1}, respectively (Chap. II). Irradiation with faster (1 MeV) electrons generated in a Van de Graaff accelerator was characterized by a coefficient $B \sim 7 \times 10^{-7}$ rad^{-1} (Chap. IV). The H^3 and Sr^{90}–Y^{90} sources produced electrons whose average energies in matter were about 4-5 and 700-800 keV, respectively, i.e., they were somewhat lower than the energies of the electrons emitted in β decay. These results indicated that the efficiency of electron irradiation increases monotonically with increasing electron energy.

SUMMARY OF RESULTS

1. It was found that the quenching of the photoluminescence of several molecular crystals irradiated with electrons did not obey the standard hyperbolic formula A (C) = $(1 + BC)^{-1}$. Deviations from this formula were manifested in the following observations: a) the reduction in

the integral (measured over the whole spectrum) luminescence efficiency agreed only roughly with this formula; b) the dependence of the luminescence quenching on the impurity concentration varied along the spectrum and this led to a gradual deformation of the luminescence spectra when the concentration of the quenching impurity was increased. Similar deviations were observed also for quenching impurities which were not generated by irradiation.

2. Two different approaches were used in the interpretation of the observed relationships governing the luminescence quenching: a) an allowance was made for the fact that the probability of the capture of an excitation migrating in a crystal by a quenching impurity should be generally a nonlinear function of the concentration of this impurity; b) an allowance for the fact that the process of excitation energy transfer ending by radiative de-excitation should be generally of multistep nature. In the first approach, (case a) a calculation was made of the probability of capture of a migrating excitation by a quenching impurity on the assumption that this impurity was distributed homogeneously over a crystal. The calculation was performed by the Wigner-Seitz cellular method (a solution was obtained of the transport equation). The results of this calculation were compared with the experimental data and the parameters occurring in the theoretical formulas were estimated. The interpretation postulating multistep energy transfer (case b) was based on the possibility of localization of free excitons and of the emission of intrinsic and defect luminescence. Solutions of the appropriate balance equations yielded expressions which explained qualitatively all the experimentally observed features of luminescence quenching. Solutions of the balance equations in the transient case led to the conclusion that the average de-excitation (decay) time τ should reach saturation at high values of the impurity concentration C and this should be accompanied by a change to an exponential luminescence decay. These conclusions were supported by fluorometric measurements. A comparison of the results deduced by these two approaches indicated that they represented effectively the same mechanism of energy transfer from free to localized excitons, in agreement with the known spectral distribution of the electronic (intrinsic), vibronic, and defect luminescence.

3. The results of experiments involving irradiation of thin anthracene single crystals with tritium β rays were used to estimate the diffusion length l of excitons in anthracene at right-angles to the crystallographic plane ab. The value obtained (1.3×10^{-5} cm) was in satisfactory agreement with the results obtained by other methods.

4. An analysis of the results obtained employing tritium targets and of the measurements of the dependences of τ on the thickness of anthracene crystals made it necessary to modify some features of the simple picture of the reabsorption of luminescence in anthracene single crystals. The results obtained could be explained only if it was assumed that: 1) the effective absorption coefficient of luminescence in anthracene does not exceed $(1-2) \times 10^3$ cm^{-1}; 2) single crystals a few microns thick luminesce almost homogeneously because of the internal reflection of luminescence from the various faces.

5. The dependence of the quantum efficiency of luminescence of anthracene crystals on the wavelength of the exciting light was determined. An investigation was made of the possible causes of the anticorrelation of the absorption and the excitation spectra (Wright effect). It was found that nonradiative annihilation of excitons on the surface of a crystal and the reabsorption of luminescence failed to explain qualitatively the experimental data on the Wright effect. This effect was probably due to a relatively deep penetration of quenching centers into a crystal and an approximately exponential distribution of these centers below the surface. The effective depth of penetration of these quenching centers was found to be about 3×10^{-4} cm.

6. Other results obtained in the present investigation were mainly related to the methods used in the various investigations. They included the method for irradiation of single crystals and for the exact determination of the energy absorbed; the use of α rays in measurements of the thickness of thin single crystals; the determination of the quenching efficiency of electrons

of various energies (the efficiency increased monotonically with the energy). Moreover, it was found that one could use either a strongly inhomogeneous (soft β rays from tritium sources) or a homogeneous (a high-energy electron beam) ionizing radiation to introduce quenching impurities and to study energy transfer processes in molecular crystals.

The author is grateful to N. D. Zhevandrov for directing this investigation and to M. D. Galanin for his continuous interest, discussion of the results, and valuable advice. Thanks are also due to V. S. Vavilov for allowing the use of a Van de Graaff accelerator in electron-irradiation experiments and to S. I. Vintovkin for this constant help in the work carried out using this accelerator.

LITERATURE CITED

1. Ya. I. Frenkel (J. Frenkel), Phys. Rev., 37:17, 1276 (1931).
2. A. S. Davydov, Theory of Molecular Excitons, McGraw-Hill, New York (1962).
3. A. S. Davydov, in: Collection in Honor of S. I. Vavilov [in Russian], Izd. AN SSSR, Moscow (1952).
4. A. S. Davydov, Usp. Fiz. Nauk, 82:393 (1964).
5. A. S. Davydov, Tr. Inst. Fiz. Akad. Nauk Ukr. SSR, 3:36 (1952).
6. A. S. Davydov and A. F. Lubchenko, Zh. Eksp. Teor. Fiz., 35:1499 (1958).
7. É. I. Rashba, Opt. Spektrosk., 2:75 (1957); 2:568 (1957).
8. Y. Toyozawa, Progr. Theor. Phys., 20:53 (1958).
9. R. S. Knox, Theory of Excitons, Suppl. No. 5 to Solid State Physics, Academic Press, New York (1963).
10. M. Trlifaj, Czech. J. Phys., 9:4 (1959).
11. M. T. Shpak, Thesis [in Russian], Institute of Physics of the Ukrainian Academy of Sciences, Kiev (1964).
12. M. D. Galanin and Z. A. Chizhikova, Opt. Spektrosk., 1:175 (1956).
13. M. D. Galanin, Tr. Inst. Fiz. Akad. Nauk SSSR, 5:339 (1950).
14. G. T. Wright, Proc. Phys. Soc. London, B68:241 (1955).
15. J. B. Birks, Scintillation Counters, Pergamon Press, London (1953).
16. V. M. Agranovich, Izv. Akad. Nauk SSSR, Ser. Fiz., 23:40 (1959).
17. V. M. Agranovich, Opt. Spektrosk., 3:84 (1957).
18. M. D. Galanin, Yu. V. Konobeev, and Z. A. Chizhikova, Opt. Spektrosk., 13:386 (1962).
19. V. M. Agranovich and Yu. V. Konobeev, Phys. Status Solidi, 27:435 (1968).
20. Y. Toyozawa, Progr. Theor. Phys., Suppl. No. 12, p. 111 (1959).
21. A. N. Faidysh, Opt. Spektrosk., 4:597 (1958).
22. A. N. Faidysh and V. M. Agranovich, Opt. Spektrosk., 1:903 (1956).
23. I. Ya. Kucherov and A. N. Faidysh, Izv. Akad. Nauk SSSR, Ser. Fiz., 22:29 (1958).
24. O. Simpson, Proc. Roy. Soc. London A, 238:402 (1957).
25. Sh. D. Khan-Magometova and G. V. Radzievskii, Opt. Spektrosk., 16:842 (1964).
26. A. N. Faidysh, Vestn. Kiev. Univ., Ser. Fiz. No. 3(1), p. 53 (1960).
27. V. L. Zima, V. M. Korsunskii, and A. N. Faidysh, Izv. Akad. Nauk SSSR, Ser. Fiz., 27:519 (1963).
28. V. M. Korsunskii and A. N. Faidysh, Opt. Spektrosk., Suppl. No. 1 (Luminescence), p. 119 (1963).
29. V. L. Zima and A. N. Faidysh, Opt. Spektrosk., 20:1022 (1966).
30. M. von Smoluhowski, Phys. Z., 17:557, 585 (1916).
31. M. von Smoluhowski, Z. Physik. Chem. (Leipzig), 92:129 (1917).
32. S. I. Vavilov (Wawilow), Z. Phys., 53:665 (1929).
33. V. Ya. Sveshnikov (B. Sveshnikoff), Acta Physicochim. USSR, 3:257, 268 (1935).

34. B. Ya. Sveshnikov, A. S. Selivanenko, V. I. Shirokov, and L. A. Kiyanskaya, Opt. Spektrosk., 14:45 (1963).
35. S. F. Kilin, M. S. Mikhelashvili, and I. M. Rozman, Opt. Spektrosk., 16:1063 (1964).
36. L. A. Kiyanskaya and B. Ya. Sveshnikov, Opt. Spektrosk., Suppl. No. 1 (Luminescence), p. 60 (1963).
37. M. D. Galanin, Zh. Eksp. Teor. Fiz., 28:485 (1955).
38. T. Förster, Ann. Phys. (Leipzig), 2:55 (1948); Z. Naturforsch., 4a:321 (1949).
39. Sh. D. Khan-Magometova, Opt. Spektrosk., 27:61 (1969).
40. Sh. D. Khan-Magometova, Izv. Akad. Nauk SSSR, Ser. Fiz., 32:1353 (1968).
41. V. M. Agranovich, Opt. Spektrosk., 4:586 (1958).
42. V. M. Agranovich and Yu. V. Konobeev, Opt. Spektrosk., 6:242 (1959).
43. V. M. Agranovich, I. Ya. Kucherov, and A. N. Faidysh, Ukr. Fiz. Zh., 2:61 (1957).
44. V. M. Agranovich and Yu. V. Konobeev, Fiz. Tverd. Tela, 5:1372 (1963).
45. I. Ya. Kucherov, A. N. Faidysh, and Z. N. Fesenko, Opt. Spektrosk., 2:462 (1957).
46. A. S. Selivanenko, Opt. Spektrosk., 4:122 (1958).
47. Yu. M. Popov and A. S. Selivanenko, Opt. Spektrosk., 9:260 (1960).
48. A. S. Selivanenko, Fiz. Tverd. Tela, 3:1009 (1961).
49. A. A. Kazzaz and A. B. Zahlan, Phys. Rev., 124:90 (1961).
50. A. Pröpstl and H. C. Wolf, Z. Naturforsch., 18a:724, 822 (1963).
51. P. Avakian and H. C. Wolf, Z. Phys., 165:439 (1961).
52. M. Trlifaj, Czech. J. Phys., 6:533 (1956).
53. M. Trlifaj, Czech. J. Phys., 8:510 (1958).
54. V. M. Korsunskii and A. N. Faidysh, Ukr. Fiz. Zh., 8:677 (1963).
55. B. J. Mulder, Philips Res. Rep., 21:283 (1966).
56. V. L. Zima, Thesis, [in Russian], Kiev State University (1965).
57. Sh. D. Khan-Magometova, Opt. Spektrosk., 25:373 (1968).
58. D. P. Craig, J. Chem. Phys., 41:4000 (1964).
59. V. K. Gorshkov and N. D. Zhevandrov, Zh. Prikl. Spektrosk., 6:267 (1967).
60. A. N. Faidysh and I. Ya. Kucherov, Ukr. Fiz. Zh., 2:68 (1957).
61. L. E. Lyons and J. W. White, J. Chem. Phys., 29:447 (1958).
62. C. W. Reed and F. R. Lipsett, J. Mol. Spectrosc., 11:139 (1963).
63. G. Gallus and H. C. Wolf, Phys. Status Solidi, 16:277 (1966).
64. V. V. Eremenko and V. S. Medvedev, Fiz. Tverd. Tela, 2:1572 (1960).
65. G. T. Wright, Phys. Rev., 100:587 (1955).
66. N. D. Zhevandrov, Tr. Fiz. Inst. Akad. Nauk SSSR, 25:3 (1964).
67. H. B. Rosenstock and J. H. Schulman, J. Chem. Phys., 30:116 (1959).
68. J. B. Birks and F. A. Black, Proc. Phys. Soc. London, A64:511 (1951).
69. J. H. Schulman, H. W. Etzel, and J. G. Allard, J. Appl. Phys., 28:792 (1957).
70. H. B. Rosenstock, J. Chem. Phys., 48:532 (1968).
71. P. Alexander, Nuclear Radiations and Life [Russian translation], Atomizdat, Moscow (1959).
72. J. Schulman and W. Shurcliff, Nucleonics, 11:10 (1953).
73. F. H. Attix, Nucleonics, 17(10):60 (1959).
74. F. A. Black, Phil. Mag., 44:263 (1953).
75. Sh. D. Khan-Magometova, N. D. Zhevandrov, and V. I. Gribkov, Izv. Akad. Nauk SSSR, Ser. Fiz., 24:561 (1960).
76. A. N. Faidysh, Thesis [in Russian], Kiev (1964).
77. F. R. Lipsett, J. Chem. Phys., 26:1444 (1957).
78. Sh. D. Khan-Magometova, Izv. Akad. Nauk SSSR, Ser. Fiz., 29:1321 (1965).
79. G. B. Radzievskii, Problems in Dosimetry and Radiation Protection [in Russian], No. 3, Atomizdat, Moscow (1964).

80. G. B. Radzievskii, Prib. Tekh. Eksp., No. 1, p. 70 (1970).
81. A. N. Faidysh, Opt. Spektrosk., 4:525 (1958).
82. Z. A. Chizhikova, Tr. Fiz. Inst. Akad. Nauk SSSR, 15:178 (1961).
83. M. S. Brodin and A. F. Prikhot'ko, Proc. Tenth All-Union Conf. on Spectroscopy [in Russian], Vol. 1, Lvov University (1957), p. 16.
84. I. M. Rozman, Thesis [in Russian], Sukhumi (1956).
85. Yu. V. Konobeev, in: Solid-State Spectroscopy [in Russian], No. 4, Nauka, Leningrad (1969), p. 209.
86. V. M. Agranovich and Yu. V. Konobeev, Opt. Spektrosk., 11:369 (1961).
87. Yu. V. Konobeev, Opt. Spektrosk., 11:504 (1961).
88. A. M. Samson, Opt. Spektrosk., 8:89 (1960).
89. I. Nakada, J. Phys. Soc. Jap., 20:346 (1965).
90. V. I. Shirokov, Izv. Akad. Nauk SSSR, Ser. Fiz., 20:605 (1956).
91. G. V. Radzievskii and D. P. Osanov, Radiobiologiya, 6:298 (1966).
92. A. T. Nelms, Circ. Nat. Bur. Stand., No. 577, p. 1 (1956).
93. C. F. Sharn, J. Chem. Phys., 34:240 (1961).
94. S. Glasstone and M. C. Edlund, The Elements of Nuclear Reactor Theory, MacMillan, London (1953).
95. R. A. Wijsman, Bull. Math. Biophys., 14:121 (1952).
96. N. D. Zhevandrov, Izv. Akad. Nauk SSSR, Ser. Fiz., 22:1332 (1958).
97. N. D. Zhevandrov, V. I. Gribkov, and V. N. Varfolomeeva, Izv. Akad. Nauk SSSR, Ser. Fiz., 23:57 (1959).
98. V. I. Gribkov, N. D. Zhevandrov, and Sh. D. Khan-Magometova, Izv. Akad. Nauk SSSR, Ser. Fiz., 24:740 (1960).
99. N. D. Zhevandrov, Izv. Akad. Nauk SSSR, Ser. Fiz., 26:67 (1962).
100. M. T. Shpak, A. V. Solov'ev, and N. I. Sheremet, Opt. Spektrosk., 13:694 (1962).
101. V. L. Broude, E. F. Sheka, M. G. Shpak, and L. G. Shpakovskaya, Opt. Spektrosk., Suppl. No. 1 (Luminescence), p. 98 (1963).
102. V. L. Broude, E. F. Sheka, and M. G. Shpak, Izv. Akad. Nauk SSSR, Ser. Fiz., 27:596 (1963).
103. M. T. Shpak and N. I. Sheremet, Opt. Spektrosk., 17:694 (1964).
104. N. N. Malykhina and M. T. Shpak, Opt. Spektrosk., 17:235 (1964).
105. F. A. Bovey, The Effects of Ionizing Radiation on Natural and Synthetic High Polymers, Interscience, New York (1958).
106. P. Alexander and A. Charlesby, Nature, 173:578 (1954).
107. A. M. Bonch-Bruevich, V. P. Kovalev, L. M. Belyaev, and G. S. Belikova, Opt. Spektrosk., 11:623 (1961).
108. V. P. Kovalev, V. K. Dobrokhotova, Yu. V. Naboikin, and L. S. Kukushkin, Izv. Akad. Nauk SSSR, Ser. Fiz., 27:524 (1963).
109. L. S. Kukushkin, Thesis [in Russian], Kharkov (1963).
110. L. A. Limareva, A. S. Cherkasov, and V. I. Shirkov, Opt. Spektrosk., 25:249 (1968).
111. M. D. Galanin and Z. A. Chizhikov, Zh. Eksp. Teor. Fiz., 26:624 (1954).
112. N. D. Zhevandrov, Izv. Akad. Nauk SSSR, Ser. Fiz., 20:553 (1956).
113. N. D. Zhevandrov, Dokl. Akad. Nauk SSSR, Ser. Fiz., 83:677 (1952).
114. M. Pope and J. Burgos, Mol. Cryst., 3:215 (1967).